Surface Science

Foundations of Catalysis and Nanoscience

Kurt W. Kolasinski

Queen Mary, University of London, UK

JOHN WILEY & SONS, LTD

Other Wiley Editorial Offices

John Wiley & Sons, Inc., 605 Third Avenue,
New York, NY 10158-0012, USA

Wiley-VCH Verlag GmbH, Pappelallee 3,
D-69469 Weinheim, Germany

John Wiley & Sons (Australia) Ltd, 33 Park Road, Milton,
Queensland 4064, Australia

John Wiley & Sons (Asia) Pte Ltd, 2 Clementi Loop #02-01,
Jin Xing Distripark, Singapore 0512

John Wiley & Sons (Canada) Ltd, 22 Worcester Road,
Rexdale, Ontario M9W 1L1, Canada

Library of Congress Cataloging-in-Publication Data
Kolasinski, Kurt.
  Surface science : foundations of catalysis and naoscience / Kurt Kolasinski.
    p. cm.
  Includes bibliographical references and index.
  ISBN 0-471-49244-2 (acid-free paper) — ISBN 0-471-49245-0 (pbk. : acid-free paper)
  1. Surface chemistry.   2. Surfaces (Physics)   3. Catalysis.   I. Title.

  QD506 .K587 2001
  541.3′3 — dc21   2001026861

British Library Cataloguing in Publication Data
A catalogue record for this book is available from the British Library

ISBN 0-471-49244 2 (Hardback)
     0-471-49245 0 (Paperback)

Typeset in 10½/12½ Times by Techset Composition Ltd.
Printed and bound in Great Britain by Antony Rowe Ltd. Chippenham, Wiltshire.
This book is printed on acid-free paper responsibly manufactured from sustainable forestry,
in which at least two trees are planted for each one used for paper production.

Contents

Acknowledgements

This book has been written over the past year and a half but its genesis dates back to 1987 when I began typing up my course notes from a lecture series given by Robert Madix and Michel Boudart at Stanford University. Though this book is not based on those notes, that is when I first started to think about what I would put into my book on surface science. This book could not have been completed without the pedagogical inspiration given to me by John Yates, Richard Zare and Gehard Ertl. I have also benefited greatly from many discussions – some specific to this book and many down the pub – with George Darling, Eckart Hasselbrink, Andrew Hodgson, Steve Holloway and Mats Persson. While the most insightful aspects of this book spring from them, any errors are solely down to me. I would also like to acknowledge George Darling, Gehard Ertl, Peter Maitlis, Hari Manoharan, T. C. Shen, Hajime Takano, David Walton and Anja Wellner for providing original figures and Peter Maitlis and Jon Preece for discussions on Chapters 5 and 6, respectively.

Had I not become a surface scientist, I would have likely become a historian. Therefore, please excuse the interloping flights of fancy in my exposition. I hope they provide occasional levity without detracting from the clarity of the arguments.

Kurt W. Kolasinski
Birmingham
March 2001

Introduction

When I was an undergraduate in Pittsburgh, determined to learn about surface science, John Yates pushed a copy of Robert and McKee's *Chemistry of the Metal–Gas Interface* [1] into my hands and said 'Read this'. It was very good advice and this book remains a good starting point for surface chemistry. But since the early 1980s, the field of surface science has changed dramatically. With the discovery by Binnig and Rohrer [2, 3] of scanning tunnelling microscopy (STM) in 1983, surface science changed indelibly. Thereafter it was possible to image almost routinely surfaces and surface bound species with atomic scale resolution. Not long afterward, Eigler and Schweizer [4] demonstrated that matter could be manipulated on an atom by atom basis. Furthermore, with the inexorable march of smaller, faster, cheaper, better in the semiconductor device industry, technology was marching closer and closer to surfaces. The STM has allowed us to visualize quantum mechanics as never before. As an example, we show two images of a Si(1 0 0) surface. In one case, Figure 1(a), a bonding state is imaged. In Figure 1(b) an antibonding state is shown. Just as expected, the antibonding state exhibits a node between the atoms whereas the bonding state exhibits enhanced electron density between the atoms.

Surface science had always been about nanoscale science, even though it was never phrased that way. Catalysis has been the traditional realm of surface chemistry. With the advent of nanotechnology, it became apparent that the control of matter on the molecular scale at surfaces was of much wider relevance. This book is an attempt, from the point of

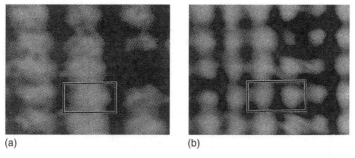

(a)   (b)

Figure 1 (a) Bonding and (b) antibonding electronic states on the Si(1 0 0) surface as imaged by scanning tunnelling microscope. Reproduced with permission from R. J. Hamers, P. Avouris and F. Bozso, *Phys. Rev. Lett.* **59** (1987) 2071. ©1987 by the American Physical Society.

view of a dynamicist, to approach surface science as the underpinning science of both heterogeneous catalysis and nanotechnology.

Heterogeneous Catalysis

One of the great motivations for studying chemical reactions on surfaces is the will to understand heterogeneous catalytic reactions. Heterogeneous catalysis is the basis of the chemical industry. Heterogeneous catalysis is involved in literally billions of pounds worth of economic activity. The chemical industry would not exist, as we know it today, if it were not for the successful implementation of heterogeneous catalysis. The aim of this book is not to review catalytic chemistry but to understand why catalytic activity occurs and how we can control it.

First we should define what we mean by catalysis and a catalyst. The term catalysis (from the Greek $\lambda\nu\sigma\iota\zeta$ and $\kappa\alpha\tau\alpha$, roughly 'wholly loosening') was coined by Berzelius in 1836 [5]. Armstrong proposed the word *catalyst* in 1885. A catalyst is an active chemical spectator. It takes part in a reaction but is not consumed. A catalyst produces its effect by changing activation barriers as shown in Figure 2. By lowering the height of an activation barrier, a catalyst speeds up a reaction. It does not, however, change the properties of the equilibrated state. Remember that whereas the kinetics of a reaction is determined by the relative heights of activation barriers (in combination with Arrhenius pre-exponential factors), the equilibrium constants are determined by the relative energetic positions of the initial and final states.

It is important to remember that the acceleration of reactions is not the only key factor in catalytic activity. If catalysts only accelerated reactions, they would not be nearly as

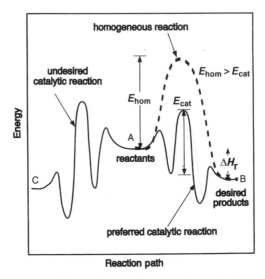

Figure 2   Activation energies and their relationship to an active and selective catalyst. (a) Reactants; (b) desired product; (c) undesired product; E_{hom}, activation barrier for the homogeneous reaction; E_{cat}, activation barrier with use of catalyst; ΔH_r, change in enthalpy of reactants compared with product.

important or as effective as they actually are. Catalysts can be designed not only to accelerate reactions: the best of them can also perform this task *selectively*. In other words, it is important for catalysts to speed up the right reactions, not simply every reaction. This is also illustrated in Figure 2 wherein the activation barrier for the desired product B is decreased more than the barrier for the undesired product C.

Why Surfaces?

Heterogeneous reactions occur in systems in which two or more phases are present, for instance, solids and liquids, or gases and solids. The reactions occur at the interface between these phases. The interfaces are where the two phases meet. Liquid–solid and gas–solid interfaces are of particular interest because the surface of a solid gives us a place to deposit and immobilize a catalytic substance. By immobilizing the catalyst we can ensure that it is not washed away and lost in the stream of products that are made. Very often catalysts take the form of small particles (the active agent) attached to the surfaces of high surface area solids (the substrate).

However, surfaces are of particular interest not only because they are where phases meet and because they give us a place to put catalysts. The surface of a solid is inherently different than the rest of the solid (the bulk). The bonding at the surface is different than that in the bulk; therefore, we should expect the chemistry of the surface to be unique. Surface atoms simply cannot satisfy their bonding requirements in the same way as bulk atoms. Therefore, surface atoms will always want to react in some way, either with each other or with foreign atoms, to satisfy their bonding requirements.

Where are Heterogeneous Reactions Important?

To illustrate a variety of topics in heterogeneous catalysis, I will make reference to a list of catalytic reaction systems that I label the (unofficial) Industrial Chemistry Hall of Fame. These reactions are selected not only because they demonstrate a variety of important chemical concepts but also because they have been of particular importance both historically and politically.

Haber–Bosch process

$$N_2 + 3H_2 \rightarrow 2NH_3 \tag{1}$$

Nitrogen fertilizers underpin modern agriculture. The inexpensive production of fertilizers would not be possible without the Haber–Bosch process. Ammonia synthesis is almost exclusively performed over an alkali-metal-promoted iron catalyst. It is a structure-sensitive reaction. Already a number of questions arise. Why an iron catalyst? What do we mean by 'promoted' and why does an alkali metal act as a promoter? What is a structure-sensitive reaction? By the end of this book all of the answers should be clear.

Fischer–Tropsch chemistry

$H_2 + CO \rightarrow$ methanol or liquid fuels or other hydrocarbons (HCs) and oxygenates

Fischer–Tropsch chemistry transforms synthesis gas ($H_2 + CO$) into useful fuels and intermediate chemicals. It is the basis of the synthetic fuels industry and has been important in sustaining economies that were shut off from crude oil, two examples of which were Germany in the 1930s and 1940s and, more recently, South Africa. It represents a method of transforming either natural gas or coal into more useful chemical intermediates and fuels. Interest in Fischer–Tropsch chemistry is rising again because, in addition to these feedstocks, biomass may also be used to produce synthesis gas.

Fischer–Tropsch reactions are often carried out over iron or cobtalt catalysts. Obviously, selectivity is a major concern because numerous products are possible but only a select few are desired for any particular application.

Threeway catalyst

$$NO_x, \ CO \text{ and } HC \rightarrow H_2O + CO_2 + N_2 \tag{2}$$

Catalysis is not always about creating the right molecule. It can equally well be important to destroy the right molecules. Increasing automobile use has led to increasing concerns about pollution. The catalytic conversion of noxious exhaust gasses to more benign chemicals has made a massive contribution to the reduction of automotive pollution. The threeway catalyst is composed of platinum, rhodium and palladium. Lead rapidly poisons the catalyst. How does this poisoning (loss of reactivity) occur?

Semiconductor Processing and Nanotechnology

The above is the traditional realm of heterogeneous catalytic chemistry. However, modern surface science is composed of other areas as well and has become particularly important to the world of microtechnology and nanotechnology [6]. The latest PowerPC[TM] and Pentium[®] processors now incorporate critical dimensions of only 180 nm in width, and this will soon drop to 130 nm. The thickness of insulating oxide layers is now only 5–10 atomic layers. Obviously, there is a need to understand materials properties and chemical reactivity at the molecular level if semiconductor processing is to continue to advance to even smaller dimensions. It has already been established that surface cleanliness is one of the major factors affecting device reliability. Eventually, however, the engineers will run out of 'room at the bottom'. Furthermore, as length scales shrink, the effects of quantum mechanics inevitably become of paramount importance. This has led to the thought that a whole new device world may exist, which is ruled by quantum mechanical effects. Devices such as a single-electron transistor have been built. Continued fabrication and study of such devices requires an understanding of atomic Lego[®]s – the construction of structures on an atom-by-atom basis.

Figure 3 shows images of some devices and structures that have been crafted at surfaces. Not only electronic devices are of interest. Microelectromechanical systems (MEMS) are provoking increasing interest. The first commercial example is the accelerometer that triggers airbags in cars. These structures are made by a series of surface etching and growth reactions.

The ultimate control of growth and etching would be to perform these one atom at a time. Figure 4 demonstrates how hydrogen atoms can be removed one by one from a silicon surface. The uncovered atoms are subsequently covered with oxygen and then etched. In Figure 4(b) we see a structure built out of xenon atoms. There are numerous ways to create structures at surfaces. We will investigate several of these in which the architect must actively pattern the substrate. We will also investigate self-assembled structures, that is, structures that form spontaneously without the need to push around the atoms or molecules that compose the structure.

Other Areas of Relevance

Surface science touches on a vast array of applications and basic science. The fields of corrosion, adhesion and tribology are all closely related to interfacial properties. The importance of heterogeneous processes in atmospheric and interstellar chemistry has now been realized. Virtually all of the molecular hydrogen that exists in the interstellar medium had to be formed on the surfaces of grains and dust particles. The role of surface chemistry in the formation of the over 100 other molecules that have been detected in outer space remains an open question [7, 8]. Many electrochemical reactions occur heterogeneously. Our understanding of charge transfer at interfaces and the effects of surface structure and adsorbed species remains in a rudimentary state.

Structure of the Book

The aim of this book is to provide an understanding of chemical transformations and the formation of structures at surfaces. In essence, our objective is to understand Chapters 5 and 6. To do this we need (a) to assemble the appropriate vocabulary and (b) to gain a familiarity with an arsenal of tools and a set of principles that guide our thinking and aid interpretation and prediction. Chapter 1 introduces us to the structure (geometric, electronic and vibrational) of surfaces and adsorbates. This gives us a picture of what surfaces look like and how they compare with molecules and bulk materials. Chapter 2 introduces the techniques with which we look at surfaces. We quickly learn that surfaces present some unique experimental difficulties. This chapter might be skipped in a first introduction to surface science. However, some of the techniques are themselves methods for surface modification. In addition, a deeper insight into surface processes is gained by understanding the manner in which data are obtained. Finally, a proper reading of the literature cannot be made without an appreciation of the capabilities and limitations of the experimental techniques.

After these foundations have been set in the first two chapters, the next two chapters deliver dynamical, thermodynamic and kinetic principles. These principles allow us to

100 μm

(a)

(b)

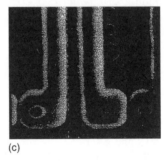

(c)

Figure 3 Examples of devices and structures that can be made at surfaces: (a) a three-level suspended tungsten stage with folded springs and comb-type actuators; (b) letters with vertical walls over 200 μm high, fabricated by ultraviolet lithography; (c) a CMOS circuit fabricated with X-ray lithography to achieve a critical dimension of 20 nm. Part (a) reproduced with permission from K. Deguchi, K. Miyoshi, H. Ban, T. Matsuda, T. Ohno and Y. Kada, *J. Vac. Sci. Technol. B* **13** (1995) 3040. Copyright 1995 by American Vacuum Society. Part (b) reproduced with permission from W. Hofmann, L.-Y. Chen and N. C. MacDonald, *J. Vac. Sci. Technol. B* **13** (1995) 2701. ©1995 by American Vacuum Society. Part (c) reproduced with permission from K. Y. Lee, N. LaBianca, S. A. Rishton, S. Zolgharnain, J. D. Gelorme, J. Shaw and T. H.-P. Chang, *J. Vac. Sci. Technol. B* **13** (1995) 3012. ©1995 by American Vacuum Society.

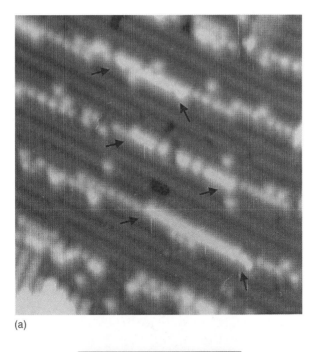

(a)

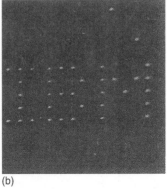

(b)

Figure 4 Examples of surface manipulation with atomic-scale resolution: (a) nanolithography can be performed on a hydrogen-terminated silicon surface by using a scanning tunnelling microscope tip to remove hydrogen atoms one at a time from the surface; (b) individual xenon atoms can be moved with precision by a scanning tunnelling microscope tip to write on the surface. Part (a) reprinted with permission from T.-C. Shen, C. Wang, G. C. Abeln, J. R. Tucker, J. W. Lyding, P. Avouris and R. E. Walkup, *Science* **268** (1995) 1590. ©1995 American Association for the Advancement of Science. Part (b) reproduced with permission from D. M. Eigler and E. K. Scwheizer, *Nature* **344** (2000) 524. ©2000 Macmillan Magazines Ltd.

understand how and why chemical transformations occur at surfaces. They deliver the mental tools required to interpret the data encountered in catalysis and growth studies.

Each chapter builds from simple to more advanced principles. Each chapter is sprinkled with 'advanced topics'. The advanced topics serve two purposes. First they provide material beyond the introductory level, which is set aside so as not to interrupt the flow of the introductory material. Second, they highlight some frontier areas. The frontiers are often too complex to explain in depth at the introductory level; nonetheless, they are included to provide a taste of the exciting possibilities of what can be done with surface science. Each chapter is also accompanied by exercises. The exercises act not only to demonstrate concepts arising in the text, but also as extensions to the text. They should be considered an integral part of the whole.

References

[1] M. W. Roberts and C. S. McKee, *Chemistry of the Metal–Gas Interface* (Clarendon Press, Oxford, 1978).

[2] G. Binnig and H. Rohrer, *Rev. Mod. Phys.* **71** (1999) S324.

[3] G. Binnig, H. Rohrer, C. Gerber and E. Weibel, *Phys. Rev. Lett.* **49** (1982) 57.

[4] D. M. Eigler and E. K. Schweizer, *Nature* **344** (1990) 524.

[5] K. J. Laidler, *The World of Physical Chemistry* (Oxford University Press, Oxford, 1993).

[6] P. Moriarty, *Rep. Prog. Phys.* **64** (2001) 297.

[7] J. S. Mathis, *Rep. Prog. Phys.* **56** (1993) 605.

[8] G. Winnewisser and E. Herbst, *Rep. Prog. Phys.* **56** (1993) 1209.

1

Surface Structure

To start our discussion of surface chemical processes, we build up from the surface itself. We need to understand the structure of clean and adsorbate-covered surfaces and use this as a foundation for understanding surface chemical problems. We will use our knowledge of surface structure to develop a new strand of chemical intuition that will allow us to know when we can apply things that we have learned from reaction dynamics in other phases and when we need to develop something completely different to understand reactivity in the adsorbed phase.

What do we mean by surface structure? There are two inseparable aspects to structure: electronic structure and geometric structure. The two aspects of structure are inherently coupled and we should never forget this point. Nonetheless, it is pedagogically helpful to separate these two aspects when we attack them experimentally and in the ways that we conceive of them.

When we speak of structure in surface science we can further subdivide the discussion into that of the clean surface, the surface in the presence of an adsorbate (substrate structure) and that of the adsorbate (adsorbate structure or overlayer structure). That is, we frequently refer to the structure of the first few layers of the substrate with and without an adsorbed layer on top of it. We can in addition speak of the structure of the adsorbed layer itself. Adsorbate structure not only refers to how the adsorbed molecules are bound with respect to the substrate atoms but also how they are bound with respect to one another.

1.1 Clean Surface Structure

1.1.1 *Ideal flat surfaces*

We are most concerned with transition metal and semiconductor surfaces. First, we consider the type of surface we obtain by truncating the bulk structure of a perfect crystal. The most important crystallographic structures of metals are the face-centred cubic (fcc), body-centred cubic (bcc) and hexagonal close-packed (hcp) structures. Many transition metals of interest in catalysis take up fcc structures under normal conditions. Notable exceptions are iron, molybdenum and tungsten, which assume bcc structures, and cobalt and ruthenium, which assume hcp structures. The most important structure for elemental (group IV: C, Si, Ge) semiconductors is the diamond lattice, whereas compound semiconductors from groups III and V (III–V compounds: e.g. GaAs and InP) assume a zincblende structure.

A perfect crystal can be cut along any arbitrary angle. The directions in a lattice are indicated by the Miller indices. Miller indices are related to the positions of the atoms in the lattice. Directions are uniquely defined by a set of three (fcc, bcc and diamond) or four (hcp) rational numbers and are denoted by enclosing these numbers in square brackets, for example [1 0 0]; hcp surfaces can also be defined by three unique indices, and both notations are encountered, as shown in Figure 1.3. In these designations, negative numbers are indicated by an overbar. A plane of atoms is uniquely defined by the direction that is normal to the plane. To distinguish a plane from a direction, a plane is denoted by enclosing the numbers associated with the defining direction in parentheses, for example (1 0 0).

The most important planes to learn by heart are the low-index planes. Low-index planes can be thought of as the basic building blocks of surface structure as they represent some of the simplest and flattest of the fundamental planes. The low-index planes in the fcc system [e.g. (1 0 0), (1 1 0) and (1 1 1)] are shown in Figure 1.1. The low-index planes of bcc symmetry are displayed in Figure 1.2, and the more complex structures of the hcp symmetry are shown in Figure 1.3.

We concentrate our discussion on Figure 1.1. The ideal structures shown in Figure 1.1 demonstrate several interesting properties. Note that these surfaces are not perfectly

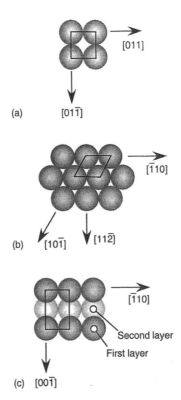

Figure 1.1 Hard-sphere representations of face-centred cubic (fcc) low-index planes: (a) fcc(1 0 0); (b) fcc(1 1 1), (c) fcc(1 1 0)

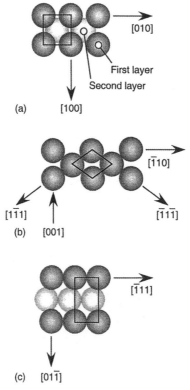

Figure 1.2 Hard-sphere representations of body-centred cubic (bcc) low-index planes: (a) bcc(1 0 0); (b) bcc(1 1 0); (c) bcc(2 1 1)

isotropic. We can pick out several sites on any of these surfaces that are geometrically unique. On the (1 0 0) surface we can identify sites of onefold (on top of and at the centre of one atom), twofold (bridging two atoms) or fourfold coordination (in the hollow between four atoms). The coordination number is equal to the number of surface atoms bound directly to the adsorbate. The (1 1 1) surface has onefold, twofold and threefold coordinated sites. Among others, the (1 1 0) surface presents two different types of twofold sites: a long bridge site between two atoms on adjacent rows and a short bridge site between two atoms in the same row. As one might expect, based on the results of coordination chemistry, the multitude of sites on these surfaces leads to heterogeneity in the interactions of molecules with the surfaces. This will be important in our discussion of adsorbate structure as well as for surface chemistry.

The formation of a surface from a bulk crystal is a stressful event. Bonds must be broken and the surface atoms no longer have their full complement of coordination partners. Therefore, the surface atoms find themselves in a higher energy situation compared with being buried in the bulk, and they must relax. Even on flat surfaces, such as the low-index planes, the top layers of a crystal react to the formation of a surface by changes in their bonding geometry. For flat surfaces, the changes in bond lengths and bond angles usually only amount to a few percent. These changes are known as

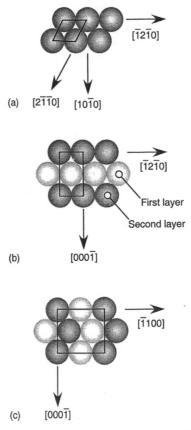

Figure 1.3 Hard-sphere representations of hexagonal close-paced (hcp) low-index planes: (a) hcp$(0\,0\,1)=(0\,0\,1)$; (b) hcp$(1\,0\,\bar{1}\,0)=(1\,0\,0)$; (c) hcp$(1\,1\,\bar{2}\,0)=(1\,1\,0)$

relaxations. Relaxations can extend several layers into the bulk. The near surface region, which has a structure different from that of the bulk, is called the selvedge. This is our first indication that bonding at surfaces is inherently different from that in the bulk, both because of changes in coordination and because of changes in structure.

1.1.2 High-index and vicinal planes

Surface structure can be made more complex either by cutting a crystal along a higher index plane or by the introduction of defects. High-index planes [surfaces $(h\,k\,l)$ with h, k, or $l > 1$ or 2] often have open structures that can expose second and even third layer atoms. The fcc$(1\,1\,0)$ surface shows how this can occur even for a low-index plane. High-index planes often have large unit cells that encompass many surface atoms. An assortment of defects is shown in Figure 1.4.

One of the most straightforward types of defects at surfaces is that introduced by cutting the crystal at an angle slightly off of the perfect $[h\,k\,l]$ direction. A small miscut

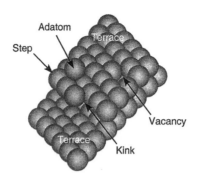

Figure 1.4 Hard-sphere representation of a variety of defect structures that can occur on single-crystal surfaces

angle leads to vicinal surfaces. Vicinal surfaces are close to but not flat low-index planes. The effect of a small miscut angle is demonstrated in Figure 1.4. Because of the small miscut angle, the surface cannot maintain the perfect $(h\,k\,l)$ structure over long distances. Atoms must come in whole units and, in order to stay as close to a low-index structure as possible while still maintaining the macroscopic surface orientation, step-like discontinuities are introduced into the surface structure. On the microscopic scale, a vicinal surface is composed of a series of terraces and steps. Therefore, vicinal surfaces are also known as stepped surfaces.

Stepped surfaces have an additional type of heterogeneity compared with flat surface, which has a direct effect on their properties [1]. They are composed of terraces of low-index planes with the same types of symmetry as normal low-index planes. In addition, they have steps. The structure of step atoms must be different from that of terrace atoms because of the different bonding that they exhibit. Step atoms generally relax more than terrace atoms. The effect of steps on electronic structure is illustrated in Figure 1.5. The electrons of the solid react to the presence of the step and attempt to minimize the energy of the defect. They do this by spreading out in a way that makes the discontinuity at the step less abrupt. This process is known as Smoluchowski smoothing [2]. Since the electronic structure of steps differs from that of terraces, we expect that their chemical reactivity is different as well. Note that the top and the bottom of a step are different, and this has implications, for instance, for diffusion of adsorbates over steps. It is often the case that diffusion in one direction is significantly easier than in the other. Furthermore,

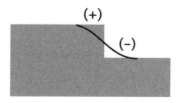

Figure 1.5 Smoluchowski smoothing: the electrons at a step attempt to smooth out the discontinuity of the step

we expect that diffusion on the terraces may differ significantly from diffusion across steps (see Section 3.2).

1.1.3   Faceted surfaces

Not all surfaces are stable. The formation of a surface is always endothermic (see Chapter 4). However, the formation of a larger surface area of low-energy (low-index) planes is sometimes favoured over the formation of a single layer of a high energy (high-index) plane. Many high-index planes are known to facet at equilibrium. Faceting is the spontaneous formation of arrays of low-index planes. Numerous systems exhibit ordered arrays of low-index facets. These have been catalogued by Shchukin and Bimberg [3] and include vicinal surfaces of Si(1 1 1), GaAs(1 0 0), Pt(1 0 0), high-index planes of Si(2 1 1) and low-index planes of TaC(1 1 0).

1.1.4   Bimetallic surfaces

A surface composed of a mixture of two metals often exhibits unique properties. The catalytic behaviour of Au + Ni surfaces, for example, will be discussed in Chapter 5. The surface alloy of $Pt_3Sn(1\ 1\ 1)$ has also attracted interest because of its unusual catalytic properties [4]. Materials containing mixtures of metals introduce several new twists into a discussion of surface structure. Consider a single crystal composed of two metals that form a true intermetallic compound. An ideal single-crystal sample would exhibit a surface structure much like that of a monometallic single crystal. The composition of the surface would depend on the bulk composition and the exposed surface plane. Several examples of this type have been observed [5], for example: $Cu_3Au(1\ 0\ 0)$; (1 0 0), (1 1 1) and (1 1 0) surfaces of Ni_3Al; (1 1 0) and (1 1 1) surfaces of NiAl; as well as $TiPt_3(1\ 0\ 0)$.

Not all combinations of metals form intermetallic compounds. Some metals have limited solubility in other metals. Furthermore, the solubility of a given metal may be different in the bulk than it is at the surface. That is, if the surface energy of one component of an alloy is significantly lower than that of the other component, the low-surface-energy species preferentially segregates to the surface. This leads to enrichment in the surface concentration as compared with the bulk concentration. Most alloys show some degree of segregation and enrichment of one component at the surface. The factors that lead to segregation are much the same as those that we will encounter in Chapter 6 when we investigate growth processes. For a binary alloy AB, the relative strengths of the A–A, A–B and B–B interactions as well as the relative sizes of A and B determine whether alloy formation is exothermic or endothermic. These relative values determine whether segregation occurs. If alloy formation is strongly exothermic, that is, if the A–B interaction is stronger than either the A–A or the B–B interactions, then there is little tendency toward segregation. The relative atomic sizes are important for determining whether lattice strain influences the energetics of segregation. In summary, surface segregation is expected unless alloy formation is highly exothermic and there is good matching of the atomic radii.

If a bimetallic surface is made not from a bulk sample but instead from the deposition of one metal on top of another, the surface structure depends sensitively on the conditions under which deposition occurs. In particular, the structure depends on whether the deposition process is kinetically or dynamically controlled. These issues will be dealt with in Chapter 6.

1.2 Reconstruction and Adsorbate Structure

1.2.1 Implications of surface heterogeneity for adsorbates

As alluded to above, the natural heterogeneity of solid surfaces has several important ramifications for adsorbates. Simply looked at from the point of view of electron density, we see that low-index planes, let alone vicinal surfaces, are not completely flat. Undulations in the surface electron density exist that reflect the symmetry of the surface atom arrangement as well as the presence of defects such as steps, missing atoms or impurities. The ability of different regions of the surface to exchange electrons with adsorbates, and thereby form chemical bonds, is strongly influenced by the coordination number of the various sites on the surface. More fundamentally, the ability of various surface sites to enter into bonding is related to the symmetry, nature and energy of the electronic states found at these sites. It is a poor approximation to think of the surface atoms of transition metals as having unsaturated valences (dangling bonds) waiting to interact with adsorbates. The electronic states at transition metal surfaces are extended, delocalized states that correlate poorly with unoccupied or partially occupied orbitals centred on a single metal atom. The concept of dangling bonds, however, is highly appropriate for covalent solids such as semiconductors. Nontransition metals, such as aluminium, can also exhibit highly localized surface electronic states.

The heterogeneity of low-index planes presents an adsorbate with a more or less regular array of sites. Similarly, the strength of the interaction varies in a more or less regular fashion that is related to the underlying periodicity of the surface atoms and the electronic states associated with them. These undulations are known as corrugation. Corrugation can refer to either geometric or electronic structure. A corrugation of zero corresponds to a completely flat surface. A high corrugation corresponds to a mountainous topology.

Since the sites at a surface exhibit different strengths of interaction with adsorbates, and since these sites are present in ordered arrays, we expect adsorbates to bind in well-defined sites. Interactions between adsorbates can enhance the order of the overlayer; indeed, these interactions can also lead to a range of phase transitions in the overlayer [6]. We will discuss the bonding of adsorbates extensively in Chapter 3 and adsorbate–adsorbate interactions in Chapter 4. The symmetry of overlayers of adsorbates may sometimes be related to the symmetry of the underlying surface. We distinguish three regimes: random adsorption, commensurate structures and incommensurate structures. Random adsorption corresponds to the lack of two-dimensional order in the overlayer, even though the adsorbates may occupy (one or more) well-defined adsorption sites. Commensurate structures are formed when the overlayer structure corresponds to the structure of the substrate in some rational fraction. Incommensurate structures are formed when the overlayer exhibits two-dimensional order; however, the periodicity of the overlayer is not related in a simple fashion to the periodicity of the substrate. A more precise and quantitative discussion of the relationship of overlayer to substrate structure is discussed in Section 2.2, which deals with low-energy electron diffraction (LEED).

1.2.2   Clean surface reconstructions

In most instances, the low-index planes of metals are stabilized by simple relaxations. Sometimes relaxation of the selvedge is not sufficient to stabilize the surface as is the case for Au(1 1 1) and Pt(1 0 0). To minimize the surface energy, the surface atoms reorganize the bonding among themselves. This leads to surfaces with periodicities that differ from the structure of the bulk-terminated surface, called reconstructions. For semiconductors it is the rule rather than the exception that the surfaces reconstruct. This can be traced back to the presence of dangling bonds on semiconductor surfaces whereas metal electrons tend to occupy delocalized states. Delocalized electrons adjust more easily to relaxations and conform readily to the geometric structure of low-index planes. Dangling bonds are high-energy entities, and solids react in extreme ways to minimize the number of dangling bonds. The step atoms on vicinal surfaces are also associated with localized electronic states, even on metals. In many cases vicinal surfaces are not sufficiently stabilized by simple relaxations and they therefore undergo faceting, as described in Section 1.1.3.

One of the most important and interesting reconstructions is that of the Si(1 0 0) surface, shown in Figure 1.6, which is the plane most commonly used in integrated circuits. The Si(1 0 0)-(2 × 1) reconstruction completely eliminates all dangling bonds from the original Si(1 0 0)-(1 × 1) surface. The complex Si(1 1 1)-(7 × 7) reconstruction reduces the number of dangling bonds from 49 to just 19. The Wood's notation used to describe reconstructions is explained in Section 2.2.

1.2.3   Adsorbate-induced reconstructions

An essential tenet of thermodynamics is that at equilibrium a system will possess the lowest possible chemical potential and that all phases present in the system have the same

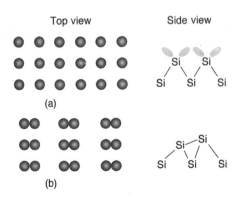

Figure 1.6 The Si(1 0 0)-(2 × 1) reconstruction: (a) unreconstructed clean Si(1 0 0)-(1 × 1); (b) reconstructed clean Si(1 0 0)-(2 × 1)

chemical potential. This is true in the real world in the absence of kinetic or dynamic constraints or, equivalently, in the limit of sufficiently high temperature and sufficiently long time. This tenet must also hold for the gas phase/adsorbed phase/substrate system and has several interesting consequences. We have already mentioned that adsorbates can form ordered structures, [see Figures 4 (in the introduction), 2.2 and 2.5]. This may appear contrary to the wishes of entropy, but if an ordered array of sites is to be maximally filled then the adsorbate must also assume an ordered structure. The only constraint on the system is that chemisorption must be sufficiently exothermic to overcome the unfavourable entropy factors.

Adsorbates not only can assume ordered structures, but also can induce reconstruction of the substrate. One way to get rid of dangling bonds is to involve them in bonding. The Si−H bond is strong and nonpolar. Hydrogen atom adsorption represents a perfect method of capping the dangling bonds of silicon surfaces. Hydrogen atom adsorption is found to lift the reconstruction of silicon surface, that is, the clean reconstructed surfaces are transformed to a new structure by the adsorption of hydrogen atoms. By adsorbing one hydrogen atom per surface silicon atom, the Si(1 0 0)-(2 × 1) asymmetric dimer structure is changed into a Si(1 0 0)-(2 × 1):H symmetric dimer structure. The Si(1 1 1) surface takes on the bulk-terminated (1 × 1) structure in the presence of chemisorbed hydrogen.

Adsorbate-induced reconstructions can have a dramatic effect on the kinetics of reactions on reconstructed surfaces. Of particular importance is the reconstruction of Pt(1 1 0). The clean surface is reconstructed into a (1 × 2) missing-row structure, a rather common type of reconstruction. However, CO adsorption leads to a lifting of the reconstruction. Adsorbate-induced reconstruction of a metal surface is associated with the formation of strong chemisorption bonds.

The surface does not present a static template of adsorption sites to an adsorbate. Somorjai [7] has collected an extensive list of clean-surface and adsorbate-induced reconstructions. When an adsorbate binds to a surface, particularly if the chemisorption interaction is strong, we need to consider whether the surface is stable compared with reconstruction [8]. For sufficiently strong interactions and high enough adsorbate concentrations we may have to consider whether the surface is stable compared with the formation of a new solid chemical compound, such as the formation of an oxide layer, or the formation of a volatile compound, as in the etching of silicon by halogen compounds or atomic hydrogen. Here we concentrate on interactions that lead to reconstruction.

The chemical potential of an adsorbate/substrate system is dependent on the temperature T, the chemical identity of the substrate S and adsorbate A, the number density of the adsorbates σ_A and surface atoms σ_S, and the structures that the adsorbate and surface assume. The gas phase is coupled, in turn, to σ_A through the pressure. Thus, we can write the chemical potential as

$$\mu(T, \sigma_S, \sigma_A) = \sum_i \sigma_{S,i}\mu_i(T) + \sum_j \sigma_{A,j}\mu_j(T) \tag{1.1}$$

In Equation (1.1) the summations are over the i and j possible structures that the surface and adsorbates, respectively, can assume. The adsorption energy can depend on the surface structure. If the difference in adsorption energy between two surface structures is sufficiently large so as to overcome the energy required to reconstruct the substrate, the surface structure can switch from one structure to the next when a critical adsorbate coverage is exceeded. Note also that Equation (1.1) is written in terms of areal densities to emphasize that variations in adsorbate concentration can lead to variations in surface structure across the surface. In other words, inhomogeneities in adsorbate coverage may lead to inhomogeneities in the surface structure. An example is the chemisorption of hydrogen on Ni(1 1 0) [9]. Up to a coverage of 1 hydrogen atom per surface nickel atom ($\equiv$1 monolayer, or 1 ML), a variety of ordered overlayer structures are formed on the unreconstructed surface. As the coverage increases further, the surface reconstructs locally into islands that contain 1.5 ML of hydrogen atoms. In these islands the rows of nickel atoms pair up to form a (1×2) structure.

Equation (1.1) indicates that the equilibrium surface structure depends both on the density of adsorbates and on the temperature. The surface temperature is important in two ways. First, the chemical potential of each surface structure will depend on temperature. Therefore, the most stable surface structure can change as a function of temperature. Second, the equilibrium adsorbate coverage is a function of the pressure and temperature. Because of this coupling of adsorbate coverage to temperature and pressure, we expect that the equilibrium surface structure can change as a function of these two variables as well as the identity of the adsorbate [8]. This can have important consequences for working catalysts because surface reactivity can change with surface structure.

An example of the restructuring of a surface and the dependence on adsorbate coverage and temperature is the H/Si(1 0 0) system (Figure 1.7) [10–13]. The clean Si(1 0 0) surface [Figure 1.6(a)] reconstructs into a (2×1) structure [Figure 1.6(b)] caused by the formation of silicon dimers on the surface. The dimers are buckled at low temperature but the rocking motion of the dimer is a low-frequency vibration, which means that at room temperature the average position of the dimers appears symmetric. When hydrogen adsorbs on the dimer in the monohydride structure (1 hydrogen atom per surface silicon atom) the dimer becomes symmetric [Figure 1.7(a)], the dimer bond expands and much of the strain in the subsurface layers is relaxed. Further increasing the hydrogen coverage by exposure of the surface to atomic hydrogen leads to the formation of dihydride units [SiH_2; Figure 1.7(b)]. These form in appreciable numbers only if the temperature during adsorption is below c. 400 K. If the surface is exposed to hydrogen at room temperature, trihydride units (SiH_3) can also form. These are a precursor to etching by the formation of SiH_4, which desorbs from the surface. Northrup [14] has shown how these changes are related to the chemical potential and lateral interactions. Neighbouring dihydride units experience repulsive interactions and are unable to assume their ideal positions. This lowers the stability of the fully covered dihydride surface such that some spontaneous formation and desorption of SiH_4 is to be expected.

The LEED pattern of a Si(1 0 0) surface exposed to large doses of hydrogen at room temperature or below has a (1×1) symmetry. This has been incorrectly interpreted as a complete coverage of the surface, with dihydrides, with the silicon atoms assuming the ideal bulk termination. Instead, the surface is rough, disordered and composed of a mixture of SiH, SiH_2 and SiH_3 units. The (1×1) pattern arises from the subsurface layers, which are also probed by LEED. If the surface is exposed to hydrogen atoms at

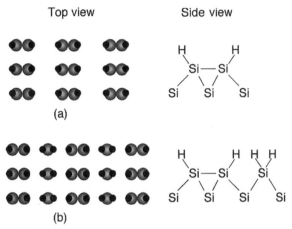

Figure 1.7   The adsorption of hydrogen onto Si(1 0 0): (a) Si(1 0 0)- $(2 \times 1):H$, $\theta(H) = 1$ ML; (b) Si(1 0 0)-$(3 \times 1):H$, $\theta(H) = 1.33$ ML. Note: these structures obtained from the adsorption of hydrogen onto Si(1 0 0) are a function of coverage. $\theta(H)$, coverage of hydrogen on Si(1 0 0); ML, monolayer, 1 ML = 1 hydrogen atom per surface silicon atom

c. 380 K, a (3×1) LEED pattern is observed. This surface is composed of an ordered structure consisting of alternating SiH and SiH_2 units [Fig. 1.7(b)]. Thus, there are 3 hydrogen atoms for every 2 silicon atoms. Heating the (3×1) surface above *c.* 600 K leads to rapid decomposition of the SiH_2 units. The hydrogen desorbs from the surface as H_2 and the surface reverts to the monohydride (2×1) structure.

The reconstructed and nonreconstructed platinum surfaces are shown in Figure 1.8. Of the clean low-index planes, only the Pt(1 1 1) surface is stable compared with reconstruction. The Pt(1 0 0) surface reconstructs into a quasi-hexangonal (hex) phase, which is *c.* 40 kJ mol^{-1} more stable than the (1×1) surface. The Pt(1 1 0) reconstructs into a (1×2) missing-row structure. These reconstructions lead to dramatic changes in the chemical reactivity, which can lead to spatiotemporal pattern formation during CO oxidation (see Chapter 5) [15]. The clean surface reconstructions can be reversibly lifted by the presence of certain adsorbates including CO and NO. This is driven by the large difference in adsorption energy between the two reconstructions. For CO the values are 155 kJ mol^{-1} and 113 kJ mol^{-1} on the (1×1) and hex phases, respectively; just large enough to overcome the energetic cost of reconstruction.

The surface structure not only effects the heat of adsorption but also can dramatically change the probability of dissociative chemisorption: O_2 dissociates with a probability of 0.3 on the Pt(1 0 0)-(1×1) surface. On the Pt(1 0 0) hex phase, the dissociation probability drops to *c.* $\sim 10^{-4}$–10^{-3}. We will investigate the implications of these changes further in Chapter 5. In Chapter 3 we discuss the dynamical factors that affect the dissociation probability.

1.2.4   Islands

A flux of gas molecules strikes a surface at random positions. If no attractive or repulsive interactions exist between the adsorbed molecules (lateral interactions), the distribution of

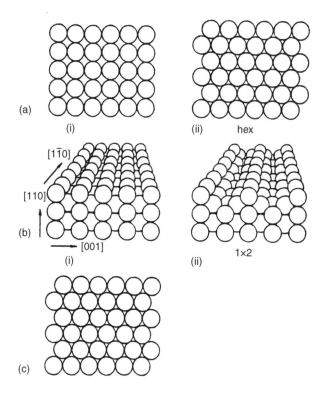

Figure 1.8 Reconstructed and nonreconstructed surfaces for three low-index planes of platinum: (a) (1 0 0) plane, (i) (1 × 1) surface, (ii) reconstructed surface [quasi-hexagonal (hex)]; (b) (1 1 0) plane, (i) (1 × 1) surface, (ii) reconstructed surface; (c) (1 × 1) surface of (1 1 1) plane [the Pt(1 1 1) surface is stable compared with reconstruction]. Reproduced with permission from R. Imbihl and G. Ertl. *Chem. Rev.* **95** (1995) 697, ©1995 American Chemical Society

adsorbates on the surface would also be random. However, if the surface temperature is high enough to allow for diffusion (see Section 3.2), then the presence of lateral interactions can lead to nonrandom distributions of the adsorbates. In particular, the adsorbates can coalesce into regions of locally high concentration separated by low-concentration or even bare regions. The regions of high coverage are known as islands. Since the coverage in islands is higher than in the surrounding regions then, according to Equation (1.1), the substrate beneath the island might reconstruct whereas regions outside of the islands might not. In some cases, such as H/Ni(1 1 0), it is the capacity of a reconstructed region to accommodate a higher coverage than an unreconstructed region that drives the formation of islands. In subsequent chapters, we shall see that islands can have important implications for surface kinetics (Chapter 4), in particular for spatiotemporal pattern formation (Section 5.8) as well as the growth of self-assembled monolayers and thin films (Chapter 6).

1.3 Band Structure of Solids

1.3.1 Bulk electronic states

A bulk solid contains numerous electrons. The electrons are classified as being either valence or core electrons. Valence electrons are the least strongly bound electrons and have the highest values of the principle quantum number. Valence electrons form delocalized electronic states that are characterized by three-dimensional wavefunctions known as Bloch waves. The energy of Bloch waves depends on the wavevector, k, of the electronic state. The wavevector describes the momentum of the electron in a particular state. Since momentum is a vector, it is characterized by both its magnitude and direction. In other words, the energy of an electron depends not only on the magnitude of its momentum but also on the direction in which the electron moves. The realm of all possible values of k is known as k space; k space is the world of the solid described in momentum space as opposed to the more familiar world of *xyz* coordinates, known as real space.

Because of the great number of electrons in a solid, there are a large number of electronic states. These states overlap to form continuous bands of electronic states, and the dependence of the energy on the momentum is known as the electronic band structure of the solid. Two bands are always formed: the valence band and the conduction band. The energetic positions of these two bands and their occupation determine the electrical properties of the solid.

The core electrons are the electrons with the lowest values of the principal quantum number and the highest binding energy. These electrons are strongly localized near the nuclei and they do not form bands. Core electrons are not easily moved from their positions near the nuclei and therefore they do not directly participate either in electrical conduction or in chemical bonding.

1.3.2 Metals, semiconductors and insulators

The simplest definition of metals, semiconductors and insulators is found in Figure 1.9. In a metal the valence band and the conduction band overlap. There is no energy gap

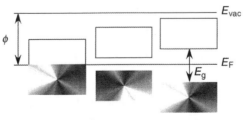

Metal   Semiconductor   Insulator

Figure 1.9 Fermi energies, vacuum energies and work functions in a metal, a semiconductor and an insulator. The presence and size of a gap between electronic states at the Ferm energy, E_F, determines whether a material is a metal, semiconductor or an insulator. E_g, band gap; ϕ, the work function, equal to the difference between E_F and the vacuum energy, E_{vac}

between these bands and the conduction band is not fully occupied. The energy of the highest occupied electronic state (at 0 K) is known as the Fermi energy, E_F. At 0 K, all states below the Fermi level (a hypothetical energy level at energy E_F) are occupied and all states above it are empty. Because the conduction band is not fully occupied, electrons are readily excited from occupied to unoccupied states and the electrons in the conduction band are therefore quite mobile. Excitation of an electron from an occupied state to an unoccupied state leads to an excited electronic configuration in which an electron occupies an excited state and in which an absent electron 'occupies' the original electronic state. This absent electron is known as a hole. The hole is a pseudo-particle that acts something like the mirror image of an electron. A hole is effectively a positively charged particle that can be characterized according to its effective mass and its mobility analogous to electrons. Electron excitation always creates a hole. Therefore, we often speak of electron–hole ($e^- - h^+$) pair formation. As these particles possess opposite charges, they can interact with one another. Creation of electron–hole pairs in the conduction band of metals represents an important class of electronic excitation, with a continuous energy spectrum that starts at zero energy.

Figure 1.9 illustrates another important property of materials. The vacuum energy, E_{vac}, is defined as the energy of a material and an electron at infinite separation. The difference between E_{vac} and E_F is known as the work function, ϕ:

$$\phi = E_{vac} - E_F \tag{1.2}$$

and at 0 K it represents the minimum energy required to remove one electron from the material to infinity. For ideal semiconductors and insulators, the actually minimum ionization energy is greater than ϕ because there are no states at E_F. Instead, the highest energy electrons reside at the top of the valence band. Not apparent in Figure 1.9 is that the work function sensitively depends on the crystallographic orientation of the surface, the presence of surface defects in particular steps [2, 16] and, of course, the presence of chemical impurities on the surface.

The reason for the dependence of the work function on the surface properties can be traced back to the electron distribution at the surface (Figure 1.10). The electron density does not end abruptly at the surface. Instead, the electron density oscillates near the surface before decaying slowly into the vacuum (Friedel oscillations). This distribution of electrons creates an electrostatic dipole layer at the surface. The surface dipole contribution D is equal to the difference between the electrostatic potential far into the vacuum, $V(\infty)$, and the mean potential deep in the bulk, $V(-\infty)$:

$$D = V(\infty) - V(-\infty) \tag{1.3}$$

If we reference the electrostatic potential with respect to the mean potential in the bulk [i.e. $V(-\infty) = 0$] then the work function can be written as

$$\phi = D - E_F \tag{1.4}$$

Therefore, ϕ is determined both by a surface term D and a bulk term E_F. With this definition of the work function, changes in D arising from surface structure and adsorbates are responsible for changes in ϕ, because surface properties cannot affect E_F.

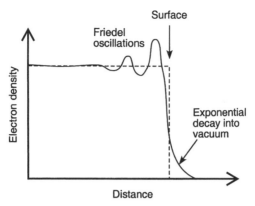

Figure 1.10 Friedel oscillations: the electron density near the surface oscillates before decaying exponentially into the vacuum

The occupation of electronic states is governed by Fermi–Dirac statistics. All particles with noninteger spin, such as electrons, which have a spin of $\frac{1}{2}$, obey the Pauli exclusion principle. This means that no more than two electrons can occupy any given electronic state. The energy of electrons in a solid depends on the availability of electronic states and the temperature. Metals sometimes exhibit regions of k-space in which no electronic states are allowed. These forbidden regions of k-space are known as partial bandgaps. At finite temperature, electrons are not confined only to states at or below the Fermi level. At equilibrium they form a reservoir with an energy equal to the chemical potential, μ. The probability of occupying allowed energy states depends on the energy of the state, E, and the temperature according to the Fermi–Dirac distribution:

$$f(E) = \{\exp[(E - \mu)/k_{\mathrm{B}}T] + 1\}^{-1} \tag{1.5}$$

where T is the temperature and k_{B} is the Boltzmann constant. The Fermi–Dirac distribution for several temperatures is drawn in Figure 1.11. At $T = 0$ K, the Fermi–Dirac function is a step function, that is $f(E) = 1$ for $E < \mu$, and $f(E) = 0$ for $E > 0$. At this temperature the Fermi energy is identical to the chemical potential:

$$\mu(T = 0 \text{ K}) \equiv E_{\mathrm{F}} \tag{1.6}$$

More generally, the chemical potential is defined as the energy at which $f(E) = 0.5$. It can be shown [17] that

$$\mu(T) \approx E_{\mathrm{F}}\left[1 - \frac{\pi^2}{12}\left(\frac{k_{\mathrm{B}}T}{E_{\mathrm{F}}}\right)^2\right] \tag{1.7}$$

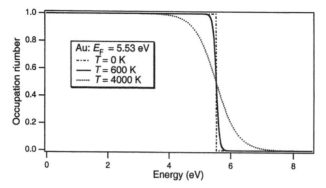

Figure 1.11 Fermi–Dirac distribution for three differet temperatures, T, for gold, E_F, Fermi energy

A direct consequence of Fermi–Dirac statistics is that the Fermi energy is not zero at 0 K. In fact, E_F is a material-dependent property that depends on the electron density, ρ, according to

$$E_F = \frac{\hbar^2}{2m_e}(3\pi^2\rho)^{2/3} \tag{1.8}$$

where $\hbar = h/2\pi$, where h is Planck's constant. E_F is of the order of several electron volts and E_F/k_B, known as the Fermi temperature, T_F, is of the order of several thousand kelvin. Consequently, the chemical potential and E_F are virtually identical unless the temperature is extremely high.

Semiconductors exhibit a complete bandgap between the valence and conduction bands. The energy of the conduction band minimum is E_C, and the energy at the valence band maximum is E_V. The magnitude of the bandgap is the difference between these two:

$$E_g = E_C - E_V \tag{1.9}$$

The Fermi level of a pure semiconductor, known as an intrinsic semiconductor, lies in the bandgap. The exact position depends on the temperature, according to

$$E_F = E_i = \frac{E_C + E_V}{2} + \frac{k_B T}{2}\ln\left(\frac{N_V}{N_C}\right) \tag{1.10}$$

where N_V and N_C are the effective densities of states of the valence and conduction bands, respectively. The densities of states can be calculated from material-specific constants and the temperature, as shown, for instance, by Sze [18]. Equation (1.10) shows that the Fermi energy of an *intrinsic* semiconductor lies near the middle of the gap. As we shall see below, this is not true for the more common case of a doped (extrinsic) semiconductor.

In the bulk of a perfect semiconductor there are no electrons at the Fermi level, even though it is energetically allowed, because there are no allowed electronic states at this energy. An equivalent statement is that in the bulk of an ideal semiconductor the density of states is zero at E_F. In any real semiconductor, defects (structural irregularities or impurities) introduce a nonzero density of states into the bandgap. These states are known as gap states. At absolute zero the valence band is completely filled and the conduction band is completely empty. At any finite temperature, some number of electrons is thermally excited into the conduction band. This number is known as the intrinsic carrier density and is given by

$$n_i = (N_C N_V)^{1/2} \exp(-E_g/2k_B T) \tag{1.11}$$

The presence of a bandgap in a semiconductor means that the electrical conductivity of a semiconductor is low. The bandgap also increases the minimum energy of electron–hole pair formation from zero to $\geq E_g$. What distinguishes a semiconductor from an insulator is that E_g in a semiconductor is sufficiently small that either thermal excitations or the presence of impurities can promote electrons into the conduction band, or holes into the valence band. Electrically active impurities are known as dopants. There are two classes of dopants. If a valence III atom, such as boron, is substituted for a silicon atom in an otherwise perfect silicon crystal, the boron atom accepts an electron from the silicon lattice. This effectively donates a hole into the valence band. The hole is mobile and leads to increased conductivity. Boron in silicon is a *p*-type dopant because it introduces a *positive* charge carrier into the silicon band structure. Boron introduces acceptor states into the silicon and the concentration of acceptor atoms is denoted N_A. On the other hand, if a valence V atom, such as phosphorus, is doped into silicon, the phosphorus atom effectively donates an electron into the conduction band. Because the resulting charge carrier is *negative*, phosphorus in silicon is known as an *n*-type dopant. Analogously, the concentration of donors is N_D.

The position of the Fermi energy in a doped semiconductor depends on the concentration and type of dopants. E_F is pushed upward by *n*-type dopants according to

$$E_F = E_i + k_B T \ln\left(\frac{N_D}{n_i}\right) \tag{1.12}$$

In a *p*-type material, E_F is pulled toward the valence band:

$$E_F = E_i - k_B T \ln\left(\frac{N_A}{n_i}\right) \tag{1.13}$$

Equations (1.12) and (1.13) hold as long as the dopant density is large compared with the n_i and density of electrically active dopants of the opposite type.

An insulator has a large bandgap. The division between a semiconductor and an insulator is somewhat arbitrary. Traditionally, a material with a bandgap larger than *c.* 3 eV has been considered to be an insulator. The push of high technology and the desire to fabricate semiconductor devices that operate at high temperatures has expanded this rule of thumb. Hence diamond, with a band gap of *c.* 5.5 eV, now represents the upper limit of wide-bandgap semiconductors.

Graphite represents one other important class of material. It does have electronic states up to the Fermi energy. However, the conduction and valence band edges, which correlate with π^* antibonding and π bonding bands formed from p_z-like orbitals, only overlap in a small region of k space. Therefore, the density of states at E_F is minimal, and graphite is considered a semimetal.

1.3.3 Energy levels at metal interfaces

An interface is generally distinguished from a surface as it is thought of as the boundary between two materials in intimate contact. At equilibrium, the chemical potential must be uniform throughout a sample. This means that the Fermi levels of two materials, which are both at equilibrium and in electrical contact, must be the same.

Figure 1.12 demonstrates what occurs when two bulk metals are brought together to form an interface. In Figure 1.12(a) we see that two isolated metals share a common vacuum level but have different Fermi levels, E_F^L and E_F^R, as determined by their work functions, ϕ_L and ϕ_R. Superscripts L and R refer to the left-hand and right-hand metals, respectively. When the two metals are connected electrically, electrons flow from the low-work-function metal to the high-work-function metal, from metal L to metal R, until the Fermi levels become equal. Consequently, metal L is slightly depleted of electrons and metal R has an excess of electrons. In other words, a dipole develops between the two metals, and with this is associated a small potential drop, the contact potential,

$$\phi_C = \phi_R - \phi_L \tag{1.14}$$

and an electrical field. The presence of the electric field is evident in Figure 1.12(b) by the sloping vacuum level. Figure 1.12 demonstrates that the bulk work functions of the two metals remain constant. Since the Fermi energies must be equal at equilibrium, the vacuum level shifts in response.

In metals, screening of free charges by valence electrons is efficient. Screening is the process by which the electrons surrounding a charge (or charge distribution such as a dipole) are polarized (redistributed) to lower the energy of the system. Screening is not very effective in insulators because the electrons are not as free to move. Therefore, when the two metals are actually brought into contact, the width of this dipole layer is only a few angstroms. The energetic separation between E_F and E_{vac} is constant after the first two or three atomic layers. At the surface, however, this separation is not constant. For the interface of two metals this means that the position of E_{vac} does not change abruptly, but continuously.

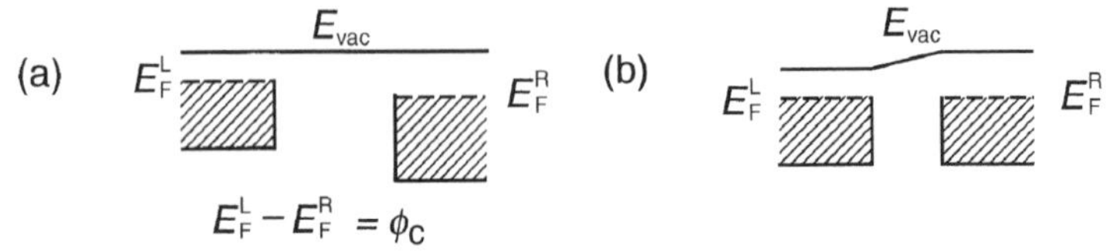

Figure 1.12 Electronic bands of metals (a) before and (b) after electrical contact. The Fermi energies of the two metals align at equilibrium when electrical contact is made. E_{vac} vacuum energy; E_F, Fermi energy; subscripts L and R refer to the left-hand and right-hand metal, respectively

The presence of a dipole layer at the surface has other implications for clean and adsorbate covered surfaces. The difference between E_F and E_{vac} in the bulk is a material-specific property. In order to remove an electron from a metal the electron must pass through the surface and into the vacuum. Therefore if two samples of the same metal have different dipole layers at the surface they exhibit different work functions. In Figure 1.5 we have seen that Smoluchowski smoothing at steps introduces dipoles into the surface. A linear relationship between step density and work-function decrease has been observed [16]. Similarly, the geometric structure of the surface determines the details of the electronic structure at the surface. Thus, the work function is found to depend both on the crystallographic orientation of the surface and on the presence of surface reconstructions.

The presence of adsorbates on the surface of a solid can introduce two distinct dipolar contributions to the work function. The first occurs if there is a charge transfer between the adsorbate and the surface. When an electropositive adsorbate such as an alkali metal forms a chemical bond with a transition metal surface the alkali metal tends to donate charge density into the metal and decreases the work function. An electronegative adsorbate, such as oxygen, sulfur or halogens, withdraws charge and increases the work function. The second contribution arises when a molecular adsorbate has an intrinsic dipole. Whether this contribution increases or decreases the work function depends on the relative orientation of the molecular dipole with respect to the surface.

1.3.4 Energy levels at metal–semiconductor interfaces

The metal–semiconductor interface is of great technological importance, not only because of the role it plays in electronic devices [18]. Many of the concepts developed here are directly applicable to charge transfer at the electrolyte–semiconductor interface as well [19, 20]. It is somewhat more complicated than the metal–metal interface, but our understanding of it can be built up from the principles that we have learned above. The situation is illustrated in Figure 1.13. Again the Fermi levels of the metal and the semiconductor must be aligned at equilibrium. Equalization of E_F is accomplished by charge transfer. The direction of charge transfer depends on the relative work functions of the metal (M) and the semiconductor (S). The case of $\phi_S > \phi_M$ for an n-type semiconductor is depicted in Figure 1.13(a), whereas the case of $\phi_S < \phi_M$ is depicted in Figure 1.13(b). Screening in a semiconductor is much less effective, which results in a near-surface region of charge density different from that of the bulk with a width of several hundred angstroms. This is called a space-charge region. In Figure 1.13(a) the near-surface region has donated charge to the metal–semiconductor interface. Because the electron density in this region is lower than in the bulk, this type of space-charge region is know as a depletion layer. In Figure 1.13(b) charge transfer has occurred in the opposite direction. The enhanced charge density in the space-charge region corresponds to an accumulation layer. The shape of the space-charge region has a strong influence on carrier transport in semiconductors and on the electrical properties of the interface. Figure 1.13(a) corresponds to an ohmic contact whereas Figure 1.13(b) demonstrates the formation of a Schottky barrier.

In the construction of Figure 1.13 we have introduced the electron affinity of the semiconductor, χ_S. This quantity as well as the band gap remain constant throughout the

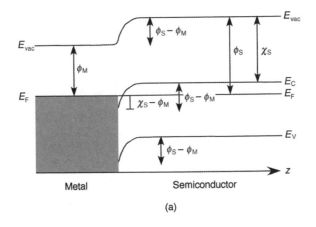

Metal   Semiconductor

(a)

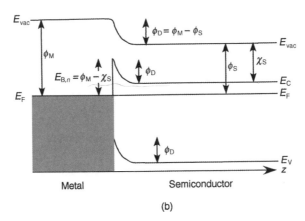

Metal   Semiconductor

(b)

Figure 1.13   Band bending in an *n*-type semiconductor at a heterojunction with a metal: (a) ohmic contact ($\phi_S > \phi_M$); (b) blocking contact (Schottky barrier, $\phi_S < \phi_M$). The energy of the bands is plotted as a function of distance z in a direction normal to the surface. ϕ_S, ϕ_M work function of the semiconductor and of the metal, respectively; E_{vac}, vacuum energy; E_C, energy of the conduction band minimum; E_F Fermi energy; E_V energy of the valence band maximum. Redrawn with permission from S. Elliott, *The Physics and Chemistry of Solids*, John Wiley, New York, 1998. ©1998 John Wiley & Sons Ltd

semiconductor. Importantly, the positions of the band edges at the surface remain constant. Thus we see that the differences $E_{vac} - E_V$ and $E_{vac} - E_C$ are constant, whereas the positions relative to E_F vary continuously throughout the space-charge region. The continuous change in E_V and E_C is called band bending.

E_{vac}, E_C and E_V all shift downward by the same amount in Figure 1.13(a). In Figure 1.13(b) the shifts are all upward, but again E_{vac}, E_C and E_V all shift by the same amount. The potential at the surface is the magnitude of the band bending and is given by

$$eV_{surf} = E_{vac}^{surf} - E_{vac}^{bulk} \qquad (1.15)$$

In Equation (1.15) E_{vac} was chosen but the shifts in E_{vac}, E_C and E_V are all the same, hence any one of these three could be used in Equation (1.15). The doping density determines the magnitude of the band bending and the depletion layer width, d. For an n-type semiconductor, the value is

$$V_{surf} = -\frac{eN_D d^2}{2\varepsilon\varepsilon_0} \tag{1.16}$$

where ε and ε_0 are the permittivities of the semiconductor and of free space, respectively. In a p-type semiconductor, the sign is reversed and N_A is substituted for N_D. From Figure 1.13 it is further apparent that

$$V_{surf} = \phi_S - \phi_M \tag{1.17}$$

for n-type ohmic contact, as in Figure 1.13(a), and

$$V_{surf} = \phi_M - \chi_S \tag{1.18}$$

for Schottky contact, as in Figure 1.13(b). V_{surf} in the case of Schottky contacts is also known as the Schottky barrier height.

Of great importance both for device applications and for electrochemistry is that the extent of band bending can be changed by application of an external bias. When a voltage V is applied across a semiconductor junction (either metal–semiconductor or electrolyte–semiconductor) it is not the chemical potential that is constant throughout the junction region but the electrochemical potential, η:

$$\eta = \mu - eV \tag{1.19}$$

as shown in Figure 1.14. In forward bias, electrons flow from the semiconductor to the metal and the barrier is reduced by eV. In reverse bias, little current flows from the metal into the semiconductor as the barrier height is increased by eV. The potential at which $\mu - eV = 0$ is known as the flatband potential.

1.3.5 Surface electronic states

All atoms in the bulk of a pure metal or elemental semiconductor are equivalent. The atoms at the surface are, by definition, different because they do not possess their full complement of bonding partners. Therefore surface atoms can be thought of as a type of impurity. Just as impurities in the bulk can have localized electronic states associated with them, so too can surface atoms. We need to distinguish two types of electronic states associated with surface atoms. An electronic state that is associated with the surface can either overlap in k space with bulk states or it can exist in a bandgap. An overlapping state is known as a surface resonance. A surface resonance exists primarily at the surface; nonetheless, it penetrates into the bulk and interacts strongly with the bulk electronic states. A true surface state is strongly localized at the surface and, because it exists in a bandgap, it does not interact strongly with bulk states. A surface state or resonance can be associated either with surface atoms of the solid (intrinsic surface state or resonance) or

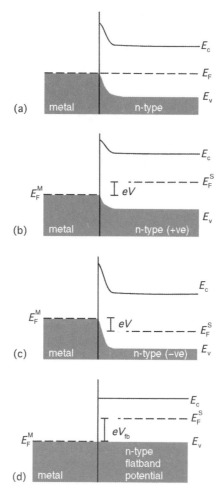

Figure 1.14 Electrochemical potential and the effects of an applied voltage on a metal–semiconductor interface. (a) No applied bias. (b) Forward bias. (c) Reverse bias. (d) Biased at the flatband potential. E_C energy of the conduction band minimum; E_F, Fermi energy; E_V, energy of the valence band maximum; superscript M and S refer to the metal and the semiconductor, respectively

with adsorbates (extrinsic surface state or resonance). Structural defects can also give rise to surface states and resonances. Surface states and surface resonances are illustrated in Figure 1.15.

Surface states play a defining role in determining the surface band structure of semiconductors. In effect, they can take the place of the metal, and the reasoning we used in Section 1.3.4 can be used to describe band bending in the presence of surface states. If the electron distribution in a semiconductor were uniform all the way to the surface, the bands would be flat. The presence of surface states means that the surface may possess a greater or lesser electron density relative to the bulk. This nonuniform electron distribution again leads to a space-charge region and band bending. Surface

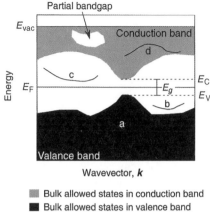

Figure 1.15 The band structure of a semiconductor, including an occupied surface resonance (a), an occupied surface state (b), a normally unoccupied surface state (c) and a normally unoccupied surface resonance (d). E_{vac} vacuum energy; E_C, energy of the conduction band minimum, E_F, Fermi energy; E_V, energy of the valence band maximum

states can either act as donor or as acceptor states. Surface states can have a strong influence on the electrical properties of devices, in particular on the behaviour of Schottky barriers [17, 18, 21].

The distinction between a resonance and a state may seem somewhat arbitrary. The difference is particularly obvious in normally unoccupied or empty states that exist above the Fermi level. Empty states can be populated by the absorption of photons with energies less than the work function. The strong interaction between bulk states and a resonance results in a short lifetime for electrons excited into the resonance. A surface state exhibits a much longer lifetime. These lifetimes have been measured directly by two-photon photoemission (see Section 2.2.3). Lifetimes depend on the specific system. For example, on Si(1 1 1)-(2 × 1) the π^* normally unoccupied surface state has a lifetime of c. 200 ps [22].

If in a metal an energy gap exists somewhere between E_F and E_{vac}, an electron excited into this gap experiences an attractive force associated with its image potential. The result is a series of bound states – image potential states [23, 24] – that are the surface analogue of Rydberg states in atomic and molecular systems. To a first approximation [24], the energies of these states form a series whose energy is given by

$$E_n = -\frac{R_\infty}{16n^2} = -\frac{0.85 \text{ eV}}{n^2} \tag{1.20}$$

converging on the energy of the vacuum level, E_{vac}. These states are bound in the z direction (normal to the surface) but are free electrons in the plane of the surface. As n increases, the wavefunction of the image states overlaps less and less with the bulk, leading to progressively less interaction between the two. The result is that the lifetime of the state increases with n. This has been measured directly by Höfer et al. [25] who have found that above Cu(1 0 0) the lifetime varies from 40 ± 6 fs to 300 ± 15 fs as n changes from 1 to 3. Similarly, the lifetime can be increased by the introduction of a spacer layer between the image state and the metal surface. Wolf, Knoesel and Hertel [26] have shown that the presence of a physisorbed xenon layer increases the $n = 1$ lifetime on Cu(1 1 1) from 10 ± 3 fs to 50 ± 10 fs.

1.4   The Vibrations of Solids

The vibrations of a solid are much more complex than the vibrations of small molecules. This arises from the many-body nature of the interactions of atoms in a solid. Analogous to electronic states, the vibrations of a solid depend not only on the movements of atoms but also on the direction in which the atoms vibrate. The multitudinous vibrations of the solid overlap to form a phonon band structure, which describes the energy of phonons as a function of a wavevector. Partial bandgaps can be found in the phonon band structure. Within the Debye model [17] there is a maximum frequency for the phonons of a solid: the Debye frequency. The Debye frequency is a measure of the rigidity of the lattice. Typical values of the Debye frequency are 14.3 meV (115 cm^{-1}) and 34.5 meV (278 cm^{-1}) for gold and tungsten, respectively, and 55.5 meV (448 cm^{-1}) for silicon. The most rigid lattice is that of diamond, which has a Debye frequency of 192 meV (1550 cm^{-1}).

From the familiar harmonic oscillator model, the energy of a vibrational normal mode in an isolated molecule is given by

$$E_v = (v + \tfrac{1}{2})\hbar\omega_0 \qquad (1.21)$$

where v is the vibrational quantum number and ω_0 is the fundamental radial frequency of the oscillator.

In a three-dimensional crystal with harmonic vibrations, this relationship needs to be modified in two important ways. First, we note that a crystal is composed of N primitive cells containing p atoms. Since each atom has three translational degrees of freedom, a total of $3pN$ vibrational degrees of freedom exist in the solid. The solutions of the wave equations for vibrational motion in a periodic solid can be decomposed in $3p$ branches. Three of these modes correspond to acoustic modes. The remaining $3(p-1)$ branches correspond to optical modes. Optical modes can be excited by the electric field of an electromagnetic wave if the excitation leads to a change in dipole moment. Acoustic and optical modes are further designated as being either transverse or longitudinal. Transverse modes represent vibrations in which the displacement is perpendicular to the direction of propagation. The vibrational displacement of longitudinal modes is parallel to the direction of propagation.

The second important modification arises from the three-dimensional structure of the solid. Whereas the vibrations of a molecule in free space do not depend on the direction of vibration, this is not the case for a vibration in an ordered lattice. This is encapsulated in the use of the wavevector of the vibration. The wavevector is given by

$$|\mathbf{k}| = \frac{2\pi}{\lambda} \tag{1.22}$$

where λ is the wavelength and $\mathbf{k}$ is a vector parallel to the direction of propagation. Consistent with this definition, we introduce the radial frequency of the vibration, ω_k,

$$\omega_k = 2\pi v_k \tag{1.23}$$

We now write the energy of a vibration of wavevector $\mathbf{k}$ in the pth branch as

$$E(\mathbf{k}, p) = [n(\mathbf{k}, p) + \tfrac{1}{2}]\hbar\omega_k(p) \tag{1.24}$$

The vibrational state of the solid is then represented by specifying the excitation numbers, $n(\mathbf{k}, p)$, for each of the $3pN$ normal modes. The total vibrational energy is thus a sum over all of the excited vibrational modes:

$$E = \sum_{\mathbf{k}, p} E(\mathbf{k}, p) \tag{1.25}$$

By direct analogy to the quantized electromagnetic field, it is conventional to describe the vibrations of a solid in terms of particle-like entities (phonons) that represent quantized elastic waves. Just as for diatomic molecules, Equations (1.24) and (1.25) show that the vibrational energy of a solid is nonvanishing at 0 K as a result of zero point motion.

The mean vibrational energy (see Exercise 1.5) is given by:

$$\langle E \rangle = \frac{1}{2}\hbar\omega_k(p) + \frac{\hbar\omega_k(p)}{\exp[\hbar\omega_k(p)/k_B T] - 1} \tag{1.26}$$

Thus, by comparison with Equation (1.24), it is confirmed that the mean phonon occupation number is

$$n(\mathbf{k}, p) \equiv \langle n(\omega_k(p), T) \rangle = \frac{1}{\exp(\hbar\omega_k(p)/k_B T) - 1} \tag{1.27}$$

and that it follows the Planck distribution law. This is expected for particles of zero (more generally, integer) spin. Accordingly, the number of phonons in any given state is unlimited and determined solely by the temperature.

Again in analogy to electronic states, there are phonon modes that are characteristic of and confined to the surface. Surface phonon modes have energies that are well defined in k space and sometimes exist in the partial bandgaps of the bulk phonons. Surface phonon modes that exist in bandgaps are true surface phonons, whereas those that overlap with bulk phonon in k space are surface phonon resonances. Furthermore, since the surface atoms are undercoordinated compared with the bulk, the surface Debye frequency is routinely much lower than the bulk Debye frequency. In particular, the vibrational amplitude perpendicular to the surface is much larger for surface than for bulk atoms.

Depending on the specific material, the root mean square vibrational amplitudes at the surface are commonly 1.4–2.6 times larger on surfaces than in the bulk.

1.5   Summary of Important Concepts

- Ideal flat surfaces are composed of regular arrays of atoms with an areal density of the order of 10^{15} cm^{-2} (10^{19} m^{-2}).

- Surfaces expose a variety of potential binding sites of different coordination numbers.

- Real surfaces will always have a number of defects (steps, kinks, missing atoms, etc.).

- Clean surfaces can exhibit relaxations or reconstructions.

  ○ Relaxations are slight changes in bond lengths and angles.

  ○ Reconstructions are changes in the periodicity of the surface compared to the bulk-terminated structure.

- Adsorbates can form ordered or random structures and may distribute themselves either homogeneously over the surface or in islands.

- Adsorbates can also lead to changes in the surface structure of the substrate, inducing either a lifting of the clean surface reconstruction or the formation of an entirely new surface reconstruction.

- The occupation of electronic states is defined by the Fermi–Dirac distribution [Equation (1.5)].

- Surface electronic states exist in a bulk bandgap.

- A surface resonance has a wavefunction that is concentrated at the surface but it does not exist in a bandgap and therefore interacts strongly with bulk states.

- Vibrations in solids are quantized and form bands analogous to the band structure of electronic states.

- The mean phonon occupation number follows the Planck distribution law [Equation (1.27)].

Exercises

1.1   The Fermi energies of caesium, silver and aluminium are 1.59, 5.49 eV and 11.7 eV, respectively. Calculate the density of the Fermi electron gas in each of these metals as well as the Fermi temperature. Calculate the difference between the chemical potential and the Fermi energy for each of these metals at their respective melting points.

1.2   Redraw Figure 1.13 for a p-type semiconductor [10, 11, 27].

1.3   Given that the partition function, q, is defined by

$$q = \sum_{i=1}^{\infty} \exp(-E_i/k_B T) \tag{1.28}$$

where E_i is the energy of the ith state. Use Equation (1.24) to show that the mean vibrational energy of a solid at equilibrium is given by Equation (1.26).

Hint: the mean energy is given by

$$\langle E \rangle = k_B T^2 \frac{\partial (\ln q)}{\partial T} \tag{1.29}$$

1.4   The Debye temperature is given by

$$\theta_D = \frac{\hbar \omega_D}{k_B} \tag{1.30}$$

and is more commonly tabulated and determined than is the Debye frequency because of its relationship to the thermodynamic properties of solids.

(a) Calculate the Debye frequencies of the elemental solids listed in Table 1.1 in Hz, meV and cm^{-1}.

(b) Calculate the mean phonon occupation number at room temperature for each of these materials at 100 K, 300 K and 1000 K.

1.5   The Debye model (see Table 1.1) can be used to calculate the mean square displacement of an oscillator in a solid. In the high-temperature limit this is given by

$$\langle u^2 \rangle = \frac{3 N \hbar^2 T}{m k_B \theta_D^2} \tag{1.31}$$

(a) Compare the root mean square displacements of platinum at 300 K to that at its melting point (2045 K). What is the fractional displacement of the metal atoms relative to the interatomic distance at the melting temperature?

(b) Compare this to the root mean square displacement of the carbon atoms at the surface of diamond at the same two temperatures.

1.6   The surface Debye temperature of Pt(1 0 0) is 110 K. Take the definition of melting to be when the fractional displacement relative to the lattice constant is equal to c. 25 percent (Lindemann criterion). What is the surface melting temperature of Pt(1 0 0)? What is the implication of a surface that melts at a lower temperature than the bulk?

1.7   The bulk terminated Si(1 0 0)-(1 × 1) surface has two dangling bonds per surface atom and is therefore unstable toward reconstruction. Approximate the dangling bonds as effectively being half-filled sp^3 orbitals. The driving force of reconstruction is the removal of dangling bonds.

(a) The stable room-temperature surface reconstructs into a (2 × 1) unit cell in which the surface atoms move closer to each other in one direction but the distance is not changed in the perpendicular direction. Discuss how the loss of one dangling bond on each silicon atom leads to the formation of a (2 × 1) unit cell.
Hint: the nearest neighbour surface silicon atoms are called dimers.

Table 1.1   Debye temperatures, θ_D, for selected elements; see Exercises 1.4 and 1.5

Element	Ag	Au	C (diamond)	C (graphite)	Pt	Si	W
θ_D (K) [17]	225	165	2230	760	240	645	400

   (b) This leaves one dangling bond per surface atom. Describe the nature of the interaction of these dangling bonds that leads to

     (i) symmetric dimers and

     (ii) tilted dimers.

   (c) Predict the effect of hydrogen adsorption on the symmetry of these two types of dimers [10, 11].

     Hint: consider first the types of bonds that sp^3 orbitals can make. Second, two equivalent dangling bonds represent two degenerate electronic states.

Further Reading

S. Elliott, *The Physics and Chemistry of Solids* (John Wiley, Chichester, Sussex, 1998).

C. Kittel, *Introduction to Solid State Physics* (John Wiley, New York, 1986).

W. A. Harrison, *Solid State Theory* (Dover Publications, New York, 1979).

H. Lüth, *Surfaces and Interfaces of Solid Materials* (Springer, Berlin, 1995).

S. R. Morrison, *Electrochemistry at Semiconductor and Oxidized Metal Electrodes* (Plenum Press, New York, 1980).

H. Over, "Crystallographic study of interaction between adspecies on metal surfaces," *Prog. Surf. Sci.* **58** (1998) 249.

M. W. Roberts and C.S. McKee, *Chemistry of the Metal–Gas Interface* (Clarendon Press, Oxford, 1978).

G. A. Somorjai, *Introduction to Surface Chemistry and Catalysis* (John Wiley, New York, 1994).

S. M. Sze, *Physics of Semiconductor Devices* (John Wiley, New York, 1981).

S. Titmuss, A. Wander and D. A. King, "Reconstruction of clean and adsorbate-covered metal surfaces," *Chem. Rev.* **96** (1996) 1291.

A. Zangwill, *Physics at Surfaces* (Cambridge University Press, Cambridge, 1988).

References

[1] H.-C. Jeong and E. D. Williams, *Surf. Sci. Rep.* **34** (1999) 171.

[2] R. Smoluchowski, *Phys. Rev.* **60** (1941) 661.

[3] V. A. Shchukin and D. Bimberg, *Rev. Mod. Phys.* **71** (1999) 1125.

[4] R. M. Watwe, R. D.Cortright, M. Mavirakis, J. K. Nørskov and J. A. Dumesic, *J. Chem. Phys.* **114** (2001) 4663.

[5] C. T. Campbell, *Annu. Rev. Phys. Chem.* **41** (1990) 775.

[6] A. Patrykiejew, S. Sokołowski and K. Binder, *Surf. Sci. Rep.* **37** (2000) 207.

[7] G. A. Somorjai, *Introduction to Surface Chemistry and Catalysis* (John Wiley, New York, 1994).

[8] G. A. Somorjai, *Annu. Rev. Phys. Chem.* **45** (1994) 721.

[9] G. Ertl, *Langmuir* **3** (1987) 4.

[10] K. W. Kolasinski, *Int. J. Mod. Phys. B* **9** (1995) 2753.

[11] R. Becker and R. Wolkow, "Semiconductor surfaces: silicon," in *Scanning Tunneling Microscopy*, eds J. A. Stroscio and W. J. Kaiser (Academic Press, Boston, MA, 1993) p. 149.

[12] Y. J. Chabal and K. Raghavachari, *Phys. Rev. Lett.* **54** (1985) 1055.

[13] J. J. Boland, *Adv. Phys.* **42** (1993) 129.

[14] J. E. Northrup, *Phys. Rev. B.* **44** (1991) 1419.

[15] R. Imbihl and G. Ertl, *Chem. Rev.* **95** (1995) 697.

[16] K. Besocke, B. Krahl-Urban and H. Wagner, *Surf. Sci.* **68** (1977) 39.

[17] S. Elliott, *The Physics and Chemistry of Solids* (John Wiley, Chichester, Sussex, 1998).

[18] S. M. Sze, *Physics of Semiconductor Devices*, 2nd edn (John Wiley, New York, 1981).

[19] S. R. Morrison, *Electrochemistry at Semiconductor and Oxidized Metal Electrodes* (Plenum, New York, 1980).

[20] H. Gerischer, *Electrochim. Acta* **35** (1990) 1677.

[21] H. Lüth, *Surfaces and Interfaces of Solid Materials*, 3rd edn (Springer, Berlin, 1995).

[22] N. J. Halas and J. Bokor, *Phys. Rev. Lett.* **62** (1989) 1679.

[23] R. Haight, *Surf. Sci. Rep.* **21** (1995) 275.

[24] W. Steinmann and T. Fauster, "Two-photon photoelectron spectroscopy of electronic states at metal surfaces", in *Laser Spectroscopy and Photochemistry on Metal Surfaces: Part I*, eds H.-L. Dai and W. Ho (World Scientific, Singapore, 1995) p. 184.

[25] U. Höfer, I. L. Shumay, C. Reuß, U. Thomann, W. Wallauer and T. Fauster, *Science* **277** (1997) 1480.

[26] M. Wolf, E. Knoesel and T. Hertel, *Phys. Rev. B* **54** (1996) 5295.

[27] K. Christmann, *Surf. Sci. Rep.* **9** (1988) 1.

2
Experimental Probes of Surface and Adsorbate Structure

In Chapter 1 we were introduced to the structural, electronic and vibrational properties of solids and their surfaces. In this chapter we investigate the techniques used to probe these properties. The emphasis here is to delve into the physical basis behind these techniques as well as the information that we can hope to gain from these techniques. We do not emphasize the instrumental side of these techniques. For more information on the instrumentation of surface science, the reader is referred to the books of Vickerman [1], Ertl and Küppers [2], Woodruff and Delchar [3], Feldman and Mayer [4] as well as the texts found in the Further Reading section at the end of this chapter. These books also introduce a number of surface-sensitive techniques that are not covered in this chapter. Experimental surface science is an instrumentally intensive experimental science and many examples of best practise and the tricks of the trade have been compiled by Yates [5].

2.1 Scanning Probe Techniques

Scanning tunnelling microscopy (STM) and the slew of scanning probe microscopy (SPM) techniques that arose in its wake represent a monumental breakthrough in surface science. In 1986 Binnig and Rohrer were awarded the Nobel Prize for the discovery of scanning tunnelling microscopy [6]. The scanning tunnelling microscope is a tool not only for detailed investigation of surface structure but also for the study of manipulations of atoms and molecules at surfaces [7, 8].

The basis of all scanning probe or proximal probe techniques is that a sharp tip is brought close to a surface. A measurement is then made of some property that depends on the distance between the tip and the surface. A variation on this theme is near-field scanning optical microscopy (NSOM), in which a small-diameter optical fibre is brought close to the surface. The diameter of the fibre is less than the wavelength of the light that is directed down the optical fibre. By working in the distance regime before the effects of diffraction have caused the light to diverge significantly (i.e. by working in the near-field regime), objects can be imaged with a resolution far below the wavelength of the light.

Numerous scanning probe techniques exist [9]. These techniques rely on the measurement of various quantities (e.g. current; van der Waals, chemical or magnetic forces;

capacitance; phonons or photons) when a probe is brought close to a surface and then scanned across it. We focus on three techniques that are widely used and illustrate the most salient aspects of scanning probe microscopy: scanning tunnelling microscopy (STM), atomic force microscopy (AFM) and near-field scanning optical microscopy (NSOM). All of these techniques are extremely versatile and can be operated under ultrahigh vacuum (UHV), at atmospheric pressure and even in solution [10].

2.1.1 Scanning tunnelling microscopy

The basic principle of scanning tunnelling microscopy is presented in Figure 2.1. A sharp conductive tip, usually a tungsten or platinum/iridium wire, is brought within a few nanometres or less of a conducting surface. A voltage difference is then applied between the tip and the surface. A measurement of the current at constant voltage (constant voltage imaging mode) or of the voltage at constant current (constant current imaging mode) is then made while the tip is scanned across the surface. As always, we set up a coordinate system in which the x and y axes lie in the plane of the surface and the z axis is directed away from the surface. Measurements of these kinds lead to the images found in Figures 2.2–2.5.

To interpret the STM images we need to understand the processes that control the flow of electrons between the tip and the surface. This is illustrated in Figure 2.6. Figure 2.6(a) represents a situation in which two metals are brought close to one another but are not connected electrically. The Fermi energies of the left-hand, E_F^L, and right-hand, E_F^R, metals have their characteristic values and, as given in Equation (1.2),

$$\phi = E_{vac} - E_F \qquad (2.1)$$

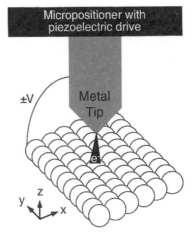

Figure 2.1 A scanning tunnelling microscope tip interacting with a surface; V, voltage; e⁻, electron flow

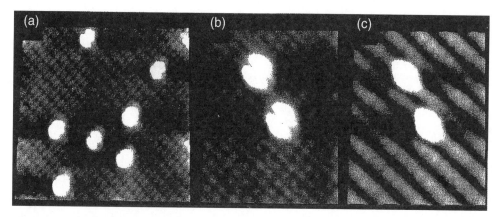

Figure 2.2   A scanning tunnelling microscope image of occupied states on a Si(1 0 0)-(2 × 1) surface nearly completely covered with adsorbed hydrogen atoms. The uncapped dangling bonds on the silicon (sites where hydrogen is not adsorbed) appear as lobes above the plane of the hydrogen-terminated sites. The rows of the (2 × 1) reconstruction are clearly visible in the hydrogen-terminated regions. Reproduced with permission from J. J. Boland, *Phys. Rev. Lett.* **65** (1990) 3325. ©1990 by the American Physical Society

The offset in the Fermi energies (contact potential) is equal to the difference in work functions of the two metals

$$\phi_c = E_F^L - E_F^R. \tag{2.2}$$

The solid line in the drawing represents the potential barrier that separates the electrons of the two metals.

In Figure 2.6(b) the metals are brought closer together and are connected electrically. The Fermi energies of the two metals line up and the vacuum level shifts. The potential between the two metals is no longer constant; therefore, an electric field exists in the vacuum between the metals. Since the two electrodes are both metals, all states up to E_F are full, and those above it are empty (ignoring thermal effects). Therefore, no current can flow between the two metals because electrons must always flow from occupied states to empty states; in Figure 2.6(b) the alignment of the Fermi levels means that the occupied states of the metals are also aligned.

Figures 2.6(c) and 2.6(d) demonstrate that not only can we turn on a current between the two metals but also we can control the direction of flow of current by adjusting the potential difference between the two. In Figure 2.6(c) the right-hand metal is biased positively with respect to the left-hand metal. This lowers E_F^R by an energy eV, where e is the elemental charge and V is the potential difference. The occupied states of the left-hand metal now lie at the same energy as some of the unoccupied states of the right-hand metal. Classically, no current can flow because of the barrier between the two. However, if the two metals are brought sufficiently close, the wavefunctions of the electrons in the two metals overlap and, by the laws of quantum mechanics, tunnelling of electrons can occur.

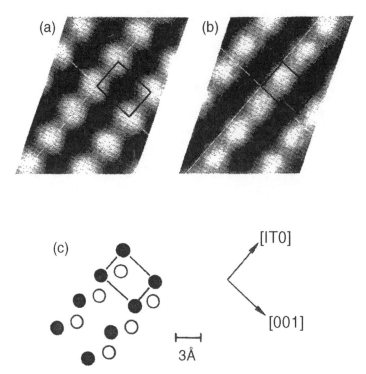

Figure 2.3 Constant-current scanning tunnelling microscopy images of the clean GaAs(1 0 0) surface: (a) the normally unoccupied states imaged at voltage $V = 1.9\,\text{V}$; (b) the normally occupied states imaged at $V = -1.9\,\text{V}$; (c) representation of the positions of the gallium ($\bullet$) and arsenic ($\bigcirc$) atoms. The rectangle is at the same position in parts (a), (b) and (c). This is an unusual example of chemically specific imaging based simply on the polarity of the tip. Reproduced with permission from R. M. Feenstra, J. A. Stroscio, J. Tersoff and A. P. Fein, *Phys. Rev. Lett.* **58** (1987) 1192. ©1987 by the American Physical Society

In a simple one-dimensional approximation [11], the tunnelling current I, then depends exponentially on the distance between the two metals, according to

$$I \propto \exp(-2\kappa d) \tag{2.3}$$

where d is the distance and where

$$\kappa^2 = \frac{2m}{\hbar^2}(eV_{\text{B}} - E) \tag{2.4}$$

in which $\hbar$ is Planck's constant divided by 2π, E is the energy of the state from which tunnelling occurs, and eV_{B} is the barrier height, which is approximately the vacuum level. κ is of the order of $1\,\text{Å}^{-1}$, hence a change in separation of just $1\,\text{Å}$ leads to an order of magnitude change in the tunnelling current. Typical tunnelling currents are of the order of nanoamps or less, and the extreme sensitivity of the tunnelling current on distance

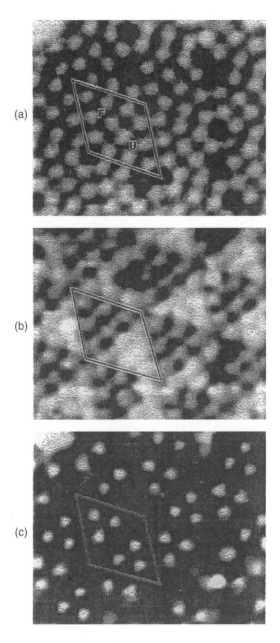

Figure 2.4 Constant-current scanning tunnelling microscopy (STM) images of the Si(1 1 1)-(7 × 7) surface (a) voltage $V = 2$ V; (b) $V = 1.45$ V; (c) $V = -1.45$ V. The quadrilateral is at the same position in parts (a), (b) and (c). Note how the apparent surface structure changes with voltage even though the surface atom positions do not change. This illustrates that STM images electronic states ("chosen" by the voltage) and not atoms directly. F and U refer to the faulted and unfaulted halves of the unit cell, which are easily distinguished in part (b). Reproducd with permission from R. J. Hamers, R. M. Tromp and J. E. Demuth, *Phys. Rev. Lett.* **56** (1986) 1972. ©1986 by the American Physical Society

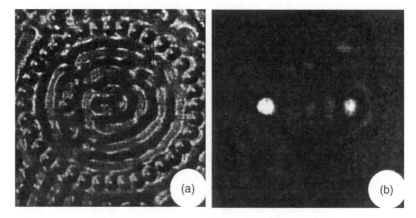

Figure 2.5 Visualization of the quantum mirage: (a) scanning tunnelling microscopy (STM) image showing cobalt atoms arranged into an ellipse on Cu(1 1 1); (b) mirage signal (under appropriate conditions a cobalt atom placed at one focus within the ellipse leads to a mirage signal, which appears as a virtual atom at the other focus). Reprinted with permission from *Nature* **403** (2000) 512, H. C. Manoharan, C. P. Lutz and D. M. Eigler. ©2000 Macmillan Magazines Limited

translates into subangstrom spatial resolution. In addition to the distance dependence, it is important to remember the requirement that electrons always flow from occupied to unoccupied states. Therefore, STM images always represent a convolution of the density of states of occupied and unoccupied electronic states between the tip and the surface. In other words, *STM does not image atoms, it images electronic states.* However, since the density of electronic states is correlated with the positions of the nuclei, STM images are always correlated with the positions of atoms. Nonetheless, great care must be taken in interpreting STM images, as one can see from the images in Figs. 2.2–2.5.

By comparing Figures 2.6(c) and 2.6(d), we see that the types of states that are imaged depend on the sign of the applied voltage. Electrons always flow towards a positive voltage and from occupied states to empty states. Therefore we can control the direction of current flow simply by setting the voltage appropriately. Let us assume that the right-hand metal is the tip and the left-hand metal in Figure 2.6 is the sample. Control of the voltage not only changes the direction of current flow but also changes whether the tip images the occupied or the unoccupied states of the sample.

STM images result from the convolution of tip and sample electronic structure as well as the distance between them. Unfortunately, the bumps in an image are not labelled with atomic symbols. For instance, an oxygen atom does not always look the same. It may appear as a protrusion at one voltage and as a depression at another. On one substrate it may be imaged at one voltage whereas a different voltage is required for a different substrate. The lack of chemical specificity in STM images represents one difficulty with the technique. Under special circumstances images can be associated with specific atoms, as in Figure 2.4. Functional group identification has also been possible in several instances [12, 13].

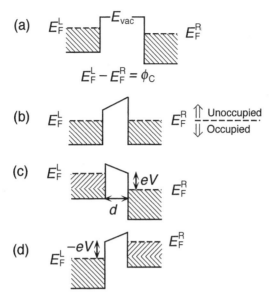

Figure 2.6   Fermi and vacuum levels for two metals separated by distance d: (a) isolated metals; (b) metals after electrical contact, in the absence of an applied bias; (c) metals after electrical contact with applied bias (biasing shifts the relative positions of the Fermi levels and makes available unoccupied states in an energy window eV into which electrons can tunnel); (d) metals after electrical contact with applied bias [the direction of tunnelling is switched compared with the case in part (c) simply by changing the sign of the applied bias]. E_F, Fermi energy; E_{vac}, vacuum energy; ϕ_c, contact potential; subscripts L and R refer to the left-hand and right-hand metal, respectively. Adapted with permission from J. Tersoff and N. D. Lang, "Theory of scanning tunnelling microscopy, in *Scanning Tunnelling Microscopy*, eds J. A. Stroscio and W. J. Kaiser (Academic Press, Boston, MA, 1993), page 1. ©1993 Academic Press

2.1.2   Scanning tunnelling spectroscopy

The scanning tunnelling microscope has the ability to deliver more information than just topography. Equations (2.3) and (2.4) show that the image depends on the voltage on the tip and not just the topography. Control of the voltage determines the electronic states from which tunnelling occurs. Therefore, STM can be used to measure the electronic structure with atomic resolution. This mode of data acquisition is known as scanning tunnelling spectroscopy (STS) [18].

The tunnelling current at bias voltage V can be shown [14] to depend on the surface density of states at energy E, $\rho_s(E)$, and the transmission probability through the barrier, $D(E)$, according to

$$I \propto \int_0^{eV} \rho_s(E)D(E)\mathrm{d}E \qquad (2.5)$$

The differential conductance is given by

$$\sigma = \frac{dI}{dV} \tag{2.6}$$

Both σ and the total conductivity I/V depend to first order on $D(E)$. Therefore, the normalized differential conductance depends only on the density of states and the tunnelling voltage:

$$\frac{dI}{dV}\left(\frac{I}{V}\right)^{-1} = \rho_s(eV)\left[\left(\frac{1}{eV}\right)\int_0^{eV} \rho_s(eV)dE\right]^{-1} \tag{2.7}$$

Simultaneous measurements of I, V and position on the surface are sometimes called current imaging tunnelling spectroscopy (CITS) and provide images of the surface as well as of the local electronic states at each position in the scan. This is illustrated in Figure 2.7.

Figure 2.7 demonstrates the ability of STS to perform atom-resolved spectroscopy. The electronic structure of a Si(1 1 1)-(7 × 7) surface depends on which atom is probed. Furthermore, as the right-hand side of Figure 2.7 shows, STS has the ability to identify how the chemisorption of an atom changes the electronic structure of the surface.

2.1.3   Atomic force microscopy

Many materials are not conductors and hence cannot be imaged by STM. Soft interfaces and biological molecules may be sensitive to electron bombardment, which would complicate STM analysis. These are some of the motivations to seek alternative approaches to SPM. These efforts were culminated in 1986 with the development of atomic force microscopy (AFM) by Binnig, Quate and Gerber [15]. AFM is also referred to as scanning force microscopy (SFM).

If a tip is brought near a surface, it experiences attractive or repulsive interactions with the surface. These interactions may be of the van der Waals type or of a chemical, magnetic or electrostatic nature, for example. These forces and their ranges are outlined in Table 2.1. The tip is attached to a nonrigid cantilever and the force experienced by the tip causes the cantilever to bend. The deflection of the cantilever can be detected by reflection of a laser beam off the cantilever. AFM can image conducting as well as insulating surfaces. In addition, it can harness a number of different forces to provide for the deflection of the cantilever – just as in STM a piezoelectric micropositioner and some type of electrical feedback mechanism for the micropositioner are integral parts of the microscope design.

As the piezoelectric drive moves a cantilever of the type shown in Figure 2.8 in free space, the motion of the tip on the cantilever exactly follows the motion of the drive. As the tip approaches a surface it experiences any one of a number of forces. The action of a force on the tip leads to a displacement of the tip in addition to the motion of the

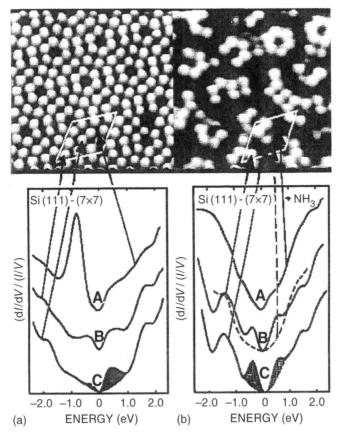

Figure 2.7 (a) Topography (top panel) and atom-resolved tunnelling spectra (bottom panel) of the unoccupied states of the clean Si(1 1 1)- (7 ×) surface; (b) topography (top panel) and atom-resolved tunnelling spectra (bottom panel) obtained after exposure of the Si(1 1 1)-(7 × 7) surface to NH_3. The curves represent spectra acquired over different sites in the reconstructed surface. Curve A, restatoms; curve B corner adatoms; curve C, middle adatoms. Negative energies correspond to occupied states; positive energies correspond to empty states. The different sites exhibit different reactivities with respect to NH_3 adsorption, with the resatoms being the most reactive and the middle adatoms being the least reactive. Reproduced with permission from J. A. Stroscio and R. M. Feenstra, "Methods of tunnelling spectroscopy", in *Scanning Tunnelling Microscopy*, eds J. A. Stroscio and W. J. Kaiser (Academic Press, Boston, MA, 1993), page 96. ©1993 Academic Press

piezoelectric drive, Δz. If the force constant, k_N, of the cantilever is known, the displacement can be translated into a force according to

$$F_N = k_N \Delta z. \tag{2.8}$$

The force curve is obtained by a measurement of the force as a function of the distance of the tip from the surface. The sensitivity of AFM is of the order of $10^{-13} - 10^{-8}$ N, which allows for direct measurements of even van der Waals interactions.

Table 2.1   Interaction forces appropriate to scanning force microscopy and their ranges

Force	Range (nm)
Electrostatic	100
Double layer in electrolyte	100
van der Waals	10
Surface-induced solvent ordering	5
Hydrogen bonding	0.2
Contact	0.1

Source: H. Takano, J. R. Kenseth, S.-S. Wong, J. C. O'Brien and M. D. Porter, *Chem. Rev.* **99** (1999) 2845.

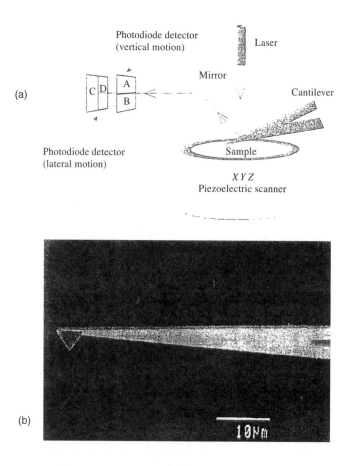

Figure 2.8   (a) Principal components of an optical lever type atomic force microscope. Detection of the reflected laser beam with a quadrant, position-sensitive photodiode facilitates the simultaneous detection of bending and torsion of the cantilever. (b) A scanning electron micrograph of the cantilever and tip of a typical atomic force microscope. Reprinted with permission from H. Takano, J. R. Kenseth, S.-S. Wong, J. C. O'Brien and M. D. Porter, *Chem. Rev.* **99** (1999) 2845. ©1999 American Chemical Society

There are several modes in which images can be acquired.

- Contact mode: the tip is advanced toward the surface until physical contact is made. The feedback circuitry is then set to maintain a constant imaging parameter. The imaging parameter is often taken to be the force, which can be measured according to Equation (2.8) by the deflection of a laser beam off the back of the cantilever. The cantilever is scanned in two dimensions and the feedback is simultaneously measured to produce a topographic image of the surface.

- Friction force mode: a lateral force applied to a tip in contact with the surface causes a twisting of the cantilever. The torsional force measured as the cantilever is scanned across the surface can be related to the frictional force between the tip and the surface.

- Tapping mode: the piezoelectric drive can be used to shake the cantilever at a frequency resonant with one of its fundamental oscillations. The tip is brought close enough to the surface such that they touch at the bottom of each oscillation. Changes either in the oscillation amplitude or in the phase can be measured as the tip is scanned across the surface to produce a topographic image.

- Noncontact mode: as in tapping mode, the piezo drives the cantilever into resonant or near-resonant oscillations. The amplitude, frequency and phase of the oscillations are then measured. This mode is particularly interesting for conducting tips, which can be biased, and for magnetic tips. This transduction mode can then be used to image magnetic and electrical forces across the surface.

Many of the forces listed in Table 2.1 are quite long range. Consequently, they tend to fall off much more slowly with distance than does tunnelling current. Therefore, the sensitivity of AFM is much less than STM. As a result, the resolution of AFM is generally less than STM, in the order of several nanometres or tens of nanometres. In exceptional cases, atomic resolution has been achieved [16, 17].

Chemical forces can be probed with AFM as well. For these purposes a tip of known chemical composition is required. Most cantilevers are made of Si_3N_4. Bare Si_3N_4 tips are unsuitable for chemical force sensing because they are easily contaminated. However, several techniques can be used to transform the tip into one with a known chemical composition. The tip can be coated with gold onto which molecules or colloidal particles can be transferred. Alkanethiol molecules and the techniques of self-assembly, treated in Chapter 6, can be used to transfer molecules terminated with a variety of functional groups to the tip. Attachment of silica spheres can also be accomplished. Silica is a versatile substrate onto which a variety of organic and biomolecules can be tethered. Takano et al. [18] have reviewed the extensive literature on tip modification.

Lieber and co-workers [19] have pursued an extremely promising variant of tip modification. They have grown carbon nanotubes on a cantilever through a catalytic process. A nanotube tip provides two advantages for AFM. The first is the known chemical composition of the tip. The second is that carbon nanotubes have high aspect ratios. This can allow for increased resolution, particularly in cases in which deep narrow crevices are present on the surface. Conventional AFM tips have a radius of 30–50 nm; a nanotube tip with a radius of only 9 nm expresses much lower convolution effects.

Once a tip has been modified in a controlled manner, the interactions of the probe molecule with surfaces or molecules immobilized on a surface can be studied. This opens up a huge range of possibilities, particularly in biochemical and biomedical applications. Since the forces measured with a chemically modified tip can be chemically specific, imaging under the influence of chemical forces can be used for compositional mapping of the substrate. By grafting one protein onto the tip and another onto a substrate is it possible to study in great detail the interaction of single binding pairs. Similarly DNA and antigen–antibody interactions can be studied [18].

2.1.4 Near-field scanning optical microscopy

Optical microscopy has a long history of providing insight into chemical, physical and biological phenomena. There is as well a familiarity and comfort in optical data as it is a human response to feel as though we understand something when we can see it. Hence wide-ranging efforts have pushed the boundaries of optical microscopy to ever-greater resolution. Conventional far-field microscopy is limited to a resolution of roughly $\lambda/2$, where λ is the wavelength of the light used for imaging. This resolution limit can be exceeded by over an order of magnitude by using so-called near-field techniques [20]. Furthermore, the use of pulsed light sources and the relative ease of performing studies as a function of wavelength mean that optical microscopic techniques lend themselves readily to time-resolved and spectrally resolved studies that can investigate dynamical behaviour and provide chemical identification.

In a series of papers written between 1928 and 1932, Synge [21–23] proposed how an aperture smaller than the wavelength of light could be used to obtain resolution far smaller than the wavelength. When light passes through an aperture it is diffracted and the light propagates away from the aperture in a highly divergent fashion. Not only does the light diverge, but also diffraction leads to an undulating transverse intensity distribution. In the far field, a distance large compared with the wavelength of light and the size of the aperture, diffraction leads to the familiar Airy disk pattern in intensity. In the near-field region, not only is the light divergent but also the intensity distribution is evolving, as shown in Figure 2.9. By working close enough to the aperture, resolution close to the size of the aperture and far below the wavelength of the light can be obtained.

The proposals of Synge were far too technically demanding to be experimentally realized in the 1930s. It was not until 1972 that the feasibility of subwavelength resolution with microwave radiation was demonstrated [24]. The advent of a device based on this effect capable of routine imaging, however, had to await several other discoveries. The widespread availability of high-quality optical fibres, visible lasers and AFM greatly aided the development of practical instruments for near-field scanning optical microscopy [NSOM, also known as scanning near-field optical microscopy (SNOM)]. Development followed from advances made by Pohl and co-workers at IBM Zürich [25] and by Lewis and co-workers at Cornell University [26].

Figure 2.10 demonstrates a variety of operating modes that can be employed to perform NSOM experiments. There are several features that most NSOM instruments have in common. They require a high-brightness light source. A laser is particularly well suited for this because of its monochromaticity, low divergence and favourable polarization

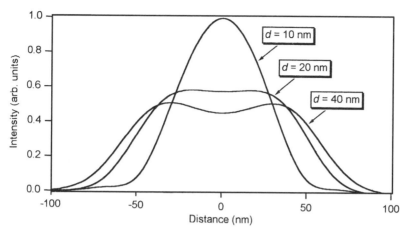

Figure 2.9 Near-field intensity distributions (arbitrary units) are shown for 532 nm light that has passed through an aperture with a diameter of 100 nm at distances d from the aperture of 10 nm, 20 nm and 40 nm. The distributions are normalized such that they all have the same integrated intensity

characteristics. Lasers can in addition be operated either with continuous-wave or pulsed radiation. The properties of laser light are particularly well suited for coupling into an optical fibre. The optical fibre plays two roles in NSOM. First, the end of an optical fibre can readily be tapered to minuscule dimensions. After tapering, the fibre is then coated with a metallic film, generally either aluminium or silver. In this way apertures with diameters of $c.$ 100 nm can be made routinely. The optical fibre also acts as a light pipe that transports the photons from the laser to the aperture. Finally, the fibre can be easily incorporated into the head of an atomic force microscope, the head being used to regulate the aperture-to-sample separation. Depending on the instrument, either the fibre or the substrate can be scanned to allow for two-dimensional imaging. The instrumentation required to scan the substrate and to obtain NSOM images is similar to that used in AFM. Indeed, often the microscope is constructed such that NSOM and AFM images are collected simultaneously.

There are several principles that can be used to collect NSOM data.

- Illumination mode: the tip is used to illuminate the sample. A transparent substrate is used so that photons can be collected below the sample. The light may either be transmitted photons from the tip, or fluorescence excited by the tip illumination.

- Collection mode: the fibre is used to gather photons and transport them to the detector. The sample is illuminated from the far field either from below or from the side.

- Reflection mode: the sample is irradiated through the optical fibre and the photons reflected by the sample are collected in the far field. This is useful for opaque samples. By using polarization tricks or wavelength separation the aperture used for illumination can also be used to collect photons.

- Photon tunnelling microscopy: total internal reflection can be used to pass light through a transparent sample. When the light reflects from the interface, an evanescent

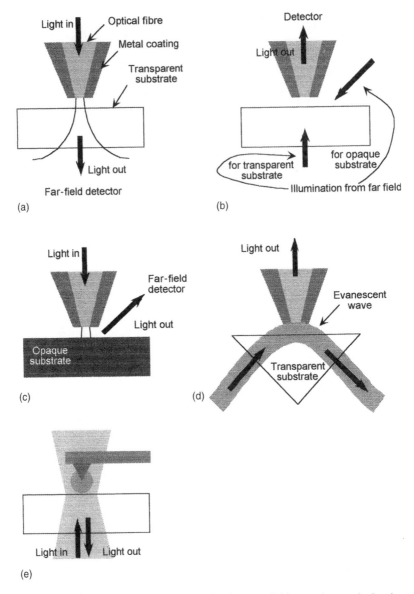

Figure 2.10 Examples of data collection modes in near-field scanning optical microscopy: (a) illumination; (b) collection; (c) reflection; (d) photon tunnelling; (e) apertureless

wave penetrates up to several hundred nanometres beyond the interface. If an aperture is brought into contact with the evanescent wave it propagates into the fibre and can then be transported to the detector. This technique is sometimes called scanning tunnelling optical microscopy (STOM) or photon scanning tunnelling microscopy (PSTM).

- Apertureless near-field microscopy: if an atomic force microscope tip is brought close to a surface and is irradiated by a far-field light source, a field enhancement occurs in a localized region between the tip and the sample. This can be used to image the surface if sophisticated filtering techniques are used to extract the signal from the large background of conventionally scattered radiation. Alternatively, a fluorescent molecule attached to the end of the tip can be excited by a laser, and its fluorescence can be used to probe the sample.

Because of the small aperture diameter, typically 80–100 nm, the optical fibre delivers only $10^{-4} - 10^{-7}$ of the light coupled into it onto the sample. This corresponds to a few tens of nanowatts of light for typical input powers. Although this may appear to be minute, it still amounts to $c.$ 100 W cm^{-2}. The ultimate resolution of NSOM is roughly 12 nm. In practice, however, a resolution of 50 nm is obtained. A number of alternative implementations of NSOM are being investigated that may be able to improve the resolution significantly [27].

Because NSOM images photons, it allows not only for nanoscale imaging but also for spectroscopic studies at high resolution without physical contact and without electron bombardment. Thus NSOM has the potential to measure the position, orientation and chemical identity of adsorbates not only on solid but also on liquid surfaces. Betzig and Chichester demonstrated single-molecule detection in 1993 [28]. With pulsed lasers, extremely high temporal resolution can be attained as well. This facilitates dynamical studies of molecular motion such as rotational reorientation and diffusion. The measurement of the time decay of fluorescence provides direct access to excited-state lifetimes.

2.2 Low-Energy Electron Diffraction

The most standard method of surface structure determination is provided by the diffraction of low-energy electrons by means of the apparatus shown in Figure 2.11 [2]. In 1927 Davidson and Germer [29] observed the angular distributions of electrons scattered from nickel and explained their data in terms of the diffraction of the electrons from crystallites in their sample. This was one of the first experimental demonstrations of the wavelike behaviour of particles as predicted by de Broglie in 1921

$$\lambda = \frac{h}{m_e v} \tag{2.9}$$

in which wavelength λ is inversely proportional to the linear momentum (a product of mass m_e and velocity v).

Electron scattering is surface sensitive if the electrons have the right energy. This is illustrated by the data in Figure 2.12. The inelastic mean free path of electrons in a solid depends on the energy of the electrons in a manner that does not depend too strongly on the chemical identity of the solid. Therefore, the curve in Figure 2.12 is known as the universal curve. For incident energies between roughly 20 eV and 500 eV, electrons interact strongly with matter and their mean free path is $c.$ 5–10 Å. Low-energy electron diffraction (LEED) measures only elastically scattered electrons, therefore, most of the electrons detected in LEED are scattered from the surface of the sample and layers deeper

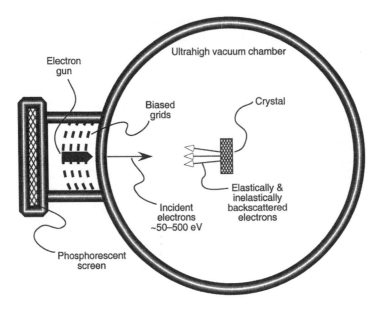

Figure 2.11   A low-energy electron diffraction chamber

than three or four atoms below the surface play virtually no role in the detected signal. To a first approximation, the patterns observed in LEED can be understood by considering only the symmetry of the surface of the sample.

The same property of electron scattering that makes it surface sensitive also makes diffraction of low-energy electrons at a surface more complicated than X-ray scattering. Low-energy electrons interact strongly with matter. This means that they have a short inelastic mean free path. This is good for surface sensitivity since no more than 3–4 layers contribute to the scattering. However, a complete description of both positions and intensities of diffraction spots requires a dynamical theory that accounts for multiple scattering from all the layers that contribute to scattering. Development of this full theory continues to be an active area of research. We do not treat it here, but more can be found

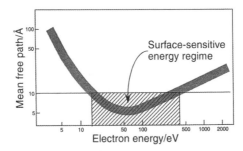

Figure 2.12   The universal curve of electron mean free path, λ, in solid matter. The shaded area approximates the bounds on measured values of λ. Data source: G. A. Somorjai, *Science* **227** (1985) 902

elsewhere [2, 31]. A simpler kinematic theory, analogous to that used in X-ray scattering, is of limited use for quantitative LEED studies. Instead, we concentrate on a simple geometric theory.

The success of the geometric theory stems from the following facts. The diffraction spot positions are determined by the space lattice (i.e. by the size and shape of the unit cell). The intensities are determined by the diffraction function (i.e. by the exact atomic coordinates within the unit cell). Further, the spot positions do not depend on the penetration depth of the electrons whereas the intensities do.

In presenting the geometric formalism of LEED theory, we follow the notation of Ertl and Küppers [2]. The basis vectors $\mathbf{a}_1$ and $\mathbf{a}_2$ describe the unit cell in real space. These vectors define the smallest parallelogram from which the structure of the surface can be reconstructed by simple translations. A reciprocal space representation of the real space lattice is described by basis vectors $\mathbf{a}_1^*$ and $\mathbf{a}_2^*$. The real and reciprocal space lattices are represented by

$$\mathbf{a}_i \cdot \mathbf{a}_j^* = \delta_{ij} \tag{2.10}$$

where $i, j = 1$ or 2, and δ_{ij} is the Kronecker delta function, where $\delta_{ij} = 0$ if $i \neq j$ and $\delta_{ij} = 1$ if $i = j$. This means that $\mathbf{a}_i^* \perp \mathbf{a}_j$ for $i \neq j$. Introducing γ and γ^*, the angles between ($\mathbf{a}_1$ and $\mathbf{a}_2$) and ($\mathbf{a}_1^*$ and $\mathbf{a}_2^*$), respectively, we have

$$\mathbf{a}_1^* = \frac{1}{\mathbf{a}_1 \sin \gamma} \tag{2.11}$$

$$\mathbf{a}_2^* = \frac{1}{\mathbf{a}_2 \sin \gamma} \tag{2.12}$$

$$\sin \gamma = \sin \gamma^* \tag{2.13}$$

The inverse relationship between real and reciprocal space means that a long vector in real space corresponds to a short vector in reciprocal space.

The need for the reciprocal space description is made evident by Figure 2.13. This figure shows that an image of the diffracted electrons corresponds to a reciprocal space image of the lattice from which the electrons diffracted. Hence, by uncovering the relationship between a reciprocal space image and the real space lattice, we can use LEED patterns to investigate the surface structure.

Advanced topic: LEED structure determination

We represent a surface overlayer by the basis vectors $\mathbf{b}_1$ and $\mathbf{b}_2$. The substrate and overlayer lattices are related by

$$\mathbf{b} = \mathbf{M} \cdot \mathbf{a} = \begin{pmatrix} m_{11} & m_{12} \\ m_{21} & m_{22} \end{pmatrix} \cdot \mathbf{a} \tag{2.14}$$

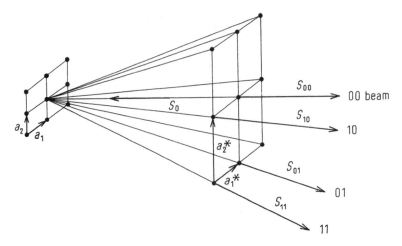

Figure 2.13 The principal of diffraction pattern formation in a low-energy electron diffraction experiment. The incident electron beam approaches along S_0 (s_0 in text). The specular beam exits along S_{00}. Reproduced with permission from G. Ertl and J. Küppers, *Low Energy Electrons and Surface Energy* 2nd edition (VCH, Weinheim, 1985). ©1985 John Wiley & Sons Limited

A similar relationship holds for the reciprocal space representation:

$$\mathbf{b^*} = \mathbf{M^*} \cdot \mathbf{a^*} = \begin{pmatrix} m_{11}^* & m_{12}^* \\ m_{21}^* & m_{22}^* \end{pmatrix} \cdot \mathbf{a^*} \tag{2.15}$$

Thus the basis vectors are related by

$$\mathbf{b}_1 = m_{11}\mathbf{a}_1 + m_{12}\mathbf{a}_2 \tag{2.16}$$

$$\mathbf{b}_2 = m_{21}\mathbf{a}_1 + m_{22}\mathbf{a}_2 \tag{2.17}$$

$$\mathbf{b}_1^* = m_{11}^*\mathbf{a}_1^* + m_{12}^*\mathbf{a}_2^* \tag{2.18}$$

$$\mathbf{b}_2^* = m_{21}^*\mathbf{a}_1^* + m_{22}^*\mathbf{a}_2^* \tag{2.19}$$

It can be shown that $\mathbf{M^*}$ is the inverse transpose of $\mathbf{M}$, which yields

$$m_{11} = \frac{1}{\det \mathbf{M^*}} m_{22}^* \tag{2.20}$$

$$m_{12} = \frac{1}{\det \mathbf{M^*}} m_{21}^* \tag{2.21}$$

$$m_{21} = \frac{1}{\det \mathbf{M^*}} m_{12}^* \tag{2.22}$$

$$m_{22} = \frac{1}{\det \mathbf{M^*}} m_{11}^* \tag{2.23}$$

The determinant of $\mathbf{M^*}$ is defined as

$$\det \mathbf{M^*} = m_{11}^* m_{22}^* - m_{21}^* m_{12}^* \tag{2.24}$$

Experimentally, the challenge is to determine the elements of **M** from the diffraction pattern measured on the LEED screen.

The diffraction condition from a one-dimensional lattice of periodicity a leads to constructive interference at angles φ when

$$a \sin \varphi = n\lambda \qquad (2.25)$$

for an electron with wavelength λ incident at normal incidence; n is an integer denoting the diffraction order. The wavelength of the electron is given by the de Broglie relationship [Equation (2.9)]. The Bragg condition of Equation (2.25) needs to be generalized to two dimensions. This leads to the Laue conditions

$$\mathbf{a}_1 \cdot (\mathbf{s} - \mathbf{s}_0) = h_1 \lambda \qquad (2.26)$$
$$\mathbf{a}_2 \cdot (\mathbf{s} - \mathbf{s}_0) = h_2 \lambda \qquad (2.27)$$

where $\mathbf{s}_0$ defines the direction of the incident beam (generally along the surface normal) and $\mathbf{s}$ defines the direction of the diffracted beam intensity maxima; h_1 and h_2 are integers and they are used to identify the diffraction reflexes that appear in the LEED pattern. The specular reflex at (0, 0) is used as the origin and arises from electrons that are elastically scattered without diffraction.

The determination of a real space structure from a reciprocal space image may seem rather esoteric. The relationship between the two is shown in Figure 2.14. By way of an example of pattern analysis, we will see that simple geometric theory can lead to a rapid determination of surface structure as well as other properties of the adsorbed layer.

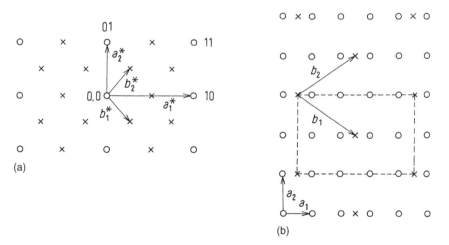

Figure 2.14 Real space and reciprocal space patterns: (a) reciprocal lattice (low-energy electron diffraction pattern) composed of substrate (normal) spots (○) and overlayer (extra) spots (×); (b) real lattice of the substrate (○) and overlayer (×). The dashed line delineates the $c(4 \times 2)$ unit cell. Reproduced with permission from G. Ertl and J. Küppers, *Low Energy Electrons and Surface Energy* 2nd edition (VCH, Weinheim, 1985). ©1985 John Wiley & Sons Limited

First, we note that the position of the (0, 0) spot does not change with electron energy (λ). This leads to easy identification of the specular reflex, as it is the only one that does not move when the electron energy is changed. There is no need to determine the absolute distance between spots. By referencing the positions of overlayer related diffraction spots to those of the substrate we can strip the analysis of a variety of experimental parameters.

There are five types of two-dimensional Bravais lattices from which all ordered surface structures can be built. These are shown in Figure 2.15. Two obvious notations emerge to describe the relative symmetry of surface layers. The most general notation uses the full matrix **M**. This notation was proposed by Park and Madden [32]. If the angle between $\mathbf{a}_1$ and $\mathbf{a}_2$ is the same as the angle between $\mathbf{b}_1$ and $\mathbf{b}_2$, a much simpler form of notation can be used [cf. Si(1 0 0)-(2 × 1)]. We express the overlayer structure in terms of $n = |\mathbf{b}_1/\mathbf{a}_1|$, $m = |\mathbf{b}_2/\mathbf{a}_2|$, and any angle of rotation, θ, between the two lattices in the form $(n \times m)R\theta°$. If $\theta = 0°$, it is excluded from the notation. In addition, a letter 'c' must be added if the overlayer corresponds to the centred rectangular lattice of Figure 2.15(c). Those who are sticklers for detail sometimes add the letter 'p' for primitive, but it is usually omitted. Most observed overlayer structures can be expressed in these terms, known as Wood's notation [33]. Several examples of ordered overlayers and the corresponding Wood's notation are given in Figure 2.16. Notice that n and m are proportional to the length of the vectors that define the parallelogram of the unit cell. Therefore, the product nm is proportional to the area of the unit cell; the area of a (2 × 2) unit cell is twice as large as that of a (2 × 1) unit cell and quadruple that of a (1 × 1) unit cell.

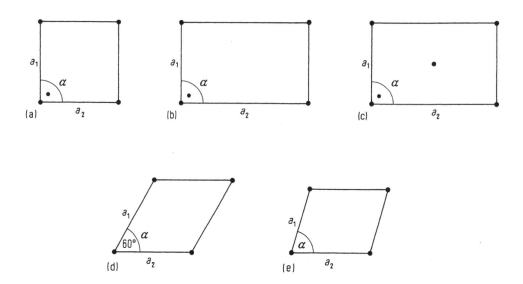

Figure 2.15   The five types of Bravais lattices: (a) square, $a_1 = a_2$, $\alpha = 90°$; (b) primitive rectangular, $a_1 \neq a_2$, $\alpha = 90°$; (c) centred rectangular, $a_1 \neq a_2$, $\alpha = 90°$; (d) hexagonal, $a_1 = a_2$, $\alpha = 60°$; (e) oblique, $a_1 \neq a_2$, $\alpha = 90°$. Reproduced with permission from G. Ertl and J. Küppers, *Low Energy Electrons and Surface Energy* 2nd edition (VCH, Weinheim, 1985). ©1985 John Wiley & Sons Limited

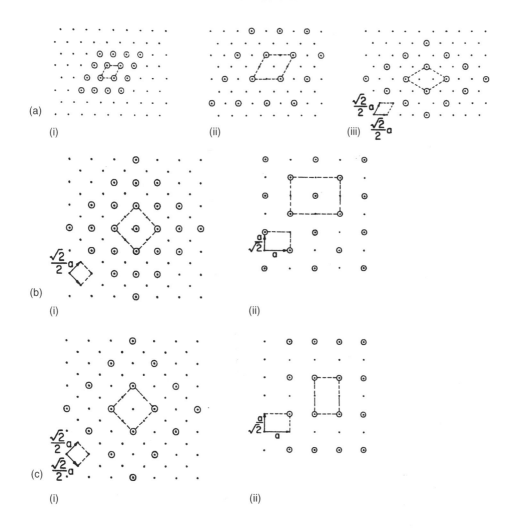

Figure 2.16   Some commonly observed adsorbate structures on low-index face-centred cubic (fcc) planes (a) fcc(1 1 1), (i) (1 × 2), (ii) (2 × 2), (iii) ($\sqrt{3}$ × $\sqrt{3}$); (b) fcc(1 0 0), (i) (2 × 2), (ii) c(2 × 2); (c) fcc(1 1 0), (i) (2 × 2), (ii) c(2 × 2). Reproduced with permission from G. A. Somorjai, *Introduction to Surface Chemistry and Catalysis* (John Wiley, New York, 1994). ©1994 John Wiley & Sons Limited

If all of the elements of **M** are integers, the overlayer forms a simple structure. A simple structure is commensurate with the substrate and all of the adsorbates occupy identical adsorption sites. If the elements are rational numbers, the overlayer forms a coincidence lattice, which is an incommensurate structure. If the elements of **M** are irrational, no common periodicity exists between the overlayer and substrate lattices. This structure type is known as an incoherent structure. The overlayer is also incommensurate with the substrate. With increasingly large values of the elements of **M**, the distinction between coincidence lattices and incoherent structures is lost.

A number of other factors affect the appearance of the diffraction pattern, namely the size of the spots and the contrast between the reflex maxima and the surrounding background. Spots are observed in the LEED pattern only if the surface is ordered in two dimensions. Streaking or the broadening of spots in one direction is indicative of the loss of order in one dimension. A disordered region results in a diffuse background.

The width of a reflex is related to the monochromaticity of the electron beam and the size of the ordered region. The width of a reflex can be expressed in terms of the angular divergence of the beam, $\delta\varphi$. It can be shown [2] that in the absence of any instrumental broadening

$$\delta\varphi = \frac{\lambda}{2d\cos\varphi} \tag{2.28}$$

where d is the diameter of the ordered region (e.g. an island). Thus, the larger the ordered region, the sharper the spot in the LEED pattern.

A spread in the electron wavelength affects the pattern in two ways. Equation (2.28) shows that $\delta\varphi$ is directly proportional to any uncertainty in λ. In addition, the diffraction pattern represents a reciprocal space image of the surface over the coherence length of the electron beam. If the electron gun were a monochromatic point source, the coherence length would be the size of the incident electron beam. Since it is not, the radiation is not completely in phase over the full extent of the beam. The amount that the radiation can be out of phase and still give coherent scattering puts a limit on the coherence length. For conventional electron beams the coherence length is ≤ 10 nm. In other words, LEED is not sensitive to disorder on length scales greater than 10 nm.

Increasing surface temperature leads to an increase in the vibrational motion of the surface layer, which results in increased diffuse scattering (increased background) and a decrease in the intensity at the reflex maximum. Similar to the case of X-ray scattering, the decrease in intensity can be described by an effective Debye–Waller factor [34]. The intensity drops exponentially with temperature according to

$$I = I_0 \exp(-2M) \tag{2.29}$$

where

$$2M = \frac{12h^2}{mk_B}\left(\frac{\cos\varphi}{\lambda}\right)\frac{T_s}{\theta_D^2} \tag{2.30}$$

In Equation (2.30), k_B is the Boltzmann constant, θ_D is the surface Debye temperature (assuming that scattering occurs only from the surface layer) and T_s is the surface temperature. Since the inelastic mean free path changes with electron energy, the value of θ_D changes with electron energy, converging on the bulk value for high energies. If the temperature becomes high enough, an overlayer of adsorbates can become delocalized as a result of diffusion, that is, it can enter a two-dimensional gas phase. Such a transition from an ordered phase to a delocalized phase would be evident in the LEED pattern as it would gradually transform from a sharp pattern to a diffuse pattern.

2.3 Electron Spectroscopy

We will study three types of electron spectroscopy in which an excitation leads to the ejection of an electron from the solid. The energy and possibly angular distribution of the ejected electrons are analysed. The three most prolific forms of electron spectroscopy are illustrated in Figure 2.17. X-rays can be used to excite photoemission. This technique is called X-ray photoelectron spectroscopy (XPS) or, especially in the older and analytical literature, electron spectroscopy for chemical analysis (ESCA). Ultraviolet light, especially that of rare-gas discharge lamps, can induce photoemission as well: ultraviolet photoemission spectroscopy (UPS). Auger electron spectroscopy (AES) involves excitation either with X-rays or, more usually, with electrons of similar energy. All of these techniques as well as LEED and other aspects of the interactions of low-energy electrons with surfaces are treated in the classic text by Ertl and Küppers [2].

2.3.1 X-ray photoelectron spectroscopy

Deep-core electrons have binding energies corresponding to the energies of photons that lie in the X-ray region. When a solid absorbs a photon with an energy in excess of the binding energy of an electron, a photoelectron is emitted and the kinetic energy of the photoelectron is related to the energy of the photon. Deep-core electrons do not participate in bonding, and their energies are characteristic of the atom from which they originate. To a first approximation, the energy of core electrons does not depend on the environment of the atom. Therefore, XPS is particularly useful for elemental analysis of a sample. Not only can XPS identify the composition of a sample but also it can be used to determine the composition *quantitatively*. The modern application of XPS owes much of its development to Siegbahn and co-workers in Uppsala, Sweden. Kai Siegbahn, whose father Manne became a Nobel laureate in 1924, was awarded the 1981 Nobel Prize for his contributions.

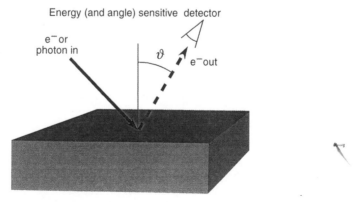

Figure 2.17 The light or electron path in X-ray photoelectron spectroscopy, ultraviolet photoemission spectroscopy and Auger electron spectroscopy

When the energies of core levels are investigated in more detail it is found that small but easily detected shifts do occur. These shifts, known as chemical shifts, depend on the bonding environment around the atom and in particular on the oxidation state of the atom. To understand chemical shifts we first need to understand the relationship between the photon energy, hv (where v is the frequency), and the electron binding energy, E_B, which is given by the Einstein equation:

$$E_B = hv - E_K \tag{2.31}$$

where E_K is the electron kinetic energy and the binding energy is referenced to the vacuum level. Because reproducibly clean surfaces of gold are easily prepared and maintained, the photoemission peaks of gold are conventionally used as a standard to calibrate the energy scale.

The binding energy of an electron is equal to the energy difference between the initial and final states of the atom. That is, the binding energy is equal to the difference in energies between the atom with n electrons and the ion with $n - 1$ electrons:

$$E_B(k) = E_f(n - 1) - E_i(n) \tag{2.32}$$

where $E_f(n - 1)$ is the final state energy, and $E_i(n)$ is the initial state energy. If there were no rearrangement of all of the spectator electrons, the binding energy would equal the negative of the orbital energy of the initial state of the electron, $-\varepsilon_k$. This approximation is known as Koopmans' theorem:

$$E_B(k) = -\varepsilon_k \tag{2.33}$$

in which the binding energy is referenced to the vacuum level.

Koopmans' theorem is based on an effective one-electron approximation in which the initial-state wavefunction, $\psi_i(n)$, can be described by a product of the single-particle wavefunction of the electron to be removed, φ_j, and the $n - 1$ electron wavefunction of the resulting ion, $\Psi_j(n - 1)$,

$$\psi_i(n) = \varphi_j \Psi_j(n - 1) \tag{2.34}$$

Within this picture the final-state wavefunction of the ion is simply (for the one-electron approximation):

$$\psi_f(n - 1) = \Psi_j(n - 1) \tag{2.35}$$

Consequently, the energies required for Equation (2.32) are

$$E_i(n) = \langle \psi_i(n) | \mathscr{H} | \psi_i(n) \rangle \tag{2.36}$$

and

$$E_f(n - 1) = \langle \Psi_j(n - 1) | \mathscr{H} | \Psi_j(n - 1) \rangle \tag{2.37}$$

Substitution of Equations (2.36) and (2.37) into Equation (2.32) demonstrates that only one peak is expected in the photoelectron spectrum.

The electrons are not frozen. The final state achieved by removing one electron corresponds to an ionic state in which a hole exists in place of the ejected photoelectron. This does not correspond to the ground state of the final ionic state. The remaining electrons relax and thereby lower the energy of the final state. This type of relaxation occurs regardless of the phase (gaseous, liquid or solid) in which the atom exists. If the atom is located in a solid, then in addition to this atomic relaxation there can be extra-atomic relaxation. In other words, not only the electrons in the atom can relax but also the electron on neighbouring atoms can relax in response to the ionization event.

These considerations lead us to reconsider our description of the final-state wavefunction. We now write a final-state wavefunction in terms of the eigenstates of the ion:

$$\psi_f(n-1) = u_k \Phi_{jl}(n-1) \tag{2.38}$$

where u_k is the wavefunction of the excited electron with momentum k, and Φ_{jl} are the wavefunctions of the l ionic states that have a hole in the jth orbital. The relaxation of the ion core in the final state means that the wavefunction $\Psi_j(n-1)$ is not an eigenstate of the ion. Therefore, there is not a unique $\Psi_j(n-1)$ that corresponds to $\Phi_{jl}(n-1)$. Instead, the $\Psi_j(n-1)$ must be projected onto the true eigenstates of the ion. This results in the formation of one or more possible final-state wavefunctions such that the final-state energy required for Equation (2.32) is

$$E_{fl}(n-1) = \langle \Phi_{jl}(n-1) | \mathcal{H} | \Phi_{jl}(n-1) \rangle \tag{2.39}$$

which has between 1 and l solutions. That there is more than one solution to Equation (2.39) means that more than one peak appears in the spectrum.

The one-electron or Hartree–Fock picture of Equation (2.33) also neglects relativistic and electron correlation effects. Both of these tend to increase the electron binding energy; hence, a more accurate expression of the binding energy is

$$E_B(k) = -\varepsilon_k - \delta\varepsilon_{relax} + \delta\varepsilon_{rel} + \delta\varepsilon_{corr} \tag{2.40}$$

where δ_{relax}, δ_{rel}, and δ_{corr} correspond to the correction factors for relaxation, relativistic effect, and electron correlation, respectively.

Recall that the energy of an electronic state is determined not only by the electronic configuration but also by angular momentum coupling. In the 'LS' coupling scheme, the total angular momentum is given by the vector sum $j = l + s$. Any state with an orbital angular momentum $l > 0$ and one unpaired electron is split into two states, a doublet, corresponding to $j = l \pm \frac{1}{2}$. More complex multiplet splittings in the initial state arise from states with higher total spins. Two conventions are encountered for the naming of electronic states. The first is the familiar convention for electronic states nL_j (n = principle quantum number, L = s, p, d, f, ... corresponding to $l = 0, 1, 2, 3, ...$). In X-ray spectroscopy, a nomenclature based on the shell is commonly used. The shell derives its name from n according to K, L, M, N, ... for $n = 1, 2, 3, 4, ...$ A subscript further designates the subshell. The number starts at 1 for the lowest (l, j) state and continues in unit steps up to the highest (l, j) state. Thus the shell corresponding to $l = 2$ has levels L_1, L_2 and L_3 corresponding to 2s, $2p_{\frac{1}{2}}$ and $2p_{\frac{3}{2}}$.

From the above discussion it is clear that initial-state and final-state effects influence the binding energy. Initial-state effects are caused by chemical bonding, which influences the electronic configuration in and around the atom. Thus, the energetic shift caused by initial state effects is known as a chemical shift, ΔE_b. To a first approximation, all core levels in an atom shift to the same extent. For most samples it is a good approximation to assume that the chemical shift is completely a result of initial-state effects and that, in particular, the relaxation energy does not depend on the chemical environment. This is obviously an approximation but it is a good general rule.

The chemical shift depends on the oxidation state of the atom. Generally, as the oxidation state increases the binding energy increases. The greater the electron-withdrawing power of the substituents bound to an atom, the higher the binding energy. This can be understood on the basis of simple electrostatics. The first ionization energy of an atom is always lower than the second ionization energy. Similarly, the higher the effective positive charge on the atom, the higher the binding energy of the photoelectron.

Most of the atomic relaxation results from rearrangement of outer-shell electrons. Inner-shell electrons of higher binding energy make a small contribution. The nature of a material's conductivity determines the nature of extra-atomic relaxation. In a conducting material such as a metal, valence electrons are free to move from one atom to a neighbour to screen the hole created by photoionization. In an insulator, the electrons do not possess such mobility. They react by being polarized by the core hole. Hence, the magnitude of the extra-atomic relaxation in metals (as much as 5–10 eV) is greater than that of insulators.

Several other final-state effects result in what are known as satellite features. Satellite features arise from multiplet splitting, shake-up events and vibrational fine structure. Multiplet splittings result from spin–spin interactions when unpaired electrons are present in the outer shells of the atom. The unpaired electron remaining in the ionized orbital interacts with any other unpaired electrons in the atom or molecule. The energy of the states formed depends on whether the spins are aligned parallel or antiparallel to one another. In a shake-up event, the outgoing photoelectron excites a valence electron to a previously unoccupied state. This unoccupied state may be either a discrete electron state, such as a π^* or σ^* state, or, especially in metals, is better thought of as electron–hole pair formation. By energy conservation, the photoelectron must give up some of its kinetic energy in order to excite the shakeup transition; hence, shake-up features always lie on the high-binding-energy (lower-kinetic-energy) side of a direct photoemission transition. Occasionally, the valence electron is excited above the vacuum level. Such a double ionization event is known as shake-off. Shake-off events generally do not exhibit distinct peaks whereas shake-up transitions involving excitation to discrete states do. An excellent discussion of final-state effects can be found in Shirley [35].

The width of photoemission peaks is determined to a large extent by the lifetime of the core hole (homogeneous broadening) and instrumental resolution. However, sample inhomogeneity and satellite features can also lead to peak broadening, especially if the instrumental resolution is insufficient to resolve the latter from the main photoemission feature. The instrumental resolution is determined not only by the resolution of the energy analyser but also by the wavelength spread in the incident X-ray beam. Through the

Heisenberg uncertainty relation, the intrinsic peak width, Γ, is inversely related to the core hole lifetime, τ, by

$$\Gamma = \frac{h}{\tau} \tag{2.41}$$

where h is the Planck constant. The lifetime generally decreases the deeper the core hole, because core holes are filled by higher lying electrons and the deeper the core hole the more de-excitation channels there are that can fill the core hole. Analogously, for a given energy level, the lifetime decreases (the linewidth increases) as the atomic number increases. Intrinsic lifetime broadening is an example of homogeneous broadening and has a Lorenzian line shape.

2.3.1.1 Quantitative Analysis

One of the great strengths of XPS is that it can be used not only for elemental analysis but also for quantitative analysis. XPS peak areas are proportional to the amount of material present because the photoionization cross-section of core levels is virtually independent of the chemical environment surrounding the element. However, since the inelastic mean free path of electrons is limited, the detection volume is limited to a region near the surface. The distribution of the analyte in the near surface region also affects the measured intensity.

The area under an XPS peak must be measured after a suitable background subtraction has been performed. In order to relate the peak area to the atomic concentration, we need to consider the region that is sampled. X-rays interact weakly with matter and penetrate several micrometres into the bulk. The photoelectrons, however, have a comparatively short inelastic mean free path, as can be seen in Figure 2.10. In addition, it depends on the kinetic energy of the electron ejected from the atom A and the material through which it passes; hence, we denote the mean free path by $\lambda_M(E_A)$.

The measured intensity is compared with that of a known standard, and a ratioing procedure is used to eliminate the necessity to evaluate a number of instrumental factors. It must be borne in mind that the best standard is one that has exactly the same concentration and matrix – in short, one that is identical to the sample. This is, of course, impractical. Therefore, the proper choice of the standard is important and the approximations introduced by using one must be considered. Appropriate standards are the signal from a known coverage, as determined for example by LEED measurements, or that from a pure bulk sample. One should try to match both the concentration and the spatial distribution of the standard as closely as possible to those of the sample to be analysed. Here, we consider only the most common case. For other distributions, the reader is referred to Ertl and Küppers [2].

Adsorbate A occupies only surface sites with a coverage θ_A The coverage θ_A represents the covered fraction of B, the substrate. The uncovered portion is then $(1 - \theta_A)$. The signal from A is not attenuated by electron scattering, hence

$$I_A = \theta_A I_A^0 \tag{2.42}$$

where I_A^0 is the signal from a known or standard coverage of A. Equation (2.42) can be used to determine the absolute coverage at an arbitrary coverage if the absolute coverage is known for I_A^0. If not then Equation (2.42) can only be used to determine a relative coverage compared with the coverage at I_A^0. The absolute coverage can be found at any arbitrary coverage if we use a ratio of XPS signals. The signal from B given by the sum of the bare surface contribution and that of the covered portion is

$$I_B = I_B^0 \left[1 - \theta_A + \theta_A \exp\left(-\frac{a_A \cos \vartheta}{\lambda_A} \right) \right] \tag{2.43}$$

where I_B^0 is the signal from a clean B substrate. Therefore,

$$\frac{I_A I_B^0}{I_B I_A^0} = \frac{\theta_A}{1 - \theta_A + \theta_A \exp\left(-\dfrac{a_A \cos \vartheta}{\lambda_A} \right)}. \tag{2.44}$$

2.3.2   Ultraviolet photoelectron spectroscopy

Ultraviolet photons can excite photoemission from valence levels. In contrast to XPS, ultraviolet photoelectron spectroscopy is difficult to use quantitatively. However, since valence electrons are involved in chemical bonding, UPS is particularly well-suited to the study of bonding at surfaces. UPS readily provides measurements of the work function and band structure of the solid, the surface and adsorbed layers.

UPS is a relatively new technique. This is largely for instrumental reasons. Eastman and Cashion [36] were the first to integrate a differentially pumped helium lamp with a UHV chamber. A helium lamp provides light at either 40.82 or 21.21 eV. Atomic emission lines are extremely sharp, which allows for a high theoretical resolution. The resolution is, in practice, limited by the energy resolution of the electron spectrometer. Other light sources covering the range of roughly 10–100 eV have found use, particularly synchrotrons and the laser-based technique of high harmonic generation [37, 38]. Lasers have brought about the advent of multiphoton photoelectron spectroscopy [39–41].

Much of what we have learned above regarding XPS is applicable to UPS as well. Koopmans's theorem is again the starting point, in which the first approximation is that the negative of the orbital energy is equated to the measured binding energy. The binding energy, also called the ionization potential I_p, is the difference of the initial-state and final-state energies. Relaxation effects are again important but tend to be smaller than those found in XPS. UPS can be performed with such high resolution that small shifts and satellites are easily detected. These satellites can in principle include vibrational structure. Normal resolution is insufficient for most vibrational structure but high-resolution studies have observed such structure [42]. A further difficulty with observing vibrational structure arises from the natural linewidth of adsorbed molecules. Chemisorbed species are strongly coupled to the substrate. The adsorbate–substrate interaction leads to excited state lifetimes that are generally on the order of 10^{-15} s. This translates into a natural linewidth of c. 1 eV, which is much too broad to observe vibrational transitions. Physisorbed species are less strongly coupled, and this has allowed for the observation of vibrational structure in exceptional cases. High-resolution studies can yield information not only on electron and hole lifetimes but also on electron–phonon interactions and defect scattering [43].

Whereas X-ray photoelectron (XP) spectra are conventionally referenced to the vacuum level, ultraviolet photoemission (UP) spectra are commonly referenced to the Fermi energy. Thus a binding energy of $E_B = 0$ corresponds to E_F in UPS. The spectrometer work function, φ_{sp}, must be accounted for when interpreting the measured electron kinetic energies, E_K, as shown in Figure 2.18:

$$E_K = h\nu - E_B - \phi_{sp} \tag{2.45}$$

The work function of the sample and thereby the absolute energy scale for a UP spectrum is easily determined. For excitation with a fixed photon energy, $h\nu$, the maximum kinetic energy of photoelectrons is given by

$$E_{K,max} = h\nu - \phi_{sp} \tag{2.46}$$

The minimum kinetic energy is given by

$$E_{K,min} = \phi - \phi_{sp} \tag{2.47}$$

The width of the spectrum,

$$\Delta E = E_{K,max} - E_{K,min} = h\nu - \phi \tag{2.48}$$

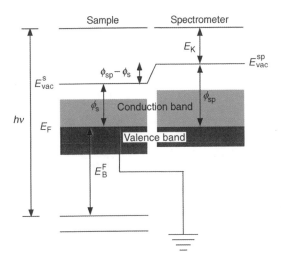

Figure 2.18 The influence of the spectrometer work function, ϕ_{sp}, on photoelectron spectra. ϕ_s, work function of the sample; E_{vac}^{sp}, E_{vac}^{s}, vacuum energy of the spectrometer and of the sample, respectively; E_F, Fermi energy; E_K, electron kinetic energy; E_B^F, binding energy; h, Planck constant; ν, frequency of incident energy. Adapted with permission from J. C. Vickerman, *Surface Analysis: The Principal Techniques* (John Wiley, Chichester, Sussex, 1997). ©1997 John Wiley & Sons

can thus be used to determine the work function of the sample:

$$\phi = hv - \Delta E \tag{2.49}$$

These relations are used to fix the energy scale, as shown in Figure 2.19. Figure 2.19 indicates the additional possibility of a signal arising from secondary electrons emanating from the spectrometer. Applying a voltage to the sample easily discriminates against the signal from 'secondaries'.

2.3.2.1 Angle-resolved ultraviolet photoemission

UPS can be used to interrogate the band structure of the solid as well as adsorbate levels. First, we note that the photon momentum is inconsequential in the photon energy range (10–100 eV) used in UPS. Therefore, the electron momentum is unchanged by photon absorption; that is, $k_i \approx k_f$. Equivalently, we state that only Franck–Condon (vertical) transitions are observed in UPS.

To obtain a measure of the band structure of a solid we need a more in-depth understanding of the photoemission process [2, 44]. The transition rate R between the initial state ψ_i and the final state ψ_f caused by a perturbation $\mathcal{H}'$ is given by Fermi's golden rule:

$$R = \frac{2\pi}{\hbar} |\langle \psi_f | \mathcal{H}' | \psi_i \rangle|^2 \delta(E_f - E_i - hv) \tag{2.50}$$

where $\delta(E_f - E_i - hv)$ is the Dirac delta function, equal to unity when $E_f - E_i - hv = 0$, and zero elsewhere. The perturbation represents the interaction of the electromagnetic

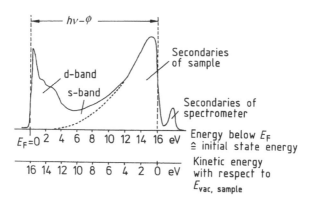

Figure 2.19   A representative ultraviolet photoemission spectrum. The relative intensities of primary and secondary electrons depend on instrumental factors. $E_{vac,sample}$, vacuum energy of the sample; E_F, Fermi energy; h, Planck constant; v, frequency of incident energy; ϕ, work function of the sample. Reproduced with permission from G. Ertl and J. Küppers, *Low Energy Electrons and Surface Energy* 2nd edition (VCH, Weinheim, 1985). ©1985 John Wiley & Sons Limited

field of the photon with the atom or molecule to be excited. As electric dipole transitions are by far the most likely to be observed, we write

$$\mathscr{H}' = \frac{e}{2mc}(\mathbf{A} \cdot \mathbf{P} + \mathbf{P} \cdot \mathbf{A}) \tag{2.51}$$

where $\mathbf{P}$ is the momentum operator $-i\hbar\nabla$ and $\mathbf{A}$ is the vector potential of the photon's electric field.

Measuring the relationship between electron energy and momentum reveals the band structure of a solid. The photoelectron kinetic energy E_K is related to the parallel and perpendicular components of momentum, $k_\parallel$ and $k_\perp$, respectively, by

$$E_K = \frac{\hbar^2}{2m}(k_\parallel^2 + k_\perp^2) \tag{2.52}$$

The momentum of the photoelectron measured in vacuum bears a direct relationship to the momentum of the electrons in the solid [44]. In order to determine the components of the momentum, we need to measure the angular distribution of photoelectrons. This is the basis of angle-resolved ultraviolet photoemission spectroscopy (ARUPS).

The differential cross-section with respect to the angle from the normal Ω must be determined. This can be written (see [44] and Exercise 2.12)

$$\frac{d\sigma}{d\Omega} \propto |\langle \psi_f | \mathbf{P} | \psi_i \rangle \cdot \mathbf{A}_0|^2 \delta(E_f - E_i - h\nu) \tag{2.53}$$

Again, we write the initial and final wavefunctions in a one-electron picture

$$\psi_i(n) = \phi_j \Psi_j(n-1) \tag{2.54}$$

$$\psi_f(n-1) = u_k \Phi_{jl}(n-1) \tag{2.55}$$

Accordingly, the differential photoionization cross-section can be written as

$$\frac{d\sigma}{d\Omega} \propto |\langle u_k | \mathbf{A} \cdot \mathbf{P} | \phi_i \rangle \langle \Phi_{jl} | \psi_i \rangle|^2 \tag{2.56}$$

Similar to what we discussed above for XPS, satellite features are expected as a result of relaxation effects in additions to the potential for vibrational fine structure. The presence of the product $\mathbf{A} \cdot \mathbf{P}$ means that the differential cross-section depends on the relative orientation between the polarization of the ultraviolet light and the transition dipole of the species to be excited. This property can be used to identify transitions and the adsorption geometry of molecular adsorbates.

A detailed interpretation of UP spectra is beyond the scope of this book but can be found in the references in the Further Reading. Figure 2.20 shows the ability of UPS to identify adsorbate electronic states and the effects of the adsorbate on the substrate. The most complete understanding of angle-resolved photoemission can only be achieved with comparisons to calculations.

UPS is surface sensitive because the low energy of the photoelectrons leads to a short inelastic mean free path. It can be used to map out the band structure of solids as well as

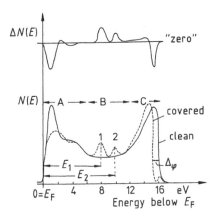

Figure 2.20 Changes in ultraviolet photoemission spectra (lower panel) on adsorption. The upper panel displays the difference spectrum. E_F, Fermi energy. $N(E)_{covered}$, $N(E)_{clear}$, count rate of photoelectrons from the adsorbate-covered and clean surfaces, respectively; $\Delta N(E) = N(E)_{covered} - N(E)_{clean}$. Reproduced with permission from G. Ertl and J. Küppers, *Low Energy Electrons and Surface Energy* 2nd edition (VCH, Weinheim, 1985). ©1985 John Wiley & Sons Limited

the electronic structure of adsorbates and the effect of adsorption on the substrate's band structure. Chemisorption involves an exchange of electrons between the adsorbate and substrate. This leads to a change in work function, which is readily detected by the change in the width of the UP spectrum. Difference spectra, the difference between spectra of adsorbate-covered and clean surfaces, clearly show the changes in the substrate as well as adsorbate induced features.

Because of the strong coupling between adsorbate and substrate, the ionization energies measured in UPS are shifted compared with those observed in the gas phase. This is illustrated in Figure 2.20 and embodied in the equation

$$E_{ad} = E_{gas} - (\phi + \Delta\phi) + E_{relax} + E_{bond\ shift} \tag{2.57}$$

where E_{ad} and E_{gas} are the ionization energies measured in the adsorbed and gas phases, respectively, E_{relax} accounts for final-state relaxation processes and $E_{bond\ shift}$ is an initial-state shift brought about by the adsorbate–substrate interaction. The remaining term accounts for changes in the work function; however, the exact form of this term remains controversial, as discussed elsewhere [2].

For chemisorbed species, $E_{bond\ shift}$ is large. The orbital energies are pinned to the Fermi level of the substrate. The magnitude of $E_{bond\ shift}$ and E_{relax} depends on the specific orbital, and not all orbitals shift to the same extent. This can lead to overlaps in UP spectra from adsorbed species that do not occur in gas-phase spectra.

Physisorbed species exhibit $E_{bond\ shift} \approx 0$ as expected from the weak coupling with the substrate. The orbitals are pinned to the vacuum level. Küppers, Wandelt and Ertl [45] exploited this by using physisorbed xenon photoemission to demonstrate that the work function is defined locally not globally. For instance, steps and defects affect the surface dipole layer in their vicinity and therefore exhibit a different work function from terraces.

Similarly, a chemically heterogeneous surface, such as a bimetallic surface, can exhibit distinct photoemission peaks associated with xenon physisorption on the two distinct metal sites on the surface.

Advanced topic: multiphoton photoemission

An extremely potent extension of photoelectron spectroscopy, especially for dynamical studies, is multiphoton photoemission (MPPE) [46]. The most common version is two-photon photoemission (2PPE). Multiphoton photoemission is the surface analogue of multiphoton ionization (MPI), which has been used to great advantage in gas phase spectroscopy. MPPE can be performed either with or without resonance enhancement. Resonance enhancement occurs when one (or more) photon(s) excites (excite) an electron into a real but normally unoccupied electronic state. Subsequently, further absorption of one (or more) photon(s) ionizes (ionize) the electron. The sum of the photon energies has to be greater than the work function. Extremely high signal to noise can be achieved by using photons that individually do not have sufficient energy to cause photoemission. This can lead to essentially background-free photoemission spectra. This is demonstrated in Figure 2.21. The photoelectron signal is resonance enhanced owing to the transient population of a normally unoccupied surface state. Figure 2.21 also demonstrates the power of polarization control to identify the nature of a transition. For instance, the diamond surface state is composed principally of sp^3 orbitals, which are aligned along the surface normal. Such normally unoccupied states are difficult to study by other spectroscopic techniques, though inverse photoemission has been profitably employed in this cause as well [47].

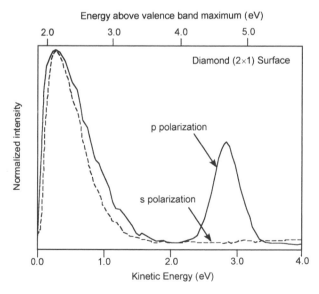

Figure 2.21   Two-photon photoemission of the reconstructed C(1 1 1)-(2 × 1) diamond surface. Owing to selection rules the normally unoccupied surface state observed at *c.* 3 eV is observed only with *p*-polarization. Reproduced with permission from G. D. Kubiak and K. W. Kolasinski, *Phys. Rev. B* **39** (1989) 1381. ©1989 by the American Physical Society

Multiphoton events are an example of nonlinear spectroscopy; that is, spectroscopy in which the signal is proportional to the light intensity raised to a power greater than one. Multiphoton transitions are extremely weak and in general they are excited only by high-intensity light sources such as lasers. The use of lasers allows for high spectral and temporal resolution. High temporal resolution is required to study electron dynamics [48]. In this way, the lifetimes of surface states and image potential states have been directly measured [38] and the electron–electron and electron–phonon scattering processes that lead to thermalization of excited electrons have been probed [49]. Two-photon photoemission has also given us a direct glimpse of the motion of desorbing caesium atoms as they leave the surface [50].

2.3.3   Auger electron spectroscopy

Another 'workhorse' technique of surface science, in particular because it can be easily incorporated into the same apparatus that is used for LEED, is Auger electron spectroscopy (AES). An Auger transition can be excited by photons, electrons or even by ion bombardment and is depicted in detail in Figure 2.22. The phenomenon was first described by Auger in 1925 [51]. Conventionally, electron bombardment at energies in the 3–5 keV range is used for surface analytical AES. Auger transitions can, however, also be simultaneously observed during XPS measurements. Like XPS, AES can be used for quantitative elemental analysis, and the measured signal is influenced by the inelastic mean free path of electrons, the atomic concentration and the concentration distribution.

Our first approximation to the photoemission process involved in XPS assumed a one-electron picture. Auger spectroscopy, in contrast, relies on the coupling between electrons. As shown in Figure 2.22, after ionization of a core level, a higher lying (core or valence) electron can fill the resulting hole in a radiationless transition. This process leaves the atom in an excited state. The excited-state energy is removed by the ejection of a second electron. Inspection of Figure 2.22 makes clear that a convenient method of labelling Auger transitions is to use the X-ray notation of the levels involved. Thus the

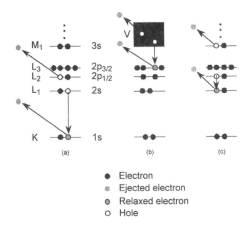

Figure 2.22   A detailed description of Auger transitions involving (a) three core levels, (b) a core level and the valence band, (c) a Coster–Kronig transition in which the initial hole is filled from the same shell

transition in Figure 2.22(a) is a KL_1L_2 transition, and that in Figure 2.22(b) is a L_3VV transition in which the V stands for a valence band electron. In some cases, more than one final state is possible and the final state is added to the notation to distinguish between these possibilities – for example, $KL_1L_3(^3P_1)$ as opposed to $KL_1L_3(^3P_2)$. Coster–Kronig transitions [Figure 2.22(c)] have particularly high rates and are therefore extremely important for determining the relative intensities of Auger transitions.

Auger spectroscopy can be used to detect virtually any element, apart from hydrogen and helium. With increasing atomic number (Z), however, the probability of radiative relaxation of the core hole, (i.e. of X-ray fluorescence) increases. Furthermore, since the incident electron energy is often only 3 keV, the Auger transitions of interest in surface spectroscopy generally have energies below 2 keV. Thus the primary excitation shell shifts upward as Z increases (e.g. K for lithium, L for sodium, M for potassium, and N for ytterbium).

The energy of an Auger transition is difficult to calculate exactly as many electron effects and final-state energies have to be considered. Fortunately, for surface analytical purposes the exact energy and line shape of Auger transitions need only be considered in the highest resolution specialist applications. The low resolution required for quantitative surface analysis also means that the small energy differences between final-state multiplets can be neglected. Auger transitions from different elements tend to be fairly widely spaced in energy and when they do overlap coincidentally there is often no easy way to use them for quantitative analysis.

The observed energy of an Auger transition is best approximated with reference to Figure 2.23. Note that the energy scale is referenced to the Fermi level. The primary electron e_p^- must have sufficient energy to ionize the core level of energy E_W. An electron at energy E_X fills the initial hole and the energy liberated in this process is transferred to an electron at E_Y, which is then ejected into the vacuum. This electron must overcome the work function of the sample, ϕ_M, to be released into the vacuum; hence the kinetic energy is reduced by this amount. However, the electron is detected by an analyser with a work function ϕ_{sp}. The vacuum level is constant and electrical contact between the sample and the analyser leads to alignment of their Fermi levels. Thus it is actually ϕ_{sp} that must be subtracted from the energy of the electron to arrive at the final kinetic energy

$$E_{WXY} = E_K - E_L - E_V - \phi_{sp} \qquad (2.58)$$

If the sample is an insulator, the Fermi levels do not automatically align. Sample charging can easily occur and special care must be taken in interpreting the spectra.

2.3.3.1   *Quantitative analysis*

The quantitative treatment of the intensity of Auger transitions is complicated not only by inelastic scattering but also by backscattering effects. For a complete discussion, see Ertl and Küppers [2] or Vickerman [1]. Approximate compositional analysis can easily be

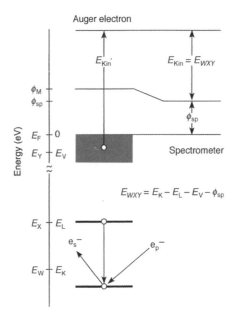

Figure 2.23   Energy levels for a KLV Auger transition, including the influence of the spectrometer work function, ϕ_{sp}, E_F, Fermi energy; E_{Kin}, kinetic energy; E_W, E_X, E_Y, energy of levels W, X and Y, respectively; E_K, E_1, E_V, energy of the K, L and valence electrons, respectively; E_{WXY}, "final" kinetic energy; ϕ_M, work function of the material. Adapted with permission from J. C. Vickerman, *Surface Analysis: The Principal Techniques* (John Wiley, Chichester Sussex, 1997). ©1997 John Wiley & Sons

obtained with use of measured relative sensitivity factors, which can be found in Davies *et al.* [52]. The mole fraction of component A in a binary mixture of A and B is given by

$$x_A = \left(\frac{I_A}{s_A}\right)\left(\frac{I_A}{s_A} + \frac{I_B}{s_B}\right)^{-1} \tag{2.59}$$

In Equation (2.59), I_A and I_B are the intensities of the peaks associated with the adsorbate and the substrate, respectively, and s_A and s_B are the relative sensitivity factors.

2.3.4   Photoelectron microscopy

Scanning electron microscopy (SEM) and transmission electron microscopy (TEM) are well-established techniques in the realm of materials science for the determination of structure. The electrons created by photoemission also lend themselves readily to the determination of the spatial information. There are, in general, three strategies that can be used to make images with use of photoemission. The first is to irradiate with a minute probe beam (either electrons or photons) and then to collect all of the photoelectrons as a function of the position of the scanned beam. In the second method, the entire surface can be illuminated but the photoelectrons are effectively imaged through a small aperture so that only a limited region of the sample is viewed. The detector is then scanned across the sample to build up an image. In the third method the entire sample is illuminated and the photoelectron distribution is imaged onto a position-sensitive detector. A photoelectron microscope can be used to measure not only the composition profile of a surface but also the geometric profile of its magnetic structure [53].

Photoelectron microscopes are useful both for spectromicroscopy and for microspectroscopy. In spectromicroscopy, micrographs (images) of electrons within a limited energy range are acquired. In other words, spectromicroscopy is energy-resolved microscopy. This is particularly important for compositional mapping. Microspectroscopy involves the acquisition of spectroscopic information (e.g. photoemission intensity as a function of kinetic energy) from a spatially resolved surface region. In this way the electronic structure of micro-scale or nano-scale structures can be measured, for example, as a function of size.

2.3.4.1   *Profiling and* xy *mapping with X-ray photoemission spectroscopy*

XPS can also be used to garner information on the geometric distribution of an element in all three dimensions. An XPS signal becomes progressively more surface sensitive for large values of the take-off angle. For take-off angles normal to the surface, roughly the first 10 nm (20–40 atomic layers) of material are probed. If information for greater depths is required, XPS can be performed with simultaneous removal of the surface layers. Removal is generally achieved by sputtering, a process in which high-energy ions are used to bombard the surface and eject material from the sample. The measurement of atomic composition as a function of depth is known as depth profiling.

The spatial resolution of XPS in the *xy* plane is impeded by the difficulty of focusing X-rays. Using focused X-rays, spatial resolutions down to *c.* 50 μm have been achieved. Alternatively, imaging of the photoelectrons with a position-sensitive detector has been used to achieve a resolution of *c.* 10 μm. The pursuit of an XPS microscope remains an active area of research.

2.3.4.2   *Depth profiling and* xy *mapping with Auger electron spectroscopy*

Depth profiling by the combination of AES and sputtering is more commonly employed than the combination with XPS. See Vickerman [1] and Hofmann [54] for a full discussion. Since high-energy electron beams can be focused to spot sizes of the order of tens of nanometres, Auger spectroscopy lends itself to high-resolution surface microscopy more readily than does XPS. High-resolution electron microscopes can be used in two modes. They can be used as a normal scanning electron microscope to image the topography of a sample. The advantage of a scanning Auger microscope is that images of elemental composition can also be obtained. This allows for correlations to be drawn between topography and elemental composition.

2.3.4.3   *Photoemission electron microscopy*

In conventional photoemission spectroscopy, the photon energy is chosen to be significantly greater than the work function of the sample. As we have learned, the work function of a heterogeneous surface is not uniform. Consider a surface that is partially covered with chemisorbed oxygen atoms. Oxygen causes a large increase in the surface dipole *in the region where it is adsorbed*. If measured by a technique that averages over a large area, this appears as a global change in the work function. However, if probed by a higher-resolution technique, we find that the work function is changed only in the parts of the surface covered with oxygen. If we choose a photon energy that is just above the work function of the clean surface, photoemission is observed only from the clean surface regions. The oxygen-covered areas are dark; that is, they do not emit electrons.

This effect can be exploited to image the chemisorption of oxygen and other adsorbates that cause large changes in the work function. This technique is known as photoemission electron microscopy (PEEM) [53, 55]. It was originally conceived in the 1930s but it was not widely exploited in surface studies until Engel developed modern instruments in the 1980s. Under favourable circumstances (i.e. when adsorbates produce sufficiently diverse work function changes to be differentiated) PEEM is capable of submicron resolution coupled with moderate temporal resolution. This combination allows PEEM to image spatiotemporal pattern formation during surface reactions with *c.* 200 nm resolution [56]. We will speak more about this topic in Chapter 5.

2.4   Vibrational Spectroscopy

Vibrational spectroscopies have long found an important place in the chemist's arsenal of analytical techniques and the same is true for those studying adsorbed systems [57, 58]. Vibrational spectroscopy is extremely useful in identifying the types of bonds that are present in a sample. Shifts in vibrational frequencies can also be used to gain insight into subtle changes in bonding. A variety of spectroscopies can give information on vibrational motions: infrared absorption [59, 60], electron energy loss [9], Raman, sum frequency generation, inelastic neutron tunnelling [58], helium scattering [61] and inelastic electron scattering [62]. Ho and co-workers have demonstrated that STM can yield vibrational spectra with high spatial resolution [63]. NSOM can also be used to obtain spatially resolved infrared and Raman spectra. Here we shall concentrate on just two of these spectroscopies – infrared absorption and electron energy loss spectroscopy.

Most surface vibrational spectroscopy is performed on the fundamental transitions, that is, on the transition from $v = 0$ to $v = 1$. The harmonic oscillator approximation is usually appropriate at this low level of excitation. The energy levels of a harmonic oscillator are equally spaced and given by

$$E_v = (v + \tfrac{1}{2})h\omega_0 \tag{2.60}$$

where v is the vibrational quantum number. The characteristic frequency of a harmonic oscillator is given by

$$\omega_0 = \frac{1}{2\pi}\left(\frac{k}{\mu}\right)^{1/2} \tag{2.61}$$

where k is the force constant of the vibration (related to the bond strength) and μ is the reduced mass

$$\mu = \frac{m_1 m_2}{m_1 + m_2} \tag{2.62}$$

of the diatomic oscillators composed of atoms of mass m_1 and mass m_2. An N-atom polyatomic molecule has $3N - 6$ (if bent) or $3N - 5$ (if linear) vibrational modes in the gas phase. Additional modes are introduced by adsorption as the translational and rotational degrees of freedom are lost for localized adsorption. These are transformed into new modes often denoted as frustrated translations and rotations. If we consider the CO molecule bound in an upright geometry, adsorption transforms the z translation into the vibration associated with the surface–CO bond. To a first approximation we can consider the various vibrational modes to be independent oscillators. The harmonic approximation breaks down for high levels of excitation because the oscillators are better described by a Morse potential than by a harmonic potential. This is especially important for low-frequency modes, such as the surface–adsorbate bond, because even moderate temperatures lead to significant populations in excited vibrational states. Anharmonicity results not only in a continuously decreasing level spacing with increasing values of v but also in couplings between the various vibrational modes. In a real (anharmonic) oscillator, overtone transitions ($\Delta v > 1$), and combination bands (simultaneous excitation of more than one vibrational mode) can be observed.

Equations (2.60) and (2.61) are written in terms of radial frequency. More commonly, vibrational transitions are categorized in terms of inverse centimetres (cm^{-1}) in infrared spectroscopy, or millielectron volts (meV) in electron energy loss spectroscopy. The inverse centimetre is the unit of wavenumber, which is denoted by $\bar{v}$ and is not to be confused with frequency (v) or angular frequency ($\omega = 2\pi v$). A very useful conversion factor is $1 \, meV = 8.065 \, cm^{-1}$.

As shown in Figure 2.24 the vibrational spectrum in the adsorbed phase differs greatly from that found in the gas phase. The loss of rotational fine structure is immediately obvious. The change in linewidth is also apparent. There are numerous mechanisms that lead to the increased linewidth in the adsorbed phase [60, 64–66]. Both homogeneous and inhomogeneous processes lead to line broadening. Inhomogeneous broadening is dominated by variations in the environment around the adsorbate. Under normal circumstances, inhomogeneous broadening is significant and leads to a Gaussian line-shape. Only the most well-ordered adsorbate structures on single-crystal surfaces exhibit spectra that are not dominated by inhomogeneous broadening. Figure 2.24 also demonstrates that more than one frequency can be associated with the same vibration in the adsorbed phase – the C–O stretch in this case – even though there is a unique value in the gas phase. The prime example of this is the ideally terminated Si(1 1 1)-(1 × 1) : H surface shown in Figure 2.25, which Chabal and co-workers have exploited to study vibrational energy transfer processes with unparalleled detail [65]. Homogeneous broadening arises both from vibrational energy transfer and dephasing events. The study of vibrational energy transfer and dephasing in adsorbates was pioneered by Cavanagh, Heilweil, Stephenson and co-workers [67–71].

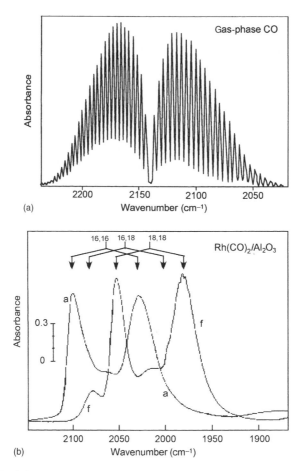

Figure 2.24   The infrared spectrum of (a) gas-phase CO compared with that of (b) CO adsorbed on dispersed rhodium clusters. The gas-phase spectrum exhibits rotational fine structure. The adsorbed CO forms a gem dicarbonyl species [Rh(CO)$_2$]. Coupling between the two adsorbed CO molecules leads to two vibrational peaks. The effect of oxygen isotopic substitution is also evident: 16, 16 refers to Rh(C^{16}O)$_2$, and so on. Adapted with permission from J. T. Yates Jr and K. Kolasinski, *J. Chem. Phys.* **79** (1983) 1026. ©1983 American Institute of Physics

Vibrational modes are named according to the type of motion involved in the vibration. There are four characteristic types of vibrations. Stretches change the bond length and are denoted with v (Greek nu, not to be confused with the vibrational quantum number v). In-plane motions that change bond angles but not bond lengths are bends. These are denoted with δ and are sometimes further subdivided between rocks, twists and wags. Out-of-plane bends are assigned the symbol γ. Torsions (τ) change the angle between two planes containing the atoms involved in the motion.

Equation (2.61) shows that the fundamental vibrational frequency depends on the reduced mass. Isotopic substitution can therefore be used as an aid in identifying vibrational transitions. The value of k depends on the vibrational potential, which

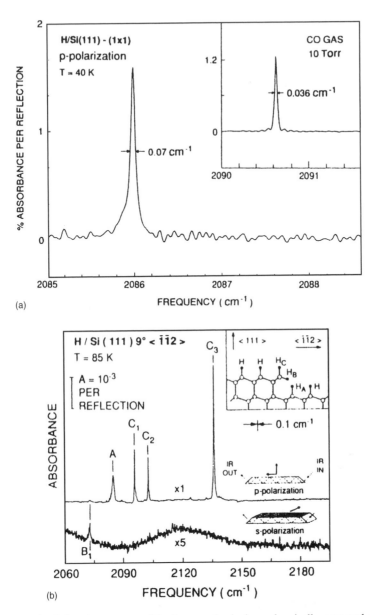

Figure 2.25 The infrared spectrum of hydrogen adsorbed on chemically prepared (a) flat and (b) stepped Si(1 1 1) surfaces. Part (a) reprinted from *Chem. Phys. Lett.* **187**, P. Jakob, Y. J. Chabal and K. Raghavachari, "Lineshape analysis of the SiH stretching mode of the ideally H-terminated Si(1 1 1) surface: the role of dynamical dipole coupling", 1991, page 325, with permission from Elsevier Science. Part (b) reproduced with permission from P. Jakob, Y. J. Chabal, K. Raghavachari and S. B. Christman, *Phys. Rev. B* **47** (1993) 6839; ©1993 by the American Physical Society

means that vibrational spectra are sensitive to changes in the vibrational potential introduced by changes in chemical bonding.

2.4.1 Infrared spectroscopy

The application of infrared (IR) spectroscopy to surfaces owes much of its early development to the work of Eichens [72, 73], Sheppard [74, 75] and Greenler [76, 77]. The absorption of infrared radiation by adsorbed species [59, 60] bears many similarities to IR absorption in other phases. The probability of a vibrational transition is proportional to the square of the transition dipole moment. The transition dipole moment is given by [59]:

$$M_{vv'} = \int_{-\infty}^{\infty} \psi(v)\mu\psi(v')d\tau \tag{2.63}$$

where $\psi(v)$ and $\psi(v')$ are the initial and final state vibrational wavefunctions, respectively, and μ is the dipole moment of the molecule. For the vibrational transition to have a nonzero probability, the integral in Equation (2.63) must be nonzero along at least one of its components (x, y or z). The z axis is conventionally chosen to lie along the surface normal. Group theory and symmetry arguments [58, 78] can be used to determine which modes are IR active, that is, capable of being observed in IR spectroscopy. This symmetry constraint depends not only on the molecule but also on the symmetry of the adsorption site [79]. This property is useful for the determination of adsorbate structure.

The strength of an absorption feature depends not only on the transition dipole moment but also on the strength of the electric field associated with the IR light incident on the sample and the orientation of the electric field vector with respect to the transition dipole. Therefore, we need to consider the electric field strength at the interface to understand what types of vibrations can be observed. Semiconductors and insulators can support electric fields both perpendicular and parallel to their surfaces. A metal, however, can support only a perpendicular electric field. This is because of the image dipole effect illustrated in Figure 2.26. Thus, on a metal, only vibrations with a component along the surface normal will be observed in IR absorption spectra. On semiconductors and insulators, the orientation of the adsorbate can be determined by measuring the intensity of spectral features as a function of the polarization angle of the incident light.

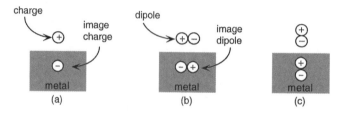

Figure 2.26 Image dipole at a metallic surface

Infrared spectroscopy is performed in a number of distinct modes:

- Transmission [Figure 2.27(a)]: this is appropriate only for transparent substrates and films that are sufficiently thin so as not to absorb too much of the incident light. It has been successfully employed in the study of dispersed metal-supported catalysts, thin films on insulating and semiconducting substrates and adsorption onto the surfaces of porous solids.
- Reflection [Figure 2.27(b)]: this is most appropriate for metal single crystals for which it is performed near grazing incidence. This type of spectroscopy is known as reflection absorption infrared spectroscopy (RAIRS) or infrared reflection absorption spectroscopy (IRAS).

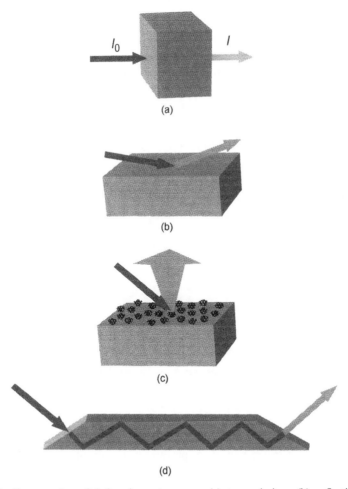

Figure 2.27   Four modes of infrared spectroscopy: (a) transmission; (b) reflection; (c) diffuse reflectance; (d) internal reflection; I_0, I, incident and transmitted light intensity, respectively

- Diffuse reflectance [Figure 2.27(c)]: Powders or rough surfaces scatter radiation diffusely rather than specularly. An absorption spectrum is measured by collecting all of the scattered radiation. This type of spectroscopy is known as diffuse reflectance infrared Fourier transform spectroscopy (DRIFTS).

- Internal reflection [Figure 2.27(d)]: infrared light can be bounced through a transparent substrate, particularly diamond, silicon, germanium or ZnSe, and absorption can occur at each surface reflection. This type of spectroscopy is known as attenuated total reflection (ATR) or multiple internal reflection (MIR) spectroscopy (ATR is the preferred abbreviation so as to avoid confusion with MIR as an abbreviation for mid-infrared). It can be used to interrogate the interface between any two phases – solid–solid, solid–liquid, solid–gas – as long as the substrate is transparent. Thus studies of semiconductor–aqueous solution interfaces become possible, as required for in situ electrochemical studies, even though the liquid is opaque. In addition to adsorbed layers on the ATR element, this technique can be used to study the surface of solids pressed against the semiconductor prism.

Other methods are discussed by Chabal [60].

Transmission is perhaps the most familiar form of infrared spectroscopy. In surface science, however, it is generally used only for high-surface-area samples and thin films because a single adsorbed monolayer adsorbs so little radiation. In transmission mode, the amount of adsorbed radiation is characterized in terms of the transmittance

$$T = \frac{I}{I_0} \tag{2.64}$$

or absorbance

$$A = -\log T = \log\left(\frac{I_0}{I}\right) \tag{2.65}$$

where I_0 is the incident intensity, and I is the transmitted intensity. In the absence of reflection and scattering, the transmittance is given by

$$T = \exp(-acl) \tag{2.66}$$

where a is the absorptivity, c the concentration, and l the path length. The absorption coefficient, α, is related to the absorptivity by

$$\alpha = ac \tag{2.67}$$

which is, in turn, related to the imaginary part, κ, of the complex refractive index, $\tilde{n}$, by

$$\alpha = \frac{4\pi\kappa}{\lambda_0} \tag{2.68}$$

where λ_0 is the vacuum wavelength of the infrared light and κ is defined through

$$\tilde{n} = n + i\kappa \tag{2.69}$$

It follows that absorbance is linearly proportional to concentration and the pathlength:

$$A = acl \tag{2.70}$$

The absorption spectra collected by diffuse reflectance are expressed in terms of the Kubelka–Munk function

$$\frac{\alpha}{S} = \frac{(1 - R_\infty)^2}{2R_\infty} \tag{2.71}$$

where S is the scattering coefficient, and R_∞ is the reflectivity of an (optically) infinitely thick sample. If the scattering coefficient is known (or assumed to be independent of wavelength) over the region of the spectrum, the Kubelka–Monk function directly transforms the measured reflectivity spectrum into the absorption spectrum.

In RAIRS and ATR, the reflection spectrum is measured for the clean substrate, $R_0(v)$, and the adsorbate-covered substrate, $R(v)$. The absorption spectrum is usually expressed in terms of the relative change in reflectivity:

$$\frac{\Delta R(v)}{R_0(v)} = \frac{R_0(v) - R(v)}{R_0(v)} \tag{2.72}$$

The relative change in reflectivity is related to the absolute coverage (molecules per unit area), σ, and the adsorbate polarizability, $\alpha(v)$ (not to be confused with the absorption coefficient) according to [60]

$$\frac{\Delta R(v)}{R_0(v)} = \frac{8\pi^2 v}{c} F(\varphi)\sigma \, \text{Im} \, \alpha(v) \tag{2.73}$$

where φ is the angle of incidence and $F(\varphi)$ contains all of the local field characteristics and the dielectric response of the system.

Equations (2.70) and (2.73) demonstrate that IR spectroscopy can be used to quantify the extent of adsorption. It is usually *assumed* that the absorbance and relative change in reflectivity are linearly proportional to adsorbate number density, as expected from Equations (2.70) and (2.73). However, this is not always so [59], in particular, deviations are to be expected if there are strong interactions between adsorbates. Although it is often a good approximation, especially at low coverage, the linear relationship between adsorbate coverage and IR absorption needs to be confirmed by corroborating measures of coverage.

Surface IR spectroscopy can be performed only if the substrate does not absorb strongly. Depending on the substrate, this leads to a cut-off in the low-frequency region of the spectrum, occurring at around $500–1000 \, \text{cm}^{-1}$. Because of this, IR spectroscopy cannot be used to interrogate low-frequency vibrations. For example, the substrate–molecular adsorbate bond generally has a very low frequency. Surface phonons also have vibrational frequencies too low to be studied by IR spectroscopy.

IR spectroscopy has two other important characteristics. The first is that it is capable of very high spectral resolution (< 0.01 cm^{-1}). This is generally much smaller than the linewidths observed for most adsorbates. High-resolution studies, however, can lead to important information regarding vibrational energy transfer and dephasing. The second characteristic is that it can be used for high-pressure, *in situ* studies. In this case, any absorption from the gas phase must be subtracted. This characteristic allows for the study of working catalysts under realistic conditions.

2.4.2 Electron energy loss spectroscopy

When electrons backscatter from a surface they can lose energy to the various degrees of freedom of the surface and adsorbed layer. Electron energy loss spectroscopy (EELS) [9] relies upon the use of a monochromatic, collimated beam of electrons that is energy analysed after it has scattered from the surface. The angular distribution of the scattered electrons contains further information. In principle, rotational, vibrational and electronic transitions can be observed in EELS. The study of vibrations by electron energy loss is often called high-resolution electron energy loss spectroscopy (HREELS) to differentiate it from the study of electronic transitions.

Figure 2.28 displays an electron energy loss (EEL) spectrum collected from a co-adsorbed layer of CO and O_2, concentrating on the vibrational part of the spectrum. By far the largest peak in the spectrum is the elastic peak. The absolute energy of the elastically scattered electrons is unimportant as far as the position of peaks in the spectrum is concerned. The centre of the elastic peak, E_0, fixes the origin of the spectrum and is conventionally set equal to zero energy. The width of the elastic peak, conventionally expressed as full width at half maximum (FWHM), is a measure of the spectral resolution.

The energy at which a peak occurs is

$$E = E_0 - \hbar\omega \tag{2.74}$$

where $\hbar\omega$ is the energy of excited transition (generally in meV), and E is the detected energy. Thus, if we take $E_0 = 0$, the energy of a peak is the energy of the vibration excited by electron scattering.

2.4.2.1 Three scattering mechanisms

The proper interpretation of EEL spectra requires a knowledge of how electrons scatter inelastically. The three mechanisms of scattering are dipole scattering, impact scattering and resonance scattering. Wave-particle duality lies behind these three scattering mechanisms.

In dipole scattering the electric field associated with the charge of a moving electron interacts with the scatterer (adsorbate or surface phonon). The electron acts as a wave and the interaction is analogous to that of a photon in IR spectroscopy. On a metal, the same selection rule is followed, that is, only vibrations associated with a dipole moment change normal to a metal surface can be observed. Just as for photons, dipole scattered electrons

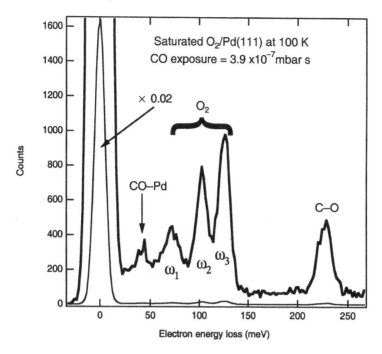

Figure 2.28 The electron energy loss spectrum of co-adsorbed $O_2 + CO$ on Pd(1 1 1). The species associated with ω_1–ω_3 are illustrated in Figure 3.8, page 94. Adapted from *Surf. Sci.* **334**, K. W. Kolasinski, F. Cemič, A. de Meijere and E. Hasselbrink, "Interactions in co-adsorbed CO + O_2/ Pd(1 1 1) layers", 1995, page 19, with permission from Elsevier Science

emerge along the specular angle. The angle of incidence equals the angle of reflection, and both the elastically scattered and inelastically scattered electrons emerge along the same propagation direction. The analysis of the spectra associated with dipole scattering, therefore, follows the same methods as RAIRS. Whereas the energetic position of the loss peaks does not depend on E_0, the intensity of the peaks does. The intensity smoothly varies with E_0, roughly following $(E_0)^{1/2}$.

In impact scattering the electron acts like a particle. The electron bounces off the scatterer, experiencing a short-range interaction and exchanging momentum. Because of momentum exchange, off-specular scattering ($\vartheta_i \neq \vartheta_f$) occurs. Thus, by measuring the angular distribution of the scattered electrons, impact and dipole scattering can be differentiated. The short-range interaction means that the selection rules valid for RAIRS are no longer appropriate for this scattering regime.

Although we may attempt to classify an electron as either a wave or a particle, an electron is, after all, an electron. In resonance scattering [80] the electron interacts with the adsorbate as only an electron can. That is, the electron actually becomes trapped in a bound (or quasi-bound) electronic state. The trapping state can be a real excited state of the isolated ion. In electron scattering such a state is called a Feshbach resonance. Alternatively, a centrifugal barrier, which arises from the angular momentum associated with the electron–molecule scattering event, can trap an electron. This is known as a shape resonance. A shape resonance is a very short-lived state (of the order of a few femtoseconds). However, even a Feshbach resonance generally has a lifetime of the order of 10^{-10}–10^{-15} s. Resonance scattering is sensitively dependent on the incident electron energy, that is, resonances in the vibrational excitation probability are observed as a function of E_0. Resonances also have characteristic angular dependencies for both their excitation and decay. The angular distribution of the electrons scattered through a resonance provides information about the symmetry of the resonance (e.g. whether a σ or a π symmetry state is involved). Because of the finite lifetime of the temporary negative ion state formed, the excitation of overtones often accompanies resonance scattering. In fact, this is one of the clearest signatures of this mechanism. It also is often the only way to study these higher-lying vibrational states.

One of the great strengths of EELS is that it can interrogate the low-frequency region (0–800 cm^{-1}) where IR cannot be used. Thus EELS can be used to investigate both the substrate–adsorbate bond and phonons. The resolution of EELS is, however, significantly lower than that of IR spectroscopy. Although it is relatively easy to obtain 10 meV resolution, it is only recently that resolution on the order of 1 meV (c. 8 cm^{-1}) has been obtained. The use of an electron beam means that EELS is constrained to use in UHV. The intensity of loss features is dependent on the adsorbate coverage and, subject to the same caveats as for RAIRS, it can be used as a quantitative measure of adsorbate coverage.

2.5 Other Surface Analytical Techniques

A large fraction of research at surfaces throughout the 1970s and 1980s was devoted to developing an 'alphabet soup' of surface-sensitive analytical techniques. An extensive collection of these acronyms has been collected by Somorjai [81]. Experimentation in surface science requires the routine intervention of a multi-technique approach and, therefore, a familiarity with a large number of techniques is virtually a must for the surface scientist.

2.6 Summary of Important Concepts

- Scanning tunnelling microscopy (STM) involves the tunnelling of electrons from occupied to unoccupied electronic states. The voltage between the tip and the surface determines the direction of current flow.

- STM images electronic states not atoms.

- Atomic force microscopy (AFM) allows for atomic scale imaging on insulating surfaces and for direct measurements of intermolecular forces.

- Near-field scanning optical microscopy (NSOM) extends optical spectroscopy to the nanoscale and even single-molecule regime.

- Low-energy (c. 20–500 eV) electrons penetrate only the first few atomic layers and can be used to investigate surface structure.

- The symmetry of low-energy electron diffraction (LEED) patterns is related to the periodicity of the substrate and adsorbate overlayer structure.

- X-ray photoelectron spectroscopy (XPS) probes the electronic states associated with core levels and is particularly well suited to quantitative elemental analysis.

- Ultraviolet photoemission spectroscopy (UPS) probes the electronic states associated with valence electrons and is particulary well suited to studying electronic changes associated with chemical bonding.

- Auger electron spectroscopy (AES) is also used for quantitative elemental analysis.

- In infrared (IR) spectroscopy at metal surfaces a strict dipole selection rule means that only vibrations with a component along the surface normal can be observed.

- In electron energy loss spectroscopy (EELS) electrons scatter through dipole, impact, and/or resonance scattering mechanisms and no strict selection rule can be assumed unless the mechanism is known.

Exercises

2.1   Lieber and co-workers [82] used AFM to measure the adhesion force arising from the contact of CH_3 groups. The adhesion force was 1.0 nN. If the tip/surface contact area was 3.1 nm^2 and the radius of a CH_3 group is $(0.2)^{1/2}$ nm, calculate the interaction force resulting from the contact of two individual CH_3 groups.

2.2   In an STM image, does an adsorbate sitting on top of a surface always look like a raised bump compared with the substrate?

2.3   Some adsorbates can be imaged in STM at low temperatures but seem to disappear at higher temperatures even though they have not desorbed from the surface. Explain.

2.4   The dimmer unit on a Si(1 0 0) surface has a bonding orbital just below E_F and an antibonding orbital just above E_F. Make a prediction about STM images that are taken at positive compared with negative voltages. Do the images look the same and, if not, how do they differ?

2.5   Use Equations (2.26) and (2.27) to show that the diffraction reflexes appear at

$$\frac{s}{\lambda} = h_1 a_1^* + h_2 a_2^* \tag{2.75}$$

Assume normal incidence of the incoming electrons.

2.6   Describe how the spots in a LEED pattern would evolve if:

   (a) incident molecules adsorbed randomly onto a surface forming an ordered overlayer only when one quarter of the substrate atoms are covered (a quarter of a monolayer);

   (b) the incident molecules form ordered islands that continually grow in size until they reach a saturation coverage of a quarter of a monolayer;

   (c) an overlayer is ordered in one direction but not in the orthogonal direction (either because of random adsorption or because of diffusion in the orthogonal direction).

2.7. Consider clean fcc(1 0 0), fcc(1 1 0) and fcc(1 1 1) surfaces. (fcc = face-centred cubic). Draw the unit cell and include the primitive lattice vectors $\mathbf{a}_1$ and $\mathbf{a}_2$. Calculate the reciprocal lattice vectors $\mathbf{a}_1^*$ and $\mathbf{a}_2^*$ and draw the LEED pattern including the reciprocal lattice vectors.

2.8. For structures (a)–(i) in Figure 2.29 determine the associated LEED patterns. Classify the structures in both Wood's notation and matrix notation.

2.9. Fractional coverage can be defined as the number of adsorbates divided by the number of surface atoms:

$$\theta = \frac{N_{ads}}{N_0} \tag{2.76}$$

For each of the structures in Exercise 2.8, calculate the coverage. Note any correlations between coverage and the LEED patterns.

2.10 Given LEED patterns (a)–(i) in Figure 2.30 obtained from adsorbate-covered face-centred cubic (fcc) substrates, determine the surface structures. Substrate reflexes are marked •; the additional adsorbate-induced reflexes are marked ×. Assume no reconstruction of the surface.

2.11 Determine all of the X-ray levels that are possible for the $n = 3$ shell.

2.12 The vector potential of an electric field can be written

$$\mathbf{A}(r, t) = \mathbf{A}_0 \exp(-i\omega t + i\mathbf{q} \cdot \mathbf{r}) \tag{2.77}$$

Use Equation (2.77) to derive Equation (2.56).

2.13 Consider the spectrum of adsorbed CO shown in Figure 2.24. CO is adsorbed as a gem-dicarbonyl on the rhodium atoms present on the Al_2O_3 substrate, as shown in Figure 2.31.

(a) If the CO molecules in the gem-dicarbonyl were independent oscillators, only one CO stretch peak would be observed. Explain why there are two peaks in the spectrum [83].

(b) Explain why substitution of ^{18}O for ^{16}O changes the positions of the bands.

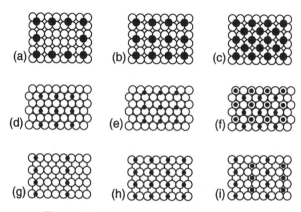

Figure 2.29   Structures (a)–(i): see Exercise 2.8

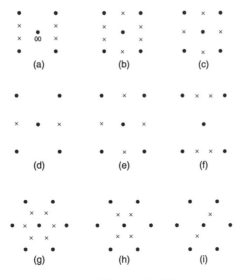

Figure 2.30   Low-energy electron diffraction (LEED) patterns (a)–(i): see Exercise 2.10

Figure 2.31   CO adsorbed as a gem-dicarbonyl on rhodium atoms present on an Al_2O_3 substrate (substrate not shown): see Exercise 2.13

Further Reading

F. Besenbacher, "Scanning tunnelling microscopy studies of metal surfaces", *Rep. Prog. Phys.* **59** (1996) 1737.

G. A. D. Briggs and A. J. Fisher, "STM experiment and atomistic modelling Hand in hand: individual molecules on semiconductor surfaces", *Surf. Sci. Rep.* **33** (1999) 1.

Y. J. Chabal, "Surface infrared spectroscopy", *Surf. Sci. Rep.* **8** (1988) 211.

D. Drakova, "Theoretical modelling of scanning tunnelling microscopy, scanning tunnelling spectroscopy and atomic force microscopy", *Rep. Prog. Phys.* **64** (2001) 205.

R. C. Dunn, "Near-field scanning optical microscopy", *Chem. Rev.* **99** (1999) 2891.

G. Ertl and J. Küppers, *Low Energy Electrons and Surface Chemistry* (VCH, Weinheim, 1985).

L. C. Feldman and J. W. Mayer, *Fundamentals of Surface and Thin Film Analysis* (North-Holland, Amsterdam, 1986).

B. Feuerbacher and R. F. Willis, "Photoemission and electron states at clean surfaces", *J. Phys. C* **9** (1976) 169.

K. Heinz, "LEED and DLEED as modern tools for quantitative surface structure determination", *Rep. Prog. Phys.* **58** (1995) 637.

H. Ibach (ed.) *Electron Spectroscopy for Surface Analysis, Topics in Current Physics Volume 4* (Springer, New York, 1977).

H. Ibach and D. L. Mills, *Electron Energy Loss Spectroscopy and Surface Vibrations* (Academic Press, New York, 1982).

F. M. Mirabella Jr (ed.), *Internal Reflection Spectroscopy: Theory and Applications* (Marcel Dekker, New York, 1992).

J. W. Niemantsverdriet, *Spectroscopy in Catalysis* (VCH, Weinheim, 1993).

E. W. Plummer and W. Eberhardt, "Angle-resolved photoemission as a tool for the study of surfaces", *Adv. Chem. Phys.* **49** (1982) 533.

H. H. Rotermund, "Imaging of dynamic processes on surfaces by light", *Surf. Sci. Rep.* **29** (1997) 265.

J. A. Stroscio and W. J. Kaiser (eds), *Scanning Tunneling Microscopy: Methods of Experimental Physics, Volume 27* (Academic Press, Boston, MA, 1993).

H. Takano, J. R. Kenseth, S.-S. Wong, J. C. O'Brien, and M. D. Porter, "Chemical and biochemical analysis using scanning force microscropy", *Chem. Rev.* **99** (1999) 2845.

S. O. Vansteenkiste, M. C. Davies, C. J. Roberts, S. J. B. Tendler, and P. M. Williams, "Scanning probe microscopy of biomedical interfaces", *Prog. Surf. Sci.* **57** (1998) 95.

J. C. Vickerman, *Surface Analysis: The Principal Techniques* (John Wiley, Chichester, Sussex, 1997).

D. P. Woodruff and T. A. Delchar, *Modern Techniques of Surface Science* (Cambridge University Press, Cambridge, 1994).

J. T. Yates Jr and T. E. Madey (eds), *Vibrational Spectroscopy of Molecules on Surfaces* (Plenum Press, New York, 1987).

References

[1] J. C. Vickerman, *Surface Analysis: The Principal Techniques* (John Wiley, Chichester, Sussex, 1997).

[2] G. Ertl and J. Küppers, *Low Energy Electrons and Surface Chemistry*, 2nd edition (VCH, Weinheim, 1985).

[3] D. P. Woodruff and T. A. Delchar, *Modern Techniques of Surface Science* (Cambridge University Press, Cambridge, 1994).

[4] L. C. Feldman and J. W. Mayer, *Fundamentals of Surface and Thin Film Analysis* (North-Holland, Amsterdam, 1986).

[5] J. T. Yates Jr, *Experimental Innovations in Surface Science: A Guide to Practical Laboratory Methods and Instruments* [AIP Press (Springer), New York, 1998].

[6] G. Binnig, H. Rohrer, C. Gerber and E. Weibel, *Phys. Rev. Lett.* **49** (1982) 57.

[7] D. M. Eigler and E. K. Schweizer, *Nature* **344** (1990) 524.

[8] P. Moriarty, *Rep. Prog. Phys.* **64** (2001) 297.

[9] H. Ibach and D. L. Mills, *Electron Energy Loss Spectroscopy and Surface Vibrations* (Academic Press, New York, 1982).

[10] K. Itaya, *Prog. Surf. Sci.* **58** (1998) 121.

[11] J. Tersoff and N. D. Lang, "Theory of scanning tunneling microscopy", in *Scanning Tunneling Microscopy*, eds J. A. Stroscio and W. J. Kaiser (Academic Press, Boston, MA, 1993), p. 1.

[12] D. M. Cyr, B. Venkataraman, G. W. Flynn, A. Black and G. M. Whitesides, *J. Phys. Chem.* **100** (1996) 13747.

[13] B. Venkataraman, G. W. Flynn, J. L. Wilbur, J. P. Folkers and G. M. Whitesides, *J. Phys. Chem.* **99** (1995) 8684.

[14] J. A. Stroscio and R. M. Feenstra, "Methods of tunneling spectroscopy", in *Scanning Tunneling Microscopy*, eds J. A. Stroscio and W. J. Kaiser (Academic Press, Boston, MA, 1993), pp. 96.

[15] G. Binnig, C. F. Quate and C. Gerber, *Phys. Rev. Lett.* **56** (1986) 930.

[16] F. Ohnesorge and G. Binnig, *Science* **260** (1993) 1451.

[17] F. J. Giessibl, *Science* **267** (1995) 68.

[18] H. Takano, J. R. Kenseth, S.-S. Wong, J. C. O'Brien and M. D. Porter, *Chem. Rev.* **99** (1999) 2845.

[19] C. L. Cheung, J. H. Hafner, T. W. Odom, K. Kim and C. M. Lieber, *Appl. Phys. Lett.* **76** (2000) 3136.

[20] J.-J. Greffet and R. Carminati, *Prog. Surf. Sci.* **56** (1997) 133.

[21] E. H. Synge, *Philos. Mag.* **6** (1928) 356.

[22] E. H. Synge, *Philos. Mag.* **11** (1931) 65.

[23] E. H. Synge, *Philos. Mag.* **13** (1932) 297.

[24] E. A. Ash and G. Nicholls, *Nature* **237** (1972) 510.

[25] D. W. Pohl, W. Denk and M. Lanz, *Appl. Phys. Lett.* **44** (1984) 651.

[26] A. Harootunian, E. Betzig, A. Lewis and M. Isaacson, *Appl. Phys. Lett.* **49** (1986) 674.

[27] R. C. Dunn, *Chem. Rev.* **99** (1999) 2891.

[28] E. Betzig and R. J. Chichester, *Science* **262** (1993) 1422.

[29] C. J. Davidson and L. H. Germer, *Phys. Rev.* **30** (1927) 705.

[30] L. de Broglie, *Phil. Mag.* **47** (1924) 446.

[31] M. A. Van Hove, *Low-energy Electron Diffraction: Experiment, Theory and Surface Structure* (Springer, Berlin, 1986).

[32] R. L. Park and H. H. Madden Jr, *Surf. Sci.* **11** (1968) 188.

[33] E. A. Wood, *J. Appl. Phys.* **35** (1964) 1306.

[34] C. Kittel, *Introduction to Solid State Physics*, (John Wiley, New York, 1986).

[35] D. A. Shirley, "Many-electron and final-state effects: beyond the one-electron picture", in *Photoemission in Solids; I, General Principles*, eds M. Cardona and L. Ley (Springer, Berlin, 1978), p. 165.

[36] D. E. Eastman and J. K. Cashion, *Phys. Rev. Lett.* **27** (1971) 1520.

[37] R. Haight and P. F. Seidler, *Appl. Phys. Lett.* **65** (1994) 517.

[38] R. Haight, *Surf. Sci. Rep.* **21** (1995) 275.

[39] J. H. Bechtel, W. L. Smith and N. Bloembergen, *Phys. Rev. B* **15** (1977) 4557.

[40] H. W. Rudolf and W. Steinmann, *Phys. Lett.* **61A** (1977) 471.

[41] H.-L. Dai and W. Ho, "Laser spectroscopy and photochemistry on metal surfaces in *Advanced Series in Physical Chemistry, Volume 5*, eds H.-L. Dai and W. Ho (World Scientific, Singapore, 1995), pp. 184–242.

[42] U. Höfer and E. Umbach, *J. Electron Spectrosc. Rel. Phenom.* **54/55** (1990) 591.

[43] R. Matzdorf, *Surf. Sci. Rep.* **30** (1997) 153.

[44] E. W. Plummer and W. Eberhardt, *Adv. Chem. Phys.* **49** (1982) 533.

[45] J. Küppers, K. Wandelt and G. Ertl, *Phys. Rev. Lett.* **43** (1979) 928.

[46] W. Steinmann and T. Fauster, "Two-photon Photoelectron Spectroscopy of Electronic States at Metal Surfaces: Part 1", in *Laser Spectroscopy and Photochemistry on Metal Surfaces. Part I*, eds H.-L. Dai and W. Ho (World Scientific, Singapore, 1995), p. 184.

[47] V. Dose, *Surf. Sci. Rep.* **5** (1985) 337.

[48] H. Petek and S. Ogawa, *Prog. Surf. Sci.* **56** (1997) 239.

[49] W. S. Fann, R. Storz, H. W. K. Tom and J. Bokor, *Phys. Rev. Lett.* **68** (1992) 2834.

[50] H. Petek, M. J. Weida, H. Nagano and S. Ogawa, *Science* **288** (2000) 1402.

[51] P. Auger, *J. Phys. Radium* **6** (1925) 205.

[52] L. E. Davies, N. C. MacDonald, P. W. Palmberg, G. E. Riach and R. E. Weber, *Handbook of Auger Electron Spectroscopy* (Perkin Elemer Corp., Physical Electronics Division, Eden Prairier, MN, 1976).

[53] W. Swiech, G. H. Fecher, C. Ziethen, O. Schmidt, G. Schönhense, K. Grzelakowski, C. M. Schneider, R. Frömter, H. P. Oepen and J. Kirschner, *J. Electron Spectrosc. Rel. Phenom.* **84** (1997) 171.

[54] S. Hofmann, *Rep. Prog. Phys.* **61** (1998) 827.

[55] H. H. Rotermund, *Surf. Sci. Rep.* **29** (1997) 265.

[56] R. Imbihl and G. Ertl, *Chem. Rev.* **95** (1995) 697.

[57] J. T. Yates Jr and T. E. Madey, "Vibrational spectroscopy of molecules on surfaces", in *Methods of Surface Characterization, Volume 1* (Plenum Press, New York, 1987).

[58] M. Pemble, "Vibrational spectroscopy from surfaces", in *Surface Analysis: The Principal Techniques*, ed. J. C. Vickerman (John Wiley, Chichester, Sussex, 1997), p. 267.

[59] B. E. Hayden, "Reflection absorption infrared spectroscopy", in *Vibrational Spectroscopy of Molecules on Surfaces*, eds J. T. Yates Jr and T. E. Madey (Plenum Press, New York, 1987), p. 267.

[60] Y. J. Chabal, *Surf. Sci. Rep.* **8** (1988) 211.

[61] A. P. Graham and J. P. Toennies, *Surf. Sci.* **428** (1999) 1.

[62] W. H. Weinberg, "Inelastic electron tunneling spectroscopy of supported homogeneous cluster compounds", in *Vibrational Spectra and Structure*, ed. J. R. Dorig (1982), Vol. 11, p. 1.

[63] B. C. Stipe, M. A. Rezaei and W. Ho, *Science* **280** (1998) 1732.

[64] A. L. Harris, K. Kuhnke, M. Morin, P. Jakob, N. J. Levinos and Y. J. Chabal, *Faraday Discuss. Chem. Soc.* **96** (1993) 217.

[65] Y. J. Chabal, A. L. Harris, K. Raghavachari and J. C. Tully, *Int. J. Mod. Phys. B* **7** (1993) 1031.

[66] M. Morin, P. Jakob, N. J. Levinos, Y. J. Chabal and A. L. Harris, *J. Chem. Phys.* **96** (1992) 6203.

[67] R. R. Cavanagh, J. D. Beckerle, M. P. Casassa, E. J. Heilweil and J. C. Stephenson, *Surf. Sci.* **269/270** (1992) 113.

[68] J. D. Beckerle, R. R. Cavanagh, M. P. Casassa, E. J. Heilweil and J. C. Stephenson, *J. Chem. Phys.* **95** (1991) 5403.

[69] J. D. Beckerle, M. P. Casassa, E. J. Heilweil, R. R. Cavanagh and J. C. Stephenson, *J. Electron Spectrosc. Rel. Phenom.* **54/55** (1990) 17.

[70] J. D. Beckerle, M. P. Casassa, R. R. Cavanagh, E. J. Heilweil and J. C. Stephenson, *J. Chem. Phys.* **90** (1989) 4619.

[71] E. J. Heilweil, M. P. Casassa, R. R. Cavanagh and J. C. Stephenson, *J. Chem. Phys.* **81** (1984) 2856.

[72] R. P. Eischens, S. A. Francis and W. A. Pliskin, *J. Phys. Chem.* **60** (1956) 194.

[73] R. P. Eischens and W. A. Pliskin, *Adv. Catal.* **10** (1958) 1.

[74] N. Sheppard and D. J. C. Yates, *Proc. R. Soc. London, Ser. A* **238** (1956) 69.

[75] N. Sheppard and T. T. Nguyen, "The vibrational spectra of carbon monoxide chemisorbed on the surfaces of metal catalysts – a suggested scheme of interpretation", in *Advances in Infrared and Raman Spectroscopy*, eds R. E. Hester and R. J. H. Clark (Heyden and Son, London, 1978), pp. 67–148.

[76] R. G. Greenler, *J. Chem. Phys.* **44** (1966) 310.

[77] R. G. Greenler, R. R. Rahn and J. P. Schwartz, *J. Catal.* **23** (1971) 42.

[78] F. A. Cotton, *Chemical Applications of Group Theory* (John Wiley, New York, 1971).

[79] N. V. Richardson and N. Sheppard, "Normal modes at surfaces", in *Vibrational Spectroscopy of Molecules on Surfaces*, ed. J. T. Yates Jr and T. E. Madey (Plenum Press, New York, 1987), p. 267.

[80] R. E. Palmer and P. J. Rous, *Rev. Mod. Phys.* **64** (1992) 383.

[81] G. A. Somorjai, *Introduction to Surface Chemistry and Catalysis* (John Wiley, New York, 1994).

[82] C. D. Frisbie, L. F. Rozsnyai, A. Noy, M. S. Wrighton and C. M. Lieber, *Science* **265** (1994) 2071.

[83] J. T. Yates Jr and K. Kolasinski, *J. Chem. Phys.* **79** (1983) 1026.

3

Chemisorption, Physisorption and Dynamics

In Chapter 1 we discussed the structure of adsorbates on surfaces. Now we turn to describing the interactions that hold adsorbates onto surfaces and the processes that get them there. In Chapter 4 we will discuss the thermodynamics and kinetics of adsorption and the adsorbed phase. In this chapter, we discuss those aspects of energetics that control the motions and binding of atoms and molecules to surfaces. In particular, we concentrate mainly on the interactions of individual molecules with clean surface. The effects of higher coverages are dealt with in Chapter 4.

3.1 Types of Interactions

When a molecule sticks to a surface, it can bind with either a chemical interaction (chemisorption) or a physical interaction (physisorption). Chemisorption involves the formation of a chemical bond between the adsorbate and the surface. Physisorption involves weaker interactions involving the polarization of the adsorbate and surface rather than electron transfer between them. Table 3.1 compares and contrasts several general aspects of chemisorption and physisorption.

Although chemisorption and physisorption may seem like nicely distinguishable categories, a more or less continuous spectrum of interaction strengths going from one to the other exists. Nonetheless, it is a useful distinction to make in most cases. At 0 K all molecular motion apart from that associated with zero point energy ceases. This means that all chemical substances condense if the temperature is low enough. Another way to interpret this is that any chemical substance becomes either chemisorbed or physisorbed if the temperature is low enough. Liquid nitrogen and liquid helium are the two most commonly used cryogenic coolants and, therefore, two important low temperatures to remember are the boiling points of N_2 (77 K) and He (4 K). Most substances stick to a surface held at liquid nitrogen (l-N_2) temperature and virtually everything sticks at liquid helium (l-He) temperature.

Chemisorption is highly directional, as are all chemical bonds. Therefore, adsorbates that are chemisorbed (chemisorbates) stick at specific sites and they exhibit a binding interaction that depends strongly on their exact position and orientation with respect to the substrate. On metals, chemisorbed atoms tend to sit in sites of the highest

Table 3.1   A comparison between chemisorption and physisorption

Chemisorption	Physisorption
Electron exchange	Polarization
Chemical bond formation	van der Waals attractions
Strong	Weak
≥ 1 eV (100 kJ mol^{-1})	≤ 0.3 eV (30 kJ mol^{-1}), stable only at cryogenic temperatures
Highly corrugated potential	Less strongly directional
Analogies with coordination chemistry	

coordination. For instance, oxygen atoms on Pt(1 1 1) bind in the face-centred cubic (fcc) threefold hollow sites with a bond energy of c. 370 kJ mol^{-1} [1]. This is not surprising since a crystal of an fcc metal is constructed by stacking the atoms one layer on top of the next by placing the atoms in the sites of highest coordination. There are, of course, exceptions. Hydrogen atoms usually bind in the sites of highest coordination but can occupy twofold sites on W(1 0 0) [2]. The diamond lattice of a semiconductor is composed of atoms stuck successively on the tetrahedrally directed dangling bonds of the surface. Thus, atoms tend to chemisorb on the highly localized dangling bonds of semiconductors in preference to sites of high coordination.

One important exception to the above rules is when an adsorbate forms such a strong bond with the substrate that it reacts to form a new compound. Oxygen forms strong bonds with a number of elements, for instance, iron, aluminium and silicon are highly susceptible to oxidation. Therefore, oxygen chemisorption can easily lead to formation not only of surface-bound oxygen atoms but also oxygen bound in subsurface sites. The bonding of oxygen atoms between and below the surface atoms of the substrate can be thought of as precursors to the formation of, for example, Fe_2O_3, Al_2O_3 and SiO_2. Hydrogen atoms, because of their small size, are also highly susceptible to the occupation of subsurface sites. Of particular interest is the H/Pd system [3, 4], in which hydrogen atoms can pack into the palladium lattice to a density greater than that of liquid H_2.

Physisorbed species (physisorbates) do not experience such strongly directional interactions. Therefore, they are more tenuously bound to specific sites and experience an attractive interaction with the surface that is much more uniform across the surface. In many cases, the interactions between physisorbates are as strong as or stronger than the interaction with the surface.

3.2   Binding Sites and Diffusion

The binding energy of an adsorbate depends on it position on the surface. This means not only that there are different binding sites at the surface but also that the sites are separated by energetic barriers, as shown in Figure 3.1. To move across the surface an adsorbate has to hop from one site to the next via pathways that traverse these barriers. Therefore, the barriers that separate the binding sites represent diffusion barriers. Figure 3.1(a) can be thought of as a one-dimensional (1D) potential energy surface (PES). Motion from one well to the next represents diffusion. Each well may also contain a number of bound vibrational states, which represent the vibrations of the adsorbate against the surface.

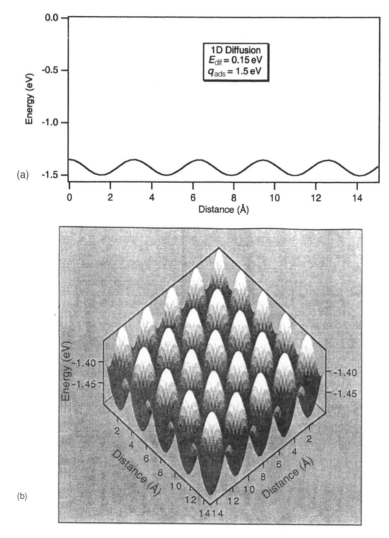

Figure 3.1 (a) One-dimensional (1D) and (b) two-dimensional representation of energy as a function of position on an ideal defect-free surface. Note how in each of these cases the interaction potential of the adsorbate is "corrugated". E_{dif}, activation energy for diffusion; q_{ads}, depth of the attractive well

Consistent with Figure 3.1, diffusion [5] is an activated process. For a uniform potential, we therefore expect diffusion to be governed by a simple Arrhenius form:

$$D = D_0 \exp\left(-\frac{E_{dif}}{RT}\right) \qquad (3.1)$$

where D is the diffusion coefficient, D_0 the diffusion prefactor, E_{dif} is the activation energy for diffusion, R is the gas constant and T is temperature.

There are two extremes of temperature for which Equation (3.1) does not describe diffusion. At very low temperatures for light adsorbates such as hydrdogen and deuterium, quantum effects can predominate [6]. In this tunnelling regime, diffusion is independent of temperature. When the temperature is sufficiently high such that $RT_s \gg E_{dif}$, the adsorbate translates freely across the surface, performing a type of Brownian motion. In other words, it is bound in the z direction but it is not bound in x and y directions. This state of free two-dimensional (2D) motion is known as a two-dimensional gas. The greater the corrugation in the adsorbate/surface potential, the larger the effective diffusion barrier experienced by the adsorbate and the higher T_s must be to form a 2D gas. Chemisorbates experience greater diffusion barriers than do physisorbates. In most cases, the T_s required to form a 2D gas is sufficiently high to engender a significant rate of desorption.

For temperatures RT_s not much greater than E_{dif}, we can interpret Equation (3.1) directly. D is related to the hopping frequency, v, the mean-square hopping length (related to the distance between sites) d^2, and the dimensionality of diffusion, b, by

$$D = \frac{vd^2}{2b} \tag{3.2}$$

For 1D diffusion $b = 1$, whereas for uniform diffusion in a plane $b = 2$. The root mean square distance, $\langle x^2 \rangle^{1/2}$, travelled by an adsorbate diffusing in one dimension in time t is given by

$$\langle x^2 \rangle^{1/2} = (2Dt)^{1/2} \tag{3.3}$$

Diffusion of a single particle in a uniform 1D potential is particularly simple; nonetheless, analysis of 1D diffusion and our knowledge of adsorbate interactions lead to several important conclusions. Figure 3.1 demonstrates that the adsorbate/surface potential is corrugated. This corrugation takes two forms: geometrical, as probed, for example by scanning tunnelling microscopy (STM), and energetic, which results in diffusion barriers. Since different binding sites have different binding energies, they also exhibit different diffusion barriers. In a 1D potential, the highest barrier determines the overall rate of diffusion.

A surface is two-dimensional and this makes diffusion over a surface somewhat more complex. For a simple uniform 2D PES, we substitute $b = 2$ into Equation (3.2) and arrive at

$$\langle x^2 \rangle^{1/2} = (4Dt)^{1/2} \tag{3.4}$$

However, the diffusion barrier is not always uniform across the surface; that is, E_{dif} depends on the direction in which the adsorbate diffuses. In this case, adsorbates diffuse anisotropically. This can lead to the formation of islands of adsorbates that do not have uniform shapes in both the x and the y directions. For instance, on surfaces that exhibit rows and troughs, such as fcc(1 1 0) or the Si(1 0 0)-(2×1) reconstruction, diffusion along the rows is much easier than across the rows. If the diffusing adsorbates tend to stop when they meet each other, they form stringlike islands, as in Figure 3.2, that are much longer in the direction parallel to the row than perpendicular to it. Swartzentruber [7], for instance, has used a scanning tunnelling microscope to track the motion of silicon atoms deposited on a Si(1 0 0)-(2×1) surface. The atoms rapidly diffuse to form dimers, called addimers. The addimers diffuse rapidly along the rows of dimers in the substrate.

Figure 3.2 Anisotropic growth of silicon islands on silicon, resulting from anisotropic diffusion. The lighter features in this scanning tunnelling microscope image are the silicon atoms that have aggregated into islands on top of the Si(1 0 0) substrate. Reprinted from *Surf. Sci.* **268**, Y.-H. Mo, J. Kleiner, M. B. Webb and M. G. Lagally, "Surface self-diffusion of Si on Si(0 0 1)", 1992, 275, with permission from Elsevier Science

Diffusion in this direction can occur either directly over the dimers or between the dimers. Although the between-dimer site is the most energetically favourable, there is a barrier between the on-top and between-dimer positions. Therefore, addimers can diffuse for long periods up and down the rows before they hop into the more stable between-dimer sites. Defects have a great influence on diffusion. Defects can facilitate the transfer from on-top to between-dimer sites. In other cases, defects may act as repulsive walls or as sinks that trap adsorbates.

A typical defect, which we have already encountered in Section 1.1.2, is a step. The electronic structure of steps differs from that of terraces. This leads to a significantly different binding potential at steps and likewise changes the diffusion barriers near steps. Frequently, as shown in Figure 3.3, diffusion in the step-down direction – that is, from one terrace, across a step, to the terrace below – is more highly activated than is diffusion on the terrace. This additional barrier height is known as the Ehrlich–Schwoebel barrier, E_s [8, 9]. For the type of potential shown in Figure 3.3, step-up diffusion is more highly activated than step-down or terrace diffusion. Consequently, only terrace and step-down diffusion are expected unless the temperature is high. Other types of defects as well as adsorbates also influence diffusion at surfaces [10]. The potentials shown in Figure 3.1 are single-adsorbate potentials. Bowker and King [11, 12] and Reed and Ehrlich [13] have shown that lateral interactions play an important role in diffusion. Whereas D is constant in the absence of lateral interactions, repulsive interactions increase the value of D, and attractive interactions decrease the value of D. Thus, the study of diffusion profiles provides one means of measuring the strength of lateral interactions.

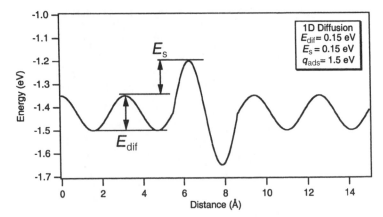

Figure 3.3   The effect of a step on diffusion energy, E_{dif} [one-dimensional (1D) diffusion]. Note that step-up diffusion is often negligible because of the increased barrier; note also the increased binding strength at the bottom of the step—a feature that is often observed. E_s, Ehrlich–Schwoebel barrier; q_{abs}, depth of the attractive well

Molecules execute the type of diffusion described above. Atoms that do not interact too strongly with the substrate diffuse likewise. However, for strongly interacting atoms, particularly for metal atoms deposited on metal substrates, another diffusion mechanism can occur. This is known as the exchange mechanism of diffusion and is depicted in Figure 3.4 This mechanism is particularly important for metal-on-metal growth systems.

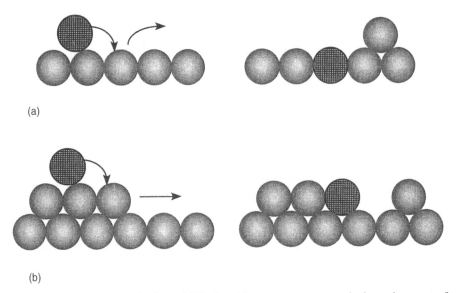

Figure 3.4   The exchange mechanism of diffusion. Mass transport occurs via the replacement of one atom with another. This can happen either (a) on a terrace or (b) at a step

3.3   Physisorption

Physisorption is a ubiquitous phenomenon. All atoms and molecules experience long-range van der Waals forces. Thus, virtually any species physisorbs if the temperature is low enough. The only exception is when the adsorbate experiences such a strong chemical attraction for the substrate that it cannot be stopped from falling into the chemisorption well; for example, rhenium adsorption on a rhenium surface. For rhenium, which has the strongest metal–metal bond, it is impossible to stop the rhenium atom from falling directly into the chemisorbed state.

Rare gases can interact only through van der Waals forces; therefore, physisorption is the only way in which they can attach to a surface. Hence, these species are bound to surfaces only at low temperatures. Some surfaces are particularly inert and therefore favour physisorption. Examples of passivated surfaces are hydrogen-terminated diamond and silicon, and sulfur-terminated rhenium as well as graphite, all of which exhibit remarkably low reactivity. These surfaces are so 'stingy' with their electrons that physisorption is the primary means of interaction between chemical species and the surface. Low surface temperature can also stabilize physisorbed states of more reactive species, especially if formation of the chemisorbed state is an activated process.

van der Waals forces exist not only between adsorbates and substrates but also among adsorbates. Because the adsorbate–surface interaction is so weak in physisorption, lateral interactions are very important for physisorbed molecules and can be as strong as the adsorbate–substrate interaction. Furthermore, physisorption is less site-specific than is chemisorption; therefore at high coverage incommensurate structures – ordered structures that lack registry with the substrate – may be formed.

Advanced topic: theoretical description of physisorption

Zaremba and Kohn [14–16] laid the foundations for a theoretical description of physisorption in the mid 1970s. The total physisorption potential between an atom and a metal surface, $V_0(z)$, can be written in terms of a short-range repulsive potential, $V_{HF}(z)$, and a long-range van der Waals attraction, $V_{corr}(z)$. As is customary, z is chosen to be the atom–surface separation along the surface normal. The repulsive term arises from the overlap of (filled) atomic orbitals with the electronic states of the surface. The approaching atom experiences a Pauli repulsion when the electrons of the surface attempt to interact with the filled atomic orbitals. If the closed-shell species is to remain closed-shell, its orbitals must be orthogonal to any metal states with which they interact. It is this orthogonalization energy that gives rise to the Pauli repulsion. This interaction can be approximated by using a Hartree–Fock treatment. The attractive van der Waals term is given by the Lifshitz potential:

$$V_{corr}(z) = \frac{C}{(z - z_0)^3} + O(z^{-5}) \qquad (3.5)$$

The terms of order z^{-5} are usually neglected and arise from multipole and other higher-order perturbations. Note that $V_{corr}(z)$ decreases as z^{-3} whereas the van der Waals interaction between two atoms decreases at z^{-6}; that is, because of the many-body surface, a physisorption interaction decays more slowly than does the van der Waals interaction between two isolated atoms. Equation (3.5) describes this interaction well only in the region in which no appreciable overlap between the atom and metal orbitals occurs. The van der Waals interaction arises from the interaction of an instantaneous dipole on the atom with an induced charge fluctuation in the metal. Therefore, the constant C is related to the polarizability of the atom, $\alpha(\omega)$, and the dielectric function of the metal, $\varepsilon(\omega)$:

$$C = \frac{1}{4\pi} \int_0^\infty \alpha(i\omega) \frac{\varepsilon(i\omega) - 1)}{\varepsilon(i\omega) + 1} \, d\omega \tag{3.6}$$

where $i = \sqrt{-1}$, and z_0 is the position to which the Lifshitz potential is referenced and is found from the weighted average of the centroid of the induced surface charge, $\bar{z}(i\omega)$:

$$z_0 = \frac{1}{2\pi C} \int_0^\infty \alpha(i\omega)\bar{z}(i\omega) \frac{\varepsilon(i\omega) - 1}{\varepsilon(i\omega) + 1} \, d\omega. \tag{3.7}$$

Equation (3.7) demonstrates that the position of the reference plane depends not only on the charge distribution but also on the dynamic screening properties of the surface.

The total potential can then be written

$$V_0(z) = V_{\mathrm{HF}}(z) + V_{\mathrm{corr}}(z) \tag{3.8}$$

Equation (3.8) succeeds in describing the physisorption potential because the equilibrium distance associated with a physisorption bond tends to be much larger than z_0. This condition ensures that the overlap of the metal and atom wavefunctions is sufficiently small to allow for the use of Equations (3.6)–(3.8).

3.4   Non-Dissociative Chemisorption

3.4.1   Theoretical treatment of chemisorption

Haber suggested that adsorption was related to 'unsaturated valence forces' in the surface of the substrate [17]. Langmuir [18, 19] subsequently formulated and confirmed the concept that chemisorption corresponds to the formation of a chemical bond between the adsorbate and the surface. This was one of his many contributions to surface science that garnered him the Nobel Prize in 1935. Theoretical approaches to the description of the adsorbate–surface bond have taken many approaches [14, 20]. All of these approaches need to contend with the problem of how an atom or molecule is attached to an essentially semi-infinite substrate. The standard methods of theoretical solid-state physics are well developed for dealing with the bulk of the substrate, and these succeed in describing extended electronic states. However, the reduction of symmetry introduced by the surface leads to the formation of localized states and computational difficulties. Furthermore, the adsorbate represents a localized impurity in this reduced-symmetry setting. Alternatively, one might approach the problem from the quantum chemical viewpoint. Quantum chemistry powerfully describes the properties of small molecules. One might then approach the surface as a cluster onto which an adsorbate is bound. The challenge is in attempting to add a sufficient number of atoms to the cluster while still explicitly treating enough electronic wavefunctions. Obviously, both approaches have their advantages and disadvantages.

Before we proceed to the specifics of chemisorption bond formation, a quick review of molecular orbital formation is in order. In Figure 3.5(a), the familiar situation of molecular orbital (MO) formation from two atomic orbitals (AOs) a and b is presented. Recall that AO interactions depend both on the energy and the symmetry of the states involved: the closer in energy the stronger the interactions, and some symmetry combinations are forbidden. Note also that the antibonding state ab* is generally more antibonding than the bonding state ab is bonding; that is, $\beta > \alpha$ in Figure 3.5(a). Consequently, if ab and ab* are both fully occupied, not only is the bond order zero, but also the overall interaction is repulsive. In the gas phase, an orbital must contain exactly zero, one or two electrons, and the orbital energy is well defined (sharp) as long as the molecule remains bound. MOs extend over the entire molecule, though in some cases the electron density is localized about a small region of the molecule or even a single atom.

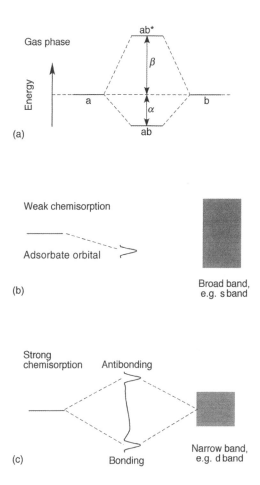

Figure 3.5 Orbital interactions: (a) gas phase; (b) weak chemisorption; (c) strong chemisorption. a, b, atomic orbitals; ab, ab*, bonding and antibonding molecular orbitals

One of the first useful approaches applied to atomic absorption at surfaces was that of the Anderson–Grimley–Newns approach [14, 21–24]. The important point arising from this approach is that the types of electronic states that arise after chemisorption depend not only on the electronic structure of the substrate and adsorbate but also on the coupling strength between the adsorbate and the substrate. The adsorbate levels may end up either inside or outside of the metal band. In the strong-coupling limit [Figure 3.5(c)], in which the adsorbate level interacts strongly with a narrow band (e.g. the d band of a transition metal) the adsorbate and metal orbitals split into bonding and antibonding combinations, one below and one above the metal band. A weak continuous part extends between these split-off states. In the weak-coupling limit [Figure 3.5(b)], in which the adsorbate level interacts with a broad band such as the s band of a transition metal, little of the metal density of states is projected onto the adsorbate. However, the adsorbate level is broadened into a Lorentzian shaped resonance centred on a narrow energy range.

The Grimley and Newns model based on the Anderson Hamiltonian is necessarily a local description of chemisorption because only the interaction of the adsorbate with its nearest neighbours is accounted for. Furthermore, electron correlation can, at best, only be included in an approximate fashion [25]. Consequently, other methods are required to obtain a quantitative understanding of chemisorption.

Figure 3.6 demonstrates the effect of surface proximity on the electronic states of an adsorbate, considered first in the weak chemisorption limit. As a molecule approaches a surface, its electronic states interact with the electronic states of the metal. This broadens the MOs (spread them out in energy) and also lowers the energy of the MOs. Whereas the energetic ordering of MOs generally is not changed, the spacing between them may change. Furthermore, orbitals that were degenerate in the gas phase may have this degeneracy lifted by the presence of a surface. The reason why MOs experience a shift and broadening is that they interact with the electrons of the substrate. If there are no substrate electrons at the energy of a MO, as for the orbital of energy E_3 in Figure 3.6, little interaction occurs and the MO remains sharp. This is particularly important for core levels, which lie below the valence band of the substrate, and for MOs that fortuitously have energies that lie in a band gap of the substrate. As metals generally do not exhibit band gaps, the valence electronic states of adsorbates usually experience strong inter-

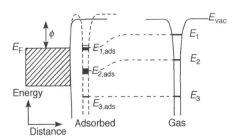

Figure 3.6 The broadening and shifting of adsorbate levels as they approach a surface. E_F, Fermi energy; E_{vac}, vacuum energy; ϕ, work function of the surface material; E_1, E_2, E_3, energies of molecular orbitals 1, 2 and 3, respectively, of the molecule far from the surface; $E_{1,ads}, E_{2,ads}, E_{3,ads}$, energies of molecular orbitals 1, 2 and 3, respectively of the adsorbed molecule; shaded area, occupied band (e.g. valence band)

actions, shifts and broadening. In the case of semiconductors and insulators where band gaps do exist, some adsorbate valence levels may show little interaction with the substrate if they fortuitously fall in a band gap.

Figure 3.6 is also instructive for the strong chemisorption limit. In this case, the bonding combination is shifted lower in energy than the initial adsorbate orbital. Broadening also occurs; indeed the broadening is generally so severe that a continuous band connects the bonding and antibonding split-off states. In this limit it is important to remember that although the final orbitals may be localized primarily on the adsorbate and correlate with specific initial adsorbate orbitals, the final orbitals are combination states that are composed of metal and adsorbate contributions. They also have a spatial extent that may include several substrate atoms.

The energy of an orbital with respect to the Fermi energy, E_F, determines the occupation of that orbital. Just as for the substrate, all electronic states of the adsorbate that lie below E_F are filled, whereas those above E_F are empty. The broadening of electronic states has an interesting effect for states near E_F. When a state lies partially below and partially above E_F, as does the highest orbital in Figure 3.6, it is partially filled. This is particularly important for the strong chemisorption limit because the bonding or antibonding combination often straddles E_F and is more or less occupied depending on its position relative to E_F. The interaction strength depends on the relative filling of the orbitals and their bonding and antibonding character. As we shall see shortly, the position of transition metal d bands relative to E_F is the most important factor in determining the filling of the bonding and antibonding combinations (for a given adsorbate orbital) and therefore the interaction strength.

The two most widely used theoretical approaches to chemisorption problems currently differ in how they set up the problem. In the repeated-slab approach, a small slice of substrate is chosen. This is several atoms wide and several layers thick. To reinstitute the symmetry properties that make bulk calculations more feasible, repeated boundary conditions are used; that is, a stack of slabs separated by a vacuum region is considered and the x and y directions fold back onto themselves. Alternatively, the cluster approach can be used to model the substrate. Increasingly, density functional theory (DFT) in combination with generalized gradient approximation (GGA) corrections is being used in both types of calculations. DFT has made great strides in recent years and has provided a method of treating larger and larger systems with increasing detail and accuracy. DFT methods attempt to model electronic states in terms of the electronic charge density, with the GGA term accounting for the correction for the shape of the electron distribution. An exchange-correlation functional is then tacked onto the charge density. Better expressions for the exchange-correlation functional remains an outstanding challenge for theoreticians. The awarding of the 1999 Nobel Prize in chemistry to Kohn marked the advances and successes of DFT theory.

One drawback of traditional DFT methods is that they are not able to calculate excited electronic states. Time-dependent DFT is, however, making rapid progress along these lines [26]. Configuration interaction (CI) calculations and other types of atomic or molecular orbital based strategies provide alternative methods to address excited states, but the cost and time required to perform these calculations increases rapidly with the number of electrons, making them difficult to perform on large systems. Currently, these methods are limited to 10–100 electrons if a high level of CI is included. This limits the usefulness of this approach for routine calculations with transition metal surfaces.

Whether a slab or cluster approach is more appropriate for a given system depends on the nature of the substrate. For semiconductors, for instance, bonding is highly localized. A cluster approach is therefore feasible, and clusters with the order of 10 atoms can provide a useful model to study the interaction of, for instance, hydrogen with silicon [27]. Metals require more atoms to be present for a fully developed band structure to evolve. The study of small metal clusters is itself an area of active research, largely because many properties as well as the reactivity of metals show cluster size dependence [28, 29]. Although interesting in their own right, these small metal clusters do not necessarily represent a good model of larger metal aggregates and single-crystal surfaces.

3.4.2   The Blyholder model of CO chemisorption on a metal

CO has long served as a model adsorbate [30] and a discussion of how CO binds with a surface illustrates a number of points that aids us in understanding nondissociative molecular chemisorption. We need to consider the electronic structure of CO in the gas phase and how the electronic structure of CO is modified by the presence of a surface.

We start the discussion of CO chemisorption with the Blyholder model [31]. As in gas-phase or coordination chemistry, the frontier orbitals – the highest occupied molecular orbital (HOMO) and the lowest unoccupied molecular orbital (LUMO) – are assumed to have the greatest effect on adsorbate/surface chemistry. Figure 3.7 depicts the frontier orbitals of CO. The HOMO of CO is the 5σ MO. The LUMO is the $2\pi^*$ MO. The former is roughly nonbonding with respect to the C–O bond, whereas the latter is antibonding with respect to C–O.

The alignment of the HOMO and LUMO with respect to the surface is also important in determining the bonding. The 5σ orbital is localized on the carbon end of the molecule. The $2\pi^*$ orbital is symmetrically distributed along the molecular axis. The combination of the energetic and orientational aspects of the molecule/surface interaction can be summarized as follows. The 5σ orbital is completely occupied as it lies below E_F. The π^* orbital is partially occupied. The 5σ orbital interacts strongly with the metallic electronic states. Effectively, the electron density of the 5σ orbital is donated to the metal, and new hybrid electronic states are formed (donation). These are predominately localized about the carbon end of the molecule; however, they also extend over several metal atoms. The $2\pi^*$ orbital accepts electron density from the metal through a process known as back donation. Again, new hybrid electronic states are constructed which are primarily localized about the CO molecule but which also extend over several substrate atoms. The overlap of the 5σ orbital with the metal states is most favourable if the molecule is oriented with the carbon end toward the surface. The overlap of the $2\pi^*$ orbital with the metal states is most favoured by a linear geometry. Consequently, we predict that the CO molecule should chemisorb carbon end down with its axis along the normal to the surface. This expectation is confirmed by experiment. CO is nearly always bound in an upright geometry. One exception is at high coverage on the Ni(1 1 0) surface where lateral interactions cause CO molecules to tilt alternately across the rows [32].

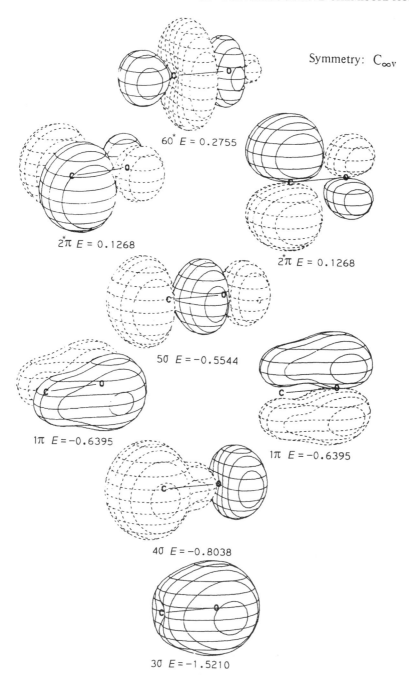

Symmetry: $C_{\infty v}$

$6\sigma^* \; E = 0.2755$

$2\pi^* \; E = 0.1268$

$2\pi^* \; E = 0.1268$

$5\sigma \; E = -0.5544$

$1\pi \; E = -0.6395$

$1\pi \; E = -0.6395$

$4\sigma \; E = -0.8038$

$3\sigma \; E = -1.5210$

Figure 3.7   The molecular orbitals of gas-phase CO. The wavefunction changes sign in going from the regions enclosed by solid lines to those enclosed by dashed lines. Energies, E, are given in atomic units (1 atomic unit $= 27.21$ eV). Orbitals with negative energies are occupied in the ground state of the neutral molecule. Reprodueed with permission from W. L. Jorgensen and L. Salem, *The Organic Chemist's Book of Orbitals* (Academic Press, New York, 1973). ©1973 Academic Press

The bonding character of the 5σ and $2\pi^*$ orbitals also tell us more about the chemisorption interaction. The hybrid orbitals formed by donation and back donation are both bonding with respect to the metal—CO (M—CO) bond (the chemisorption bond). The 5σ orbital is nonbonding in CO. Therefore, modification of this orbital does not have a strong influence on the intramolecular C—O bond. The $2\pi^*$ orbital, however, is antibonding with respect to the C—O bond; hence, the increased occupation of the $2\pi^*$ engendered by back donation leads to a weakening of the C—O bond. A weaker C—O bond leads to increased reactivity of the CO. This is one of the basic elements of catalytic activity. Not only do surfaces provide a meeting place for reactants but also they weaken (or break) bonds in the reactants, which in turn leads to a decrease in the activation barrier for the overall reaction.

The involvement of the $2\pi^*$ orbital allows us to understand the effect of chemisorption on the vibrational spectrum of CO. The decreased bond order of the CO bond caused by back donation into the $2\pi^*$ orbital results in a lower bond strength. Since the bond strength is proportional to the vibrational frequency, we observe a decrease in the vibrational frequency. The extent of this shift depends on the extent to which the $2\pi^*$ orbital is filled. Back donation into this antibonding orbital depends on the chemical identity of the metal, which determines the relative position of the d band with respect to the $2\pi^*$ orbital, as well as on the site that the CO occupies. As in coordination chemistry [33], adsorbed CO experiences greater back donation with increased coordination number of the chemisorption site. Therefore, we expect back donation to increase in the order on-top (onefold coordination) < twofold bridge < fourfold hollow on an fcc(1 0 0) surface. Correspondingly, the vibrational frequency should decrease in the order on-top > twofold bridge > fourfold hollow. These general trends are borne out by experiment.

The Blyholder model, the use of frontier orbitals and analogies to coordination chemistry have been successful in explaining the qualitative trends observed in the chemisorption of small molecules. However, when looked at in detail, the situation is more complicated than the simple picture outlined above. For instance, the adsorption energy of CO on Ni(1 0 0) is nearly twice that of CO on Cu(1 0 0); nevertheless, the C—O stretch frequency differs by less than 25 cm^{-1} [34]. In order to understand specific systems and quantitative characteristics, we need to reconsider the assumptions of the Blyholder model.

Ab initio calculations in conjunction with experimental data [34, 35] have shown that not only the 5σ and $2\pi^*$ orbitals but also the 4σ and 1π orbitals must be taken into account. In other words, the formation of bonds at the surface tends to be somewhat more complex than in the gas phase. Analysis of how the CO molecular orbitals hybridize with the electronic states of the metal shows that the σ bonding interactions slightly strengthen the C—O bond but destabilize the M—CO bond. All the σ states that contribute to bonding are initially full, and the lowest unoccupied σ state (6σ) is far too high in energy to contribute to the bonding. Therefore, the net effect of the σ states is to reorganize the electrons with respect to the gas-phase electron densities. They have little influence on the C—O bond energy but are repulsive with respect to the chemisorption bond rather than attractive as assumed in the Blyholder model. The π system, however, still acts as we expected in the Blyholder model. The π system is largely responsible for the M—CO bond, and the back donation into these hybrid orbitals is antibonding with respect to the

C−O bond. Furthermore, the energetic position of the d bands with respect to the $2\pi^*$ orbital is still an important parameter in determining the orbital populations.

Several lessons are learned from CO adsorption. The fundamental principle of chemical bonding still prevails. Orbitals combine to form bonding and antibonding combinations. The relative occupation of bonding orbitals with respect to antibonding orbitals determines relative bond strengths. The filling of orbitals that strengthen one bond, for instance the M−CO bond, generally leads to a weakening of other bonds, for example the C−O bond. As a first approximation, we should consider the interaction of frontier molecular orbitals with the electronic states of the substrate near the Fermi energy. Any hybrid orbitals that fall below E_F are occupied and those above it are unoccupied. Adsorbate-associated orbitals are broadened by their interaction with the substrate. Therefore, they may straddle E_F and become partially occupied. To understand quantitatively the bonding in an adsorbate/substrate system it may be necessary to consider other orbitals beyond the frontier orbitals.

3.4.3 Molecular oxygen chemisorption

Let us revisit the vibrational spectrum of O_2/Pd(111) (Figure 2.28, page 73), in light of our understanding of the Blyholder model. The three distinct vibrational frequencies in the O−O stretching region are indications of three distinct molecular O_2 species with different bond orders. Decreasing vibrational frequency corresponds to increased occupation of the π^*-related states and increasing M−O_2 binding. Vibrational spectroscopy alone does not allow us unambiguously to identify the geometry of the three O_2 species. However, with the aid of X-ray analysis and recourse to coordination chemistry we are able to identify the three species as those shown in Figure 3.8.

The interaction of O_2 with palladium is more complicated than that of CO with palladium. In part, this is caused by the increased reactivity of O_2 with palladium. CO does not dissociate at low temperature, whereas O_2 partially dissociates when an O_2-covered surface is heated above *c.* 180 K. This can be related to the MO structure of O_2. The two $2\pi^*$ orbitals are half-filled and degenerate in the gas phase. The degeneracy is lifted in the adsorbed phase because one orbital is oriented perpendicular and the other parallel to the surface. Both become available upon chemisorption for various amounts of back donation. In a sense, the structures depicted in Figure 3.8 are almost a snapshot of the dissociation of O_2 on Pd(111). It should come as no surprise to discover that the O_2

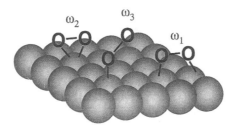

Figure 3.8 O_2/Pd(1 1 1) adsorbate structure. The labelling of the three states (ω_1−ω_3) correlates with the loss peaks observed in the electron energy loss spectrum, as shown in Figure 2.28, page 73

species that is most likely to dissociate is the ω_1 species, that is, the O_2 species with the strongest M$-O_2$ bond and the weakest O$-$O bond. Nonetheless, other simple expectations about the O_2/Pd system are not borne out [36]. The most strongly bound species (ω_1) is the last to be populated when the surface is dosed at $T_s \approx 100$ K, while the moderately bound ω_2 species is the first to be populated. Dissociation of O_2 does not occur until the temperature is raised. If adsorption is carried out at lower temperature, $T_s \approx 30$ K, a physisorbed molecular state is formed instead of the chemisorbed state [37]. This state has a vibrational frequency almost identical to that found in the gas phase, as expected for a van der Waals bound species that does not experience charge transfer into antibonding MOs from the surface.

The data for O_2/Pd(111) indicate that the predictions of simple equilibrium thermodynamics (i.e. that the system should attain the state of lowest free energy) are not borne out for the interaction of O_2 with a low-temperature palladium surface. Hence activation barriers must separate the various molecular physisorbed, molecular chemisorbed and dissociatively adsorbed states. Reaction kinetics and dynamics control the state of the system and must prevent it from attaining the state of lowest free energy. This highlights the importance of always determining whether the system under consideration has attained equilibrium or merely a steady state. Reaction kinetics and dynamics are discussed thoroughly in the following chapters.

3.4.4 The binding of Ethene

As a model of the binding of a polyatomic molecule to a surface, we consider ethene (ethylene). How would $H_2C=CH_2$ bind at a surface? In the gas phase, ethene is a planar molecule with a polarizable π-bond system connected with the C=C double bond. At cryogenic temperature, H_2CCH_2 binds weakly to the surface, lying flat on the surface as depicted in Figure 3.9(a). With a binding energy of -73 kJ mol^{-1} on Pt(111) [1] this species can be considered weakly chemisorbed. The adsorption geometry is determined by the π-bonding interaction, which finds the strongest interaction when the H_2CCH_2 lies flat rather than end-on.

A stronger interaction with the surface requires more extensive exchange of the π electrons with the surface. As π electrons are donated from H_2CCH_2 to the surface to form two σ bonds, the bond order drops to one between the carbon atoms. A planar geometry is no longer favourable for a H_2CCH_2 species. The adsorbate prefers to assume a structure closer to that of ethane. The C$-$C axis remains parallel to the surface, [Figure 3.9(b)] but the hydrogen atoms bend up away from the surface so that the bonding about the carbon atoms can assume a structure consistent with sp^3 hybridization. This electronic rearrangement occurs when π-bonded C_2H_4/Pt(111) is heated above 52 K [40]. The transformation of the π-bonded planar species to the di-σ bound chemisorbed species is mirrored in the vibrational spectrum. The C$-$H stretches are IR inactive in the π-bonded species (no component along the normal) but not the chemisorbed species [38]. The C$-$C stretch, if it could be observed, would also shift to a significantly lower frequency in the chemisorbed state.

On Pt(111), the di-σ species is stable up to c. 280 K, with a binding energy of -117 kJ mol^{-1}. Above this temperature, the molecule rearranges, and at room tempera-

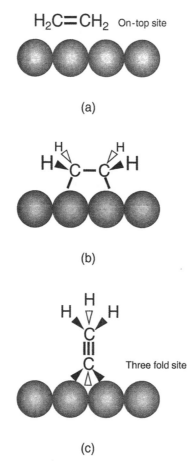

Figure 3.9   The binding of ethene at a metal surface: (a) the weakly chemisorbed π-bonded C_2H_4; (b) the di-σ-bonded chemisorbed state, (c) ethylidyne

ture the stable configuration is that of ethylidyne ($C-CH_3$), as shown in Figure 3.9(c), which requires the loss of one hydrogen atom per molecule via H_2 desorption. The migration of the hydrogen atom from one side of the molecule to the other is accompanied by the 'standing up' of the molecule. The bond breaking and formation is activated, which is why this adsorption geometry is not formed at low temperature.

The three C_2H_4-derived species exhibit not only different binding geometries and energies but also different reactivities and roles in the hydrogenation of C_2H_4 [39]. Whereas we might not expect the π-bonded species to be important at room temperature or above, high pressures (*c*. 1 atm) and the presence of H_2 in the gas phase can lead to appreciable coverages of this species. The coverage is determined by the dynamic balance between adsorption, desorption and reaction–topics that are dealt with in more detail in Chapters 4 and 5. The π-bonded species is more reactive than is the di-σ-bonded C_2H_4 toward hydrogenation, and ethylidyne does nothing more than occupy sites that could otherwise be used for reaction. The catalytic reactivity, as discussed further in Chapter 5, depends sensitively on the strength of the adsorbate–surface interaction and the balance between a strong enough interaction to activate the adsorbate but not so strong that the adsorbate prefers to remain chemisorbed rather than reacting further.

3.5   Dissociative Chemisorption: H_2 on a Simple Metal

The dissociation of H_2 on a surface is the prototypical example of dissociation at a surface [40–42]. The measurement of the dissociation probability (dissociative sticking coefficient) of H_2 and its theoretical explanation serve as the basis for a useful historical record of the advance not only of our understanding of reaction dynamics at surfaces but also as a record of the ever-more sophisticated experimental and theoretical techniques that have been brought to bear on surface reaction dynamics [43].

Figure 3.10 is an oversimplification but it is a useful starting point to understand dissociative adsorption. The real-life situation is more complicated because each molecular orbital of the molecule interacts with the surface electronic states. Each combination of molecular orbital plus surface orbital generates a bonding plus antibonding pair. The positions and widths of each of these new hybrid orbitals change as the interaction of the molecule with the surface increases. Close accounting of orbital occupation must be made not only to determine the filling of orbitals but also to keep track of to which bond an orbital is bonding or antibonding. For example, in the case of

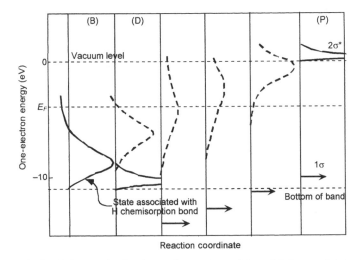

Figure 3.10   Calculated changes in the electronic structure of the orbitals associated with H_2, as the molecule approaches a magnesium surface. Moving to the left in the figure represents motion toward the surface. The molecule is physisorbed with its axis parallel to the surface in panel (P). Panel (D) corresponds to the top of the barrier before dissociation. Panel (B) represents the chemisorbed atoms. The molecular orbitals are replaced by one broad filled state associated with the binding of the atoms to the surface. Note that magnesium is an s-band metal for which d-band interactions need not be considered. Reproduced with permission from J. K. Nørskov, A. Houmøller, P. K. Johansson and B. I. Lundqvist, *Phys. Rev. Lett.* **46** (1981) 257. ©1981 by the American Physical Society

CO chemisorption, back donation into the π system is bonding with respect to the M−CO bond but antibonding with respect to the C−O bond.

As a closed-shell species approaches a surface, a repulsive interaction develops as a result of the Pauli repulsion of the electrons in the solid and molecule. A Pauli repulsion is the repulsive force that arises between filled electronic states because no more than two electrons can occupy an orbital, as predicted by the Pauli exclusion principle. Filled states must, therefore, be orthogonal, and orthogonalization costs energy. For a rare gas atom this is the end of the story because excited states of the atom lie high in energy. For H_2, however, electron density can be donated from the metal into the $2\sigma^*$ antibonding orbital. Electron transfer leads to a weakening of the H−H bond and a strengthening of the adsorbate–surface interaction. Eventually dissociation occurs if enough electron density can be transferred. Dissociation can be thought of as a concerted process in which the transfer of electron density leads to a gradual weakening of the H−H bond while M−H bonds form.

The H_2 1σ and $2\sigma^*$ orbitals shift and broaden as they approach the surface. Electron transfer from the metal to the H_2 occurs because the $2\sigma^*$ orbital drops in energy and broadens as H_2 approaches the surface. As it drops below E_F, electrons begin to populate the orbital and the H_2 bond grows progressively weaker while the M−H bonds become progressively stronger [44–46]. The process of charge transfer is illustrated in Figure 3.10.

Because the MOs involved in bond formation have a definite geometry, the orientation of the molecule with respect to the surface must be correct to facilitate orbital overlap and charge transfer. Both the final state of the dissociated molecule (two adsorbed atoms) and the orientation of the $2\sigma^*$ orbital render a parallel approach of the H_2 most favourable to charge transfer and dissociation. Furthermore, just as there is corrugation in a chemisorption potential, corrugation exists in the energetics of H_2 dissociation. In other words, there are specific sites on the surface that are more favourable than others to dissociation. The corrugation in the H_2–surface interaction exhibits both energetic and geometric components [47].

3.6   What Determines the Reactivity of Metals?

A number of new concepts have been introduced in the preceding sections. Therefore, this section is intended to act in part as a résumé, reinforcing concepts of bonding at surfaces. It also extends our understanding so that we can comprehend quantitative trends in adsorption energies and dissociation probabilities.

We already have some feeling for the reactivity of metals. Gold and silver are particularly attractive for jewellery not only because of their preciousness but also because they oxidize slowly. Iron is not very attractive for this purpose because it readily rusts. However, the oxidation behaviour of a metal is only one measure of reactivity. Platinum is also prized in jewellery, particularly in Japan, and resists oxidation as well as other chemical attack. Nonetheless, it is one of the most useful metals in catalysis. So what makes gold so noble whereas metals such as platinum and nickel are highly active catalytically?

As is to be expected, surface structure plays a role in reactivity. The presence of structural defects can lead to enhanced reactivity. For instance, C–C and C–H bonds are more likely to be broken at step and kink sites [48] on platinum-group metals than at terraces. Adsorption is generally more likely to be hindered by an energetic barrier (see Section 3.9) on close-packed metal surface – that is bcc(1 1 0), fcc(1 1 1) and hcp(0 0 1) (bcc, body-centred cubic; fcc, face-centred cubic; hcp, hexagonal close-packed) – than on more open planes. We return to issues of the surface structure dependence of reactivity in Chapter 5. Putting aside for a moment these structural factors, we concentrate below on the electronic factors that differentiate one metal from another.

In this discussion we follow the theoretical work of Hammer and Nørskov [49–51], who investigated the chemisorption of hydrogen and oxygen atoms and the dissociative chemisorption of H_2 molecules over a variety of metals. In the previous section, we considered a simple two-level picture of H_2 dissociation. We now look in more detail at how the molecular orbitals interact with the s and d bands of the metal, affecting binding energies and activation barriers.

We begin with the chemisorption of an atom, taking hydrogen as a representative example. The mixing of hydrogen and metal orbitals can be conceived of as a two-step process, as depicted in Figure 3.11. In step 1, the hydrogen 1s orbital interacts with the s band of the metal. The s band of transition metals is very broad, thus the interaction is of the weak chemisorption type. This leads to a bonding level far below the Fermi energy. Accordingly, the bonding level is filled and the overall interaction is attractive. The

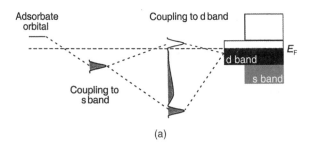

(a)

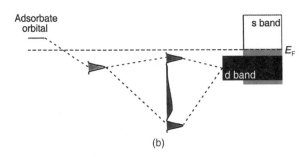

(b)

Figure 3.11   The two-step conceptualization of chemisorption bond formation on transition metal surfaces: (a) early transition metal; (b) coinage metal. E_F, Fermi energy

strength of this interaction is roughly the same for all transition metals. In step 2 we consider the interaction of the bonding level with the metal d band. This interaction leads to the formation of two levels. The first is shifted to lower energies than the original bonding state, and the second is positioned slightly higher in energy than the unperturbed metal d band. The low-energy state is bonding with respect to adsorption whereas the high-energy state is antibonding. The overall energetic effect of the resulting states depends on the coupling strength between the adsorbate s-band hybrid with the d band and the extent of filling of the antibonding state. The filling of the antibonding states is, in turn, determined by the position of the antibonding state with respect to the Fermi energy. As both the coupling strength and the Fermi energy depend on the identity of the metal, differences in step 2 explain the differences in chemisorption bond strength between metals.

Moving from left to right across a row of the transition metals, the centre of the d bands moves further below E_F. The antibonding state formed in step 2 follows the position of the d bands. For early transition metals [Figure 3.11(a)] the antibonding state lies above E_F and is not filled. Consequently, both step 1 and step 2 are attractive, and atomic chemisorption is strongly exothermic. For copper, silver and gold, however, the antibonding state lies below E_F and is filled [Figure 3.11(b)]. The resulting repulsion tends to cancel out the attraction from step 1. For both hydrogen and oxygen on copper, the overall interaction is marginally attractive, whereas for gold chemisorption is unstable. Thus, the trend across a row can be explained in terms of ε_d, the energy of the d-band centre. This correlation is confirmed in Figure 3.12 in which a linear relationship between ε_d and the oxygen atom binding energy is found.

To explain why gold is more noble than copper, the magnitude of the coupling in step 2 needs to be considered. The orthogonalization energy between adsorbate and metal d orbitals, which is repulsive, increases with increasing coupling strength. This energy increases as the d orbitals become more extended. The 5d orbitals of gold are more extended than are the 3d orbitals of copper, which renders gold less reactive than copper because of the higher energy cost of orthogonalization between the H 1s and Au 5d orbitals.

Hence, there are two important criteria effecting the strength of the chemisorption interaction: (1) the degree of filling of the antibonding adsorbate–metal d states (which correlates with ε_d) and (2) the strength of the coupling. The filling increases in going from left to right across a row of transition metals in the periodic chart and is complete for the coinage metals (copper, silver and gold). The coupling increases in going down a column in the periodic table. It also increases in going to the right across a period.

The principles used to explain atomic adsorption can also be extended to molecular adsorption. Indeed, in Chapter 5 we encounter an example (steam reforming of hydrocarbons over a nickel catalyst) in which the knowledge gained from first-principle calculations has been used to fashion a modified working industrial catalyst. The two-step process is applicable, for instance, to CO. The result is a description of the bonding that is in accord with the Blyholder model. The 5σ-derived states (bonding and antibonding combinations) are predominantly below E_F and therefore lead to a repulsive interaction. The $2\pi^*$-derived states lead to attractive interactions because the bonding combination lies below E_F and the antibonding combination is (at least partially) above E_F. Moving to

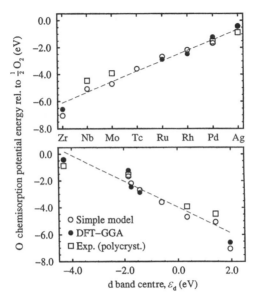

Figure 3.12 The interaction strength of chemisorbed oxygen and the way in which it varies across a row of transition metals. In the upper panel the good agreement between experimental and theoretical results is shown. In the lower panel the linear relationship between interaction strength and the energy of the d-band centre, ε_d, is demonstrated. rel., relative, Exp. (polycryst.), experimental results for polycrystalline sample; DFT, density functional theory; GGA, generalized gradient approximation. Source of data for experimental results: I. Toyoshima and G. A. Somorjai, *Catal. Rev.-Sci. Eng.* **19** (1979) 105. Figure reproduced with permission from G. Ertl, "Dynamics of reactions at surfaces", in *Advances in Catalysis, Volume 45*, eds B. C. Gates and H. Knözinger (Academic Press, Boston, MA, 2000), page 1. ©2000 Academic Press

the left in the periodic chart, the M–CO adsorption energy increases as the filling of the $2\pi^*$-metal antibonding combination rises further above E_F. However, the adsorption energy of carbon and oxygen increases at a greater rate than that experienced by the molecular adsorbate. Therefore, a crossover from molecular to dissociative adsorption occurs. This happens from cobalt to iron for the 3d transition, from ruthenium to molybdenum for 4d and from rhenium to tungsten for the 5d transition. Similar trends are observed for N_2 and NO.

In the dissociative chemisorption of H_2 both the filled σ_g orbital and the unfilled σ_u^* MOs must be considered. The σ_g orbital acts much like the H 1s orbital explained above. The σ_u^* orbital also undergoes similar hybridization, and it is the sum of both sets of interactions that determines whether an energy barrier stands in the way of dissociation. Again, the energetic positioning of the bonding and antibonding combinations with respect to the metal d band, the filling of the levels and the coupling strength are decisive. For dissociative adsorption, the strength of the σ_u^* interactions is the dominant factor that determines the height of the activation barrier. Similar trends appear for the chemisorption energy of hydrogen and the activation barrier height for H_2. Copper, silver and gold exhibit substantial barriers (>0.5 eV). The barrier height, if present at all, decreases rapidly to the left of the periodic table. For instance for nickel, the close-packed Ni(111) surface has a barrier height of only 0.1 eV, whereas other more open faces exhibit no barrier at all [52].

3.7 Atoms and Molecules Incident on a Surface

Adsorption and desorption are the simplest and most ubiquitous surface reactions. All chemistry at surfaces involves adsorption and/or desorption steps at some point. Therefore, it is natural that we should start any discussion of reaction dynamics at surfaces with a thorough investigation of these two fundamental steps. Furthermore, we will find that a great deal of what we learn by the study of adsorption and desorption will be transferable to the understanding of more complex surface processes.

3.7.1 Scattering channels

We now turn from a discussion of energetics and bonding interactions to a discussion of the dynamics of molecule–surface interactions. That is, we turn from looking at the adsorbed phase to how molecules approach and enter the adsorbed phase. A fundamental understanding of gas–solid dynamical interactions was greatly advanced by the introduction of molecular beam techniques into surface studies [53–60]. When these techniques were combined with laser spectroscopy in the groups of Ertl [61–63], Somorjai [64–66], Zare [67–69] and Kleyn, Luntz and Auerbach [70–72] we obtained our first glimpses of energy exchange between molecules and solid surface on a quantum-state-resolved level. Cavanagh and King [73, 74], Bernasek and co-workers [75–77], Haller and co-workers [78–83] and Kubiak, Sitz and Zare [84, 85] then extended such studies to molecular desorption and recombinative desorption. We are now in the position to address the question: what is the outcome of a generalized molecule–surface scattering event? To answer this question, as we shall see, we need to understand potential energy hypersurfaces and how to interpret them with the aid of microscopic reversibility.

We can classify these encounters in several ways, as illustrated in Figure 3.13. Events in which no energy is exchanged correspond to elastic scattering. Although no energy is exchanged, momentum is transferred. For the molecular beam scattering depicted in Figure 3.13, elastic scattering is characterized by scattering in which the angle of

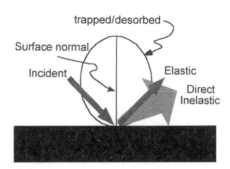

Figure 3.13 Scattering channels for a molecule incident on a surface

incidence is equal to the angle of reflection (specular scattering). The experimental and theoretical aspects of elastic scattering have received much attention in the literature [86, 87]. A special case of elastic scattering is diffraction. Diffraction is particularly important for light particles, as we have already seen for electrons (see Section 2.2), but it can also be important for heavier particles, particularly H_2 and helium. Indeed, while low-energy electron diffraction (LEED) was important in demonstrating the wavelike nature of the electron, one of the first demonstrations of the wavelike nature of an atom was provided by the experiments of Stern who observed the diffraction of helium from single-crystal salt surfaces [88–90]. Even neon can experience the effects of diffraction in scattering [91]. Diffractive scattering depends sensitively on the structure of the surface, and helium diffraction has been used as a powerful tool to examine the structure of surfaces [92] and may even find applications in microscopy [93]. A particle must lose energy to stick to a surface; therefore, an elastically scattered particle always leaves the surface and returns to the gas phase.

As far as reactions at surfaces are concerned, we are more interested in scattering events in which energy is exchanged. This is inelastic scattering. If a particle is to adsorb on a surface it must lose a sufficient amount of energy. Therefore, we differentiate between two types of inelastic scattering. The first type is direct inelastic scattering in which the particle either gains or loses energy and is returned to the gas phase. A useful example of direct inelastic scattering is the scattering of helium atoms, which can be used to investigate surface phonons, low-frequency adsorbate vibrations and diffusion [94, 95]. In Figure 3.13, the direct inelastic channel has a broader angular distribution centred about a subspecular angle owing to a range of momentum-transfer events in which, on average, the scattered molecules have lost energy. Superspecular scattering is indicative of a net gain of energy by the scattered beam. Equating subspecular (superspecular) scattering with a net loss (gain) of energy from the beam to the surface is only strictly correct for a flat surface. In this case, the parallel component of momentum is conserved and only the normal component can change. Scattering from a corrugated surface can lead to coupling between parallel and normal momentum. Nonetheless, particularly if the incident kinetic energy is not too high, conservation of parallel momentum is often observed in direct inelastic scattering.

A special case of inelastic scattering is when the particle loses enough energy to be trapped in the adsorption well at the surface. This is known as trapping or sticking. Trapping is when a molecule is transferred from the gas phase into a state that is temporarily bound at the surface. If the molecule returns to the gas phase we call the overall process trapping/desorption. Sticking is when a molecule is transferred into a bound adsorbed state. The difference between trapping and sticking becomes most distinct in the discussion of precursor-mediated adsorption (Section 3.7.5). The distinction between trapping and sticking is lost at high temperatures when surface residence times become very short even for chemisorbates. Because direct inelastic scattering and sticking lead to such different outcomes, we generally associate the term 'inelastic scattering' only with direct inelastic scattering. In contrast to elastic or direct inelastic scattering, molecules that leave the surface after trapping or sticking do so only after a residence time on the surface characteristic of the desorption kinetics. Desorption, regardless of whether it is normal thermal desorption or trapping/desorption, generally is associated with an angular distribution that is centred about the surface normal because the molecules have lost all memory of their initial conditions in the beam.

Molecular beam techniques are particularly well suited for the study of all forms of elastic and inelastic scattering, including those that lead to reactions [96–98]. Supersonic molecular beams have found wide use in chemical dynamics [99, 100]. They allow the experimenter to control the angle of incidence and energy as well as the flux of the impinging molecules. Molecular beams can be either continuous or pulsed. Pulsed molecular beams are particularly well suited to time-resolved studies of dynamics and kinetics. A typical molecular beam apparatus is sketched in Figure 3.25 (page 141). For further details on molecular beam methods consult [99] and [101]–[103].

As dynamicists studying surface reactivity we are interested in the probability with which a particle adsorbs on the surface. From the above discussion, the sticking probability on a clean surface, or initial sticking coefficient, s_0, is readily defined as

$$s_0 = \lim_{\theta \to 0} \frac{N_{\text{stick}}}{N_{\text{inc}}} = \lim_{\theta \to 0} \frac{N_{\text{stick}}}{N_{\text{el}} + N_{\text{in}} + N_{\text{stick}}} \qquad (3.9)$$

where N_{stick} is the number of particles that stick to the surface, N_{inc} is the total number incident on the surface, N_{el} is the number scattered elastically, and N_{in} is the number scattered inelastically but which do not stick. Values of s_0, even for simple molecules such as H_2 and O_2, can vary between 1 and less than 10^{-10}. This extreme range of sticking probabilities indicates that sticking is extremely sensitive to the exact form of the molecule–surface interaction potential. We will investigate the mathematical treatment of the sticking coefficient in more detail in Chapter 4, in particular, how it changes with coverage. For now, we concentrate mainly on the dynamical factors that determine the sticking coefficient on a clean surface. In the following sections we characterize the adsorption process by defining what we mean by nonactivated and activated adsorption as well as direct and precursor-mediated adsorption.

3.7.2 *Nonactivated adsorption*

One of the most illustrative ways to conceptualize the binding of an adsorbate is in terms of a 1D potential, as shown in Figure 3.14. Such diagrams were introduced by Lennard-Jones in 1932 to aid in the understanding of H_2 adsorption on metal surfaces [104], hence, the name Lennard-Jones diagrams. A molecule must lose energy to stick to a surface. Obviously, total energy is one of the decisive factors that determines whether a molecule can stick. The curve in Figure 3.14 represents the energy of a molecule with zero kinetic energy as a function of the distance from the centre of mass of the molecule to the surface. We represent a molecule with some kinetic energy by a point above the curve, and the distance between the point and the potential energy curve is equal to the kinetic energy.

Figure 3.14 represents a potential energy curve for nonactivated adsorption. In nonactivated adsorption the molecule does not encounter an energetic barrier as it moves closer to the surface, and adsorption is downhill all the way. From Figure 3.14 we can draw several important conclusions. First, the less energy a molecule has, the easier it

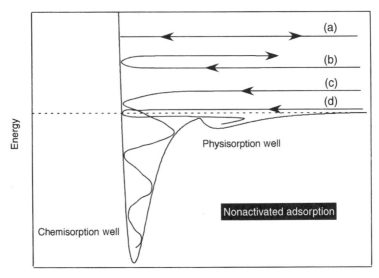

Figure 3.14 A one-dimensional representation of nonactivated adsorption: (a) elastic scattering trajectory; (b) direct inelastic scattering trajectory; (c) sticking event (chemisorption); (d) sticking event (physisorption)

is for the molecule to 'roll downhill' and stick in the chemisorption well. With increasing amounts of kinetic energy, the molecule has to lose progressively more kinetic energy when it strikes the surface. Therefore, the initial sticking coefficient decreases as the kinetic energy of the molecule increases. These intuitive expectations are true for direct nonactivated adsorption except at extremely low energies. For physisorbed systems, as the velocity of the incident molecule approaches zero, the sticking coefficient actually tends to zero. This is a quantum mechanical effect, which has been observed for neon incident on Si(100) [105]. This effect occurs only at exceptionally low energy, is not important for reactive systems and will not be considered further.

Figure 3.14 depicts four hypothetical trajectories. In trajectory a [Figure 3.14(a)], the molecule approaches the surface with a high kinetic energy. As it enters the chemisorption well, the total energy is constant but the distance between the trajectory and the potential energy curve (i.e. the kinetic energy) increases. The molecule is accelerated toward the surface by the attractive force of the chemisorption potential. The molecule then reflects from the repulsive back wall of the potential and retreats from the surface with the same total energy. This is an elastic scattering trajectory. Trajectory b [Figure 3.14(b)] approaches the surface, is reflected from the repulsive wall but, in this case, the molecule bounces off the surface with more kinetic energy than it had initially. This is a direct inelastic scattering trajectory in which the molecule gains energy. Trajectory c [Figure 3.14(b)] represents a sticking event. The molecule again approaches the surface and is scattered off the repulsive wall. Now, however, it loses sufficient energy on its first encounter with the wall that its total energy drops below the zero of the potential energy curve. The molecule then loses further energy and drops to the bottom of the chemisorption well. In trajectory d [Figure 3.14(d)], the molecule sticks but in this case it falls into the physisorption well. After sticking in the physisorption well, a barrier must be overcome to enter the chemisorbed state.

Trajectory c poses an interesting question. How long does it take the molecule to reach the bottom of the chemisorption well once it has hit the surface? Note that for the molecule to stick it need not lose *all* of its kinetic energy on the first bounce. It only has to lose *enough* energy to drop below the zero of the potential energy curve. Therefore, the molecule can retain some energy after the first encounter with the surface and it may take several bounces before the molecule fully equilibrates and falls to the bottom of the well. Indeed, from Figure 3.14 we can easily imagine a trajectory that bounces off the back wall and falls only into the chemisorption well after it samples the physisorption well. As we shall see in Section 3.7.5 and in Chapter 4, this result has important ramifications both for the dynamics and for the kinetics of adsorption.

Whereas the description of the adsorption of an atom onto a surface with a 1D potential energy curve may seem satisfactory, you should feel less confident about its ability to describe the interaction of an adsorbing molecule, especially for the case of dissociative adsorption. The interaction of a molecule is described by a multidimensional potential energy hypersurface [potential energy surface (PES); on the use of the term 'hypersurface', see the opening text of Section 3.9]. A somewhat better description is given by a 2D representation in which the molecular centre of mass–separation, z, is plotted along one axis and the intramolecular bond distance of the (diatomic) molecule, R, is plotted along the other axis. Contours denote the change in energy along these two coordinates. Such a 2D PES for nonactivated dissociative adsorption is depicted in Figure 3.15. Far from the surface (large values of z), motion along the R axis corresponds to intramolecular vibration, and motion along the z axis represents translation of the molecule with respect to the surface. At small values of z, the molecule is bound in an adsorption well. Motion along z now corresponds to the molecule–surface vibration.

In Figure 3.15, the molecule dissociates along a nonactivated pathway. On the surface, the chemisorption well is now displaced to larger values of the interatomic separation because the minimum distance between the atoms in the dissociated state must be larger than in the molecule. If the diagram were continued to larger values of R a periodic set of wells would be found corresponding to the two atoms bound at the preferred adsorption sites separated by increasing distances. The meaning of the R coordinate changes as a result of the transition from gaseous to adsorbed phase. In the gas phase, motion along R corresponds to the intramolecular vibration. In the absorbed phase, R is still the distance between adsorbed atoms but no vibrational mode can be assigned directly to motion along this coordinate.

The dashed line in Figure 3.15 represents the path of minimum energy that connect the gas-phase molecule to the adsorbed phase. This can be thought of as the reaction coordinate. Drawing the energy as a function of the position along this path results in a 1D representation of energy with position that is reminiscent of a Lennard-Jones diagram. Though some care must be taken in interpreting the energy profile along this path [106, 107], to a first approximation it is convenient and instructive to visualize the reaction coordinate in this way.

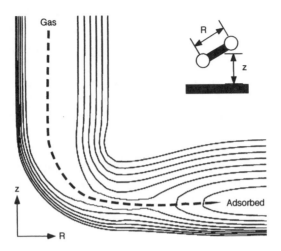

Figure 3.15   Two-dimensional potential energy surface for nonactivated dissociative adsorption; z, the distance from the molecular centre of mass to the surface (vertical axis); R, internuclear distance between the atoms of the (diatomic) molecule (horizontal axis)

3.7.3   Hard cube model

The effects of translational energy, a nonrigid surface, the relative masses of the impinging molecule and surface atom and surface temperature, T_s, upon direct nonactivated adsorption are better understood by investigating so-called hard-cube models [108–112]. The surface is modelled by a cube of mass m. The cube is confined to move along the direction normal to the surface with a velocity described by a 1D Maxwellian distribution at T_s. As the surface is assumed to be flat, the tangential momentum is conserved. The molecule with mass M and total kinetic energy E_K is assumed to be a rigid sphere (i.e. the internal degree of freedom are unaffected by the collision). Its angle of incidence is ϑ_i. The attractive part of the potential is approximated by a square well of depth ε. The attractive well accelerates the molecule as it approaches the surface. Hence, the normal velocity of the molecule at impact is

$$u_n = -\left[\frac{2}{M}(E_K \cos^2 \vartheta_i + \varepsilon)\right]^{1/2} \tag{3.10}$$

The surface atoms move with velocity v. As always, the direction away from the surface is positive. The probability distribution of surface atom velocities is

$$P(v)dv = (u_n - v)\exp(-a^2 v^2)dv \tag{3.11}$$

where

$$a^2 = \frac{m}{2k_B T_s} \tag{3.12}$$

The factor $(u_n - v)$ accounts for collisions being more probable if the cube is moving toward the atom than if it is moving away.

After collision, the parallel and normal components of velocity, v'_p and v'_n, respectively, are calculated by using the conservation of energy and linear momentum:

$$u'_p = u_p \tag{3.13}$$

and

$$u'_n = \frac{\mu - 1}{\mu + 1} u_n + \frac{2}{\mu + 1} v \tag{3.14}$$

where

$$\mu = \frac{M}{m} \tag{3.15}$$

Trapping occurs if the molecule exchanges sufficient energy to drop below the zero of energy. Stated otherwise, trapping occurs if the normal velocity of the molecule is below a critical velocity u'_c that can be contained in the well:

$$u'_c = \left(\frac{2\varepsilon}{M}\right)^{1/2} \tag{3.16}$$

This condition occurs when the velocity of the surface cube is below a critical value of

$$v_c = \frac{1}{2}\left[(1 + \mu)\left(\frac{2\varepsilon}{M}\right)^{1/2} + (1 - \mu)u_n\right] \tag{3.17}$$

The sticking coefficient is obtained by calculating the fraction of collision for which the surface atom velocity is below v_c from Equations (3.11) and (3.17):

$$s(u_n, \varepsilon, T_s) = \left[\int_{-\infty}^{v_c} (u_n - v)\exp(-a^2 v^2)dv\right]\left[\int_{-\infty}^{\infty} (u_n - v)\exp(-a^2 v^2)dv\right]^{-1} \tag{3.18}$$

Integration yields

$$s(u_n, \varepsilon, T_s) = \frac{1}{2} + \frac{1}{2}\mathrm{erf}(av_c) + \frac{\exp(-a^2 v_c^2)}{2au_n\pi^{1/2}} \tag{3.19}$$

Where erf is the error function. The results of Equation (3.19) are shown in Figure 3.16 for parameters determined by Kleyn and co-workers [109] for NO/Ag(111). We see that s drops as expected for increasing molecular velocity. There is a weak dependence of s on T_s for the low velocities typical of thermal distributions, thus the trapping probability falls slowly with increasing T_s. Furthermore, since the model involves an impulsive collision, s drops with increasing m because the collision becomes less inelastic. The effective mass of the surface cube depends on the point of impact as well as the initial velocity of the incident molecule.

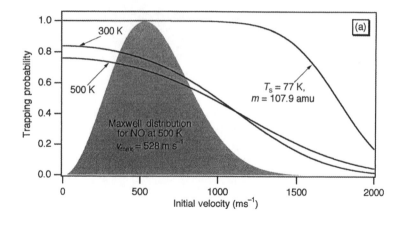

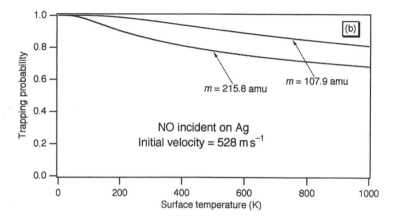

Figure 3.16   The behaviour of the trapping probability as predicted by the hard-cube model for NO incident on a silver surface: (a) the effect of surface temperature, T_s, and molecular velocity on the sticking coefficient, s (a 500 K Maxwell velocity distribution is shown to convey the relative importance of various velocities to the thermally averaged trapping probability); (b) the dependence of s on the effective surface mass and T_s [the effective surface mass is chosen to be either one (107.9 amu) or two (215.8 amu) silver atoms]. Parameters determined by Kleyn and co-workers [109]

3.7.4 Activated adsorption

Some molecules exhibit extremely low sticking coefficients that actually increase with increasing temperature, as first observed by Dew and Taylor for H_2 [113] and by Schmidt for N_2 [114]. To explain this, we need to consider the interaction of a molecule with a surface along a potential energy curve of the type shown in Figure 3.17. In this case, the molecule encounters an activation barrier as it approaches the surface and, therefore, we call this activated adsorption. Trajectories on such a potential energy curve exhibit distinctly different behaviour from that in the nonactivated case. A low-energy molecule [trajectory a, Figure 3.17(a)] reflects off the energetic barrier. In the absence of quantum

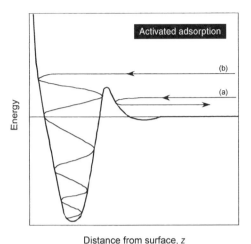

Figure 3.17 A Lennard-Jones diagram is a one-dimensional representation of the potential energy. In the case shown here, the one dimension is taken as the distance above the surface. A barrier separating the chemisorption well from the gas phase distinguishes activated adsorption. Also shown in the diagram are the energies of two hypothetical trajectories at (a) low and (b) high kinetic energy. Classically, only high energy trajectories can overcome the adsorption barrier

mechanical tunnelling, a molecule must have an energy greater than or equal to the barrier height in order to have a chance to enter the chemisorption well [trajectory b, Figure 3.17(b)]. The sticking coefficient for activated adsorption is zero for low-energy molecules. It then rises abruptly as the energy of the molecules approaches and exceeds the energy of the barrier. Classically, a step function is followed as the molecular energy exceeds the activation barrier. Quantum mechanically, tunnelling leads to a more gradual onset to barrier crossing. Other factors, discussed below, lead to a further rounding off of step-function behaviour. At sufficiently high energy, the sticking coefficient begins to decrease as the energy of the molecules becomes excessive, just as it did for nonactivated adsorption. Therefore, simply by measuring the sticking coefficient as a function of the incident kinetic energy we can distinguish activated from nonactivated adsorption. This is also true for direct adsorption. In the next section, we discuss the complications introduced by the presence of an indirect (precursor-mediated) path to the adsorbed phase.

Just as for nonactivated adsorption, a 2D representation of the PES can be drawn for activated adsorption. Three examples are drawn in Figure 3.18. These three PESs are distinguished by the position of the barrier. The barriers are characterized as early, middle or late, based upon how closely the transition state (TS), which resides at the top of the barrier, resembles the gas-phase molecule. The early barrier is in the entrance channel to adsorption, and the interatomic distance in the TS is close to the equilibrium bond distance of the molecule. The late barrier is in the exit channel for adsorption, and the TS has an elongated interatomic distance. As we shall see shortly, the position of the barrier and the shape of the PES are decisive in determining how a molecule best surmounts the activation barrier.

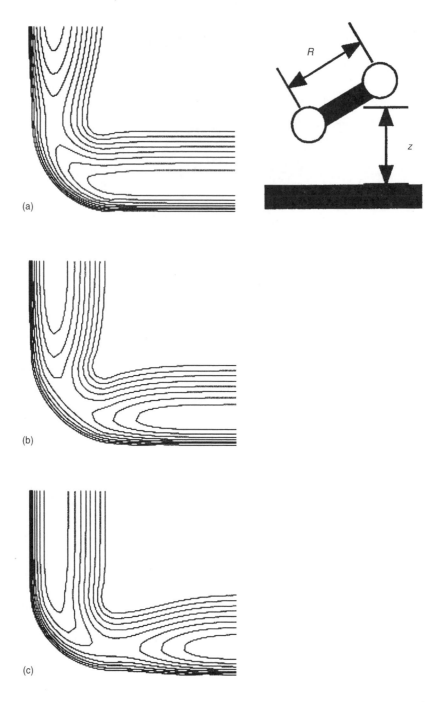

Figure 3.18 Potential energy surfaces (PESs) for activated dissociative adsorption with (a) early, (b) middle and (c) late barriers. z, the distance from the molecular centre of mass to the surface (vertical axis); R, internuclear distance between the atoms of the (diatomic) molecule (horizontal axis). George Darling is thanked for providing these very fine model PESs

3.7.5 *Direct and precursor-mediated adsorption*

The possibility of transient mobility for an adsorbing molecule brings with it two distinct types of adsorption dynamics in nonactivated adsorption. These two types of dynamics are known as direct adsorption and precursor-mediated adsorption. The adsorption mechanism has implications not only for how molecular energy affects the sticking coefficient but also for how adsorbate coverage affects the kinetics of adsorption. The effects on kinetics are discussed in Chapter 4. Table 3.2 presents a survey of some adsorption systems. A perusal of this table illustrates that a variety of adsorption dynamics are observed and that the mechanism of adsorption depends not only on the chemical identity of the molecule and surface but also on the presence of co-adsorbates, crystallographic plane, surface temperature and molecular energy.

Direct adsorption corresponds to the case in which a molecule makes the decision to stick or scatter upon its first encounter with the surface. In the extreme limit, a molecule hits the surface, loses virtually all of its energy and adsorbs at the site where it lands. In a less restrictive case, the molecule loses enough total energy to be bound but it is still able to hop one or more sites away from the point of impact. For instance, the molecule might strike the surface at an on-top site before sliding into a bridge or hollow site.

We do not expect the energy of the gas molecule to have a strong effect on the sticking coefficient in direct nonactivated adsorption. This is true at moderate energies; however, at extremes of high and low energy, the sticking coefficient is a function of energy. As already mentioned, the sticking coefficient drops at sufficiently high energies. At low molecular energies two factors conspire to increase the sticking coefficient. First, the molecule has little energy to lose. Second, a molecule may be able to stick in only one particular orientation or at one particular site. Consider, for example, NO on Pt(111), which must bind nitrogen end down. If NO approaches oxygen end first it does not stick unless the molecule is able to reorient itself before it bounces off the surface. The slower the molecule, the greater the probability that the strong attraction of the surface for the nitrogen end will pull the molecule into the proper orientation before it strikes the surface. The process of molecular reorientation is known as steering. Molecules can be steered by attractive or repulsive forces not only into the proper orientation but also to the proper surface site.

Similarly, surface temperature does not have a strong influence on the sticking coefficient in direct nonactivated adsorption (Figure 3.16). At high temperatures, adsorbed molecules desorb after a short residence time on the surface. Eventually, the residence time becomes so short that it is difficult to distinguish between desorbing molecules and those that scatter directly [115].

In precursor-mediated adsorption a molecule loses sufficient energy in the normal component of kinetic energy on the first bounce to stop it from returning to the gas phase. The molecule is trapped into a mobile precursor state. In some instances the precursor may correlate with a stable adsorbed state, such as a physisorbed state. In other cases, the precursor may be a completely dynamical state. That is, the molecule enters the chemisorption well but a bottleneck in energy transfer stops it from falling immediately to the bottom of the well. The total energy of the trapped precursor molecule may be less than that required for stable adsorption or it may be higher. In the latter case, the

Table 3.2 A survey of various adsorption systems. Unless stated to be extrinsic, precursor refers to the sticking behaviour on the clean surface

Species	Surface	Properties	Reference
Atomic:			
Xe	Pt(111)	Combination of direct and extrinsic precursor-mediated adsorption leads to s increasing with increasing coverage; nonactivated	[i]
Kr	Pt(111)	Intrinsic and extrinsic precursors; nonactivated	[ii]
Cs	W	Precursor-mediated; nonactivated	[iii]
Ir, Re, W, Pd	Ir(111)	Direct at $\theta = 0$; nonactivated	[iv]
Molecular:			
CO	Ni(100)	Direct nonactivated adsorption on clean surface	[i]
	Pt(111)	Extrinsic precursor for low E_K and low T_S; direct adsorption possible at $\theta(CO) > 0$ for higher E_K	[v, vi]
	Ni(111)	Precursor-mediated; nonactivated	[vii]
N_2	Fe(111)	Direct into first molecular state, activated transfer into second molecular state, K co-adsorption lowers barrier to transfer	[viii, ix]
NO	Pt(111)	Direct nonactivated	[x]
O_2	Ag(111)	Activation barrier between physisorbed and chemisorbed molecular states	[xi]
	Pt(111)	Precursor-mediated, switching to direct for high E_K	[xii]
	Pt(100)	s_0 two orders of magnitude lower on hex reconstruction compared with (1×1); precursor switches to direct for high E_K	[xiii]
n-alkanes	Pt(111)	Precursor-mediated; nonactivated; s increases with coverage for C_2H_6 and C_3H_8	[i]
C_6H_6	Si(111)-(7×7)	Precursor-mediated from physisorbed state to chemisorbed state	[xxvii]
Si_2H_6	Si(111)-(7×7)	Precursor-mediated; nonactivated	[xiv]
Dissociative:			
H_2	Si(111)-(7×7)	Direct activated; large barrier on clean terraces,	[xv–xviii]
	Si(100)-(2×1)	Lower barrier at steps; lower barrier in neighborhood of adsorbed H	
	Ni(997)	Precursor-mediated	[i]
	Cu(100)	Direct activated, large barrier	[xix]
	Cu(111)	Direct activated; large barrier	[xix]
	Ge(100)-(2×1)	Direct activated; large barrier	[xx]
	Pt(111)	Activated on terraces but nonactivated at steps with possible involvement of precursor	[xxi]
	W(110)	Direct activated	[xxi]
	Re(0001)	Direct activated	[xxi]
	Mo(110)	Direct nonactivated	[xxi]
	Rh(111)	Direct nonactivated	[xxi]
	Ir(111)	Direct nonactivated	[xxi]
N_2	Fe(111)	Precursor-mediated	[viii]
	W(110)	Direct activated	[i]
	W(100)	Precursor-mediated at low E_k, direct at high E_k; extrinsic precursor	[i]
	W(310)	Nonactivated	[xxi]

Table 3.2 (*continued*)

Species	Surface	Properties	Reference
O_2	Pt(111)	Precursor-mediated; barrier between molecular chemisorbed state and dissociative well	[xxii]
	W(110)	Precursor mediated for low E_K; direct activated for higher E_K	[xxiii]
	Ir(110)-(2×1)	Precursor mediated for low E_K; direct activated for higher E_K	[xxiv, xxv]
	Ge(100)-(2×1)	Precursor mediated for low E_K; direct activated for higher E_K	[xxv]
CO_2	Ni(100)	Direct activated	[i]
	Si(111)-(7×7)	Direct activated	[xxvi]
CH_4	W(110)	Direct activated	[i]
	Ni(111)	Direct activated	[i]
	Ir(110)-(2×1)	Direct activated	[i]
n-alkanes, C_2–C_4	Ni(100)	Direct activated	[i]
	Ir(110)-(2×1)	Direct activated and precursor-mediated; low T_s and E_K favour precursor, high T_s and E_K favour direct activation	[i]
SiH_4	Si(111)-(7×7)	Direct activated	[xiv]
Si_2H_6	Si(111)-(7×7)	Precursor-mediated, switching to direct activation for high E_K; dissociation suppressed by H(a)	[xiv, xxviii, xxix]
	Si(100)-(2×1)		

[i] C. R. Arumaianayagam and R. J. Madix, *Prog. Surf. Sci.* **38** (1991) 1.

[ii] A. F. Carlsson and R. J. Madix, *J. Chem. Phys.* **114** (2001) 5304.

[iii] J. B. Taylor and I. Langmuir, *Phys. Rev.* **44** (1933) 423.

[iv] S. C. Wang and G. Ehrlich, *J. Chem. Phys.* **94** (1991) 4071.

[v] C. T. Campbell, G. Ertl, H. Kuipers and J. Segner, *Surf. Sci.* **107** (1981) 207.

[vi] L. K. Verheij, J. Lux, A. B. Anton, B. Poelsema and G. Comsa, *Surf. Sci.* **182** (1987) 390.

[vii] S. L. Tang, M. B. Lee, J. D. Beckerle, M. A. Hines and S. T. Ceyer, *J. Chem. Phys.* **82** (1985) 2826.

[viii] C. T. Rettner and H. Stein, *Phys. Rev. Lett.* **59** (1987) 2768.

[ix] L. J. Whitman, C. E. Bartosch, W. Ho, G. Strasser and M. Grunze, *Phys. Rev. Lett.* **56** 1986) 1984.

[x] J. A. Serri, J. C. Tully and M. J. Cardillo, *J. Chem. Phys.* **79** (1983) 1530.

[xi] M. E. M. Spruit, E. W. Kuipers, F. H. Guezebroek and A. W. Kleyn, *Surf. Sci.* **215** (1989) 421.

[xii] K.-H. Allers, H. Pfnür, P. Feulner and D. Menzel, *Z. Phys. Chem. (Neue Folge)* **197** (1996) 253.

[xiii] X.-C. Guo, J. M. Bradley, A. Hopkinson and D. A. King, *Surf. Sci.* **310** (1994) 163.

[xiv] S. M. Gates, *Surf. Sci.* **195** (1988) 307.

[xv] K. W. Kolasinski, *Int. J. Mod. Phys. B* **9** (1995) 2753.

[xvi] M. B. Raschke and U. Höfer, *Appl. Phys. B* **68** (1999).

[xvii] M. Dürr, M. B. Raschke, E, Pehlke and U. Höfer, *Phys. Rev. Lett.* **86** (2001) 123.

[xviii] E. J. Buehler and J. J. Boland, *Science* **290** (2000) 506.

[xix] H. A. Michelsen, C. T. Rettner and D. J. Auerbach, "The adsorption of hydrogen at copper surfaces: a model system for the study of activated adsorption", in *Surface Reactions: Springer Series in Surface Sciences, Volume 34*, ed. R. J. Madix (Springer, Berlin, 1994) p. 185.

[xx] L. B. Lewis, J. Segall and K. C. Janda, *J. Chem. Phys.* **102** (1995) 7222.

[xxi] G. Ehrlich, "Activated chemisorption", in *Chemistry and Physics of Solid Surfaces VII* Volume 10, eds R. Vanselow and R. F. Howe (Springer, New York, 1988) p. 1.

[xxii] C. T. Rettner and C. B. Mullins, *J. Chem. Phys.* **94** (1991) 1626.

[xxiii] C. T. Rettner, L. A. DeLouise and D. J. Auerbach, *J. Chem. Phys.* **85** (1986) 1131.

[xxiv] R. W. Verhoef, D. Kelly and W. H. Weinberg, *Surf. Sci.* **306** (1994) L513.

[xxv] C. B. Mullins, Y. Wang and W. H. Weinberg, *J. Vac. Sci. Technol. A* **7** (1989) 2125.

[xxvi] P. W. Lorraine, B. D. Thomas, R. A. Machonkin and W. Ho, *J. Chem. Phys.* **96** (1992) 3285.

[xxvii] D. E. Brown, D. J. Moffatt and R. A. Wolkow, *Science* **279** (1998) 542.

[xxviii] S. K. Kulkarni, S. M. Gates, C. M. Greenlief and H. H. Sawin, *Surf. Sci.* **239** (1990) 13.

[xxix] J. R. Engstrom, D. A. Hansen, M. J. Furjanic and L. Q. Xia, *J. Chem. Phys.* **99** (1993)

Note: E_K, kinetic energy; s, sticking coefficient; s_0, initial sticking coefficient; T_s, surface temperature, θ, coverage; hex reconstruction, see Figure 1.8, page 12

precursor corresponds to a metastable state. The necessary criterion is that the normal component of kinetic energy is too low for the molecule to leave the surface. The precursor molecule now starts to hop from one site to the next. At each hop the molecule can either desorb into the gas phase, migrate to the next site, or become adsorbed. Desorption can occur either because of energy transfer from the surface to the precursor molecule or by intramolecular energy transfer from one molecular degree of freedom into the normal component of kinetic energy. The latter process is particularly important for metastable precursors.

The energy and temperature dependencies of precursor-mediated adsorption are more complex than for direct adsorption because multiple steps must be considered. The dynamics of trapping into the precursor state are the dynamics of direct adsorption; thus, this step has the same energy dependence as direct adsorption. The desorption rate out of the precursor state increases with increasing surface temperature. In most cases, this rate increases faster than the transfer rate from the precursor to the adsorbed phase; therefore, increased surface temperature tends to decrease the rate of adsorption in precursor-mediated adsorption. This behaviour is not, however, exclusive. If the barrier between the precursor state and the chemisorbed state has its maximum above energy $E = 0$, increasing T_s increases the rate of adsorption. This is found, for example, for the dissociative adsorption of O_2 on Pd(111) and Pt(111). For surface temperatures near l-N_2 temperature, O_2 at room temperature adsorbs into a molecular chemisorbed state and no dissociation is observed. However, if the surface temperature is raised above $c.$ 180 K, the dissociative sticking coefficient rises, as there is now competition between desorption out of the molecularly chemisorbed precursor and dissociation. In Chapter 4 we will investigate the kinetics of precursor-mediated adsorption in more detail.

A change in molecular or surface temperature can also lead to a change in the adsorption dynamics (see Table 3.2). If the energy is low, the molecule may be trapped into the physisorbed state and never move on to the chemisorbed phase. As the temperature is increased, adsorption can change from either direct or precursor-mediated adsorption into the physisorbed state to precursor-mediated adsorption into the chemisorbed phase. However, as the temperature increases further, the precursor may no longer represent an efficient mechanism for adsorption because of the high rate of desorption out of this state. At this point, the adsorption dynamics become dominated by direct adsorption. Whether such a scenario occurs is determined by the exact details of the molecule–surface interaction as is the relative efficacy of direct compared with precursor-mediated adsorption. By the end of this chapter, these factors should be clear.

3.8   Microscopic Reversibility in Ad/Desorption Phenomena

At equilibrium, the number of molecules per unit area striking a surface (the incident molecules) must be equal to the number per unit area receding from the surface (the emitted molecules) (no net mass flow). Likewise, there can be no net energy or momentum transfer. It follows that the angular distributions of the incident and emitted molecules must be identical (no flow directionality). Although these statements may seem simplistic, they are so profound that they and their consequences have been repeatedly discussed (and misinterpreted) for more than a century. Whereas any remaining questions regarding angular distributions were settled in the 1970s and 1980s, issues pertaining to energy transfer continue to remain an active area of research, particularly now that quantum-state-resolved measurements are accessible.

James Clerk Maxwell [116] was the first to attempt a description of the angular and velocity distributions of molecules emitted from a surface. Maxwell proceeded with trepidation because, as he admitted, knowledge of the nature of the gas–surface interface was essentially nonexistent in 1878. He assumed that a fraction, essentially the sticking coefficient, of the molecules strike the surface and are then emitted with angular and velocity distributions characteristic of equilibrium at the surface temperature. The remaining molecules scatter elastically from the surface. Maxwell's assumptions were well founded, and traces of them still pervade the thinking of many scientists to this day. However, as we shall see, these assumptions violate the laws of thermodynamics. First, we know that molecules also scatter inelastically. Second, and more importantly, we know that the sticking coefficient is not a constant.

Knudsen [117] made the next step forward when he formulated his law on the angular distribution of emitted molecules. According to Knudsen, the angular distribution must be proportional to $\cos \vartheta$, where ϑ is the angle measured from the surface normal. Such a distribution is often called diffuse. That Knudsen's proposition should attain the status of a 'law' is remarkable because (1) it was incorrectly derived and (2) experiments of the type he performed to prove it should exhibit deviations from a $\cos \vartheta$ distribution. Smoluchowski [118] pointed out these and other deficiencies. In particular, he noted that Knudsen assumed that the sticking coefficient was unity – an assumption that cannot be generally valid. Gaede [119] scoffed at Smoluchowski's more general derivation. Gaede used an argument based on the second law of thermodynamics and the impossibility of building a perpetual motion machine, concluding that $s = 1$ and that the $\cos \vartheta$ distribution must be observed. Millikan [120], in the course of his oil-drop experiments, 'set the cat among the pigeons' when he demonstrated that specular scattering occurs. His results were unmistakably confirmed by Stern's observation of diffraction in atomic scattering at surfaces [88–90]. Therefore, the sticking coefficient cannot, in general, be unity. Nonetheless, Millikan's attempt at an explanation [121] was at best a description of a special case if not an outright violation of the second law, as he introduced a type of direct inelastic scattering characterized by a $\cos \vartheta$ distribution and Maxwellian velocity distribution.

The situation did not improve until the landmark work of Clausing [122]. By this time, Langmuir's investigations of gas–surface interactions had brought about a revolution in our understanding of surface processes at the molecular level. Clausing drew heavily on this work. In 1916 Langmuir stated [123], 'Since evaporation and condensation are in general thermodynamically reversible phenomena, the mechanism of evaporation must be the exact reverse of that of condensation, even down to the smallest detail.' This remains a guiding principle that we use throughout the remainder of this chapter.

To analyse and describe correctly the energy and angular distributions of adsorbed, desorbed and scattered molecules, we need to apply correctly two closely related principles. These are microscopic reversibility, originally introduced by Tolman [124], and detailed balance, first proposed by Fowler and Milne [125]. These principles are formal statements of the idea expressed by Langmuir. Microscopic reversibility refers to individual trajectories whereas detailed balance refers to the rates of forward and reverse processes. That is, microscopic reversibility pertains to individual particles whereas detailed balance pertains to averages over ensembles of particles, though appropriate averaging can be used with microscopic reversibility to make predictions about rates.

Microscopic reversibility states that if we reverse individual trajectories in space and time they must follow exactly the same trajectory. In other words, the interaction of a molecule with a surface (the effective potential energy curve) is independent of the direction of propagation. The same potential energy curve (or set of curves if electronic excited states are involved) governs the interaction of a molecule–surface system regardless of whether the molecule is approaching the surface or leaving the surface. In the strictest sense, microscopic reversibility does not apply to molecule–surface interactions. This is because of the symmetry of the problem. As was first noted by Clausing [122], we cannot reverse the coordinates of all the atoms (as the molecules must always approach the surface from above), and a less restrictive form of reversibility, called reciprocity, is obeyed for molecule–surface interactions [126, 127]. This is largely a matter for purists and is commonly glossed over in the literature. The essential consequence of microscopic reversibility, however, is not lost: gas–surface interactions exhibit time-reversal symmetry such that adsorption and desorption trajectories can be related to one another.

Although the trajectories of individual particles are reversible, this does not mean that molecule–surface interactions cannot lead to irreversible changes, nor does it mean that molecules have memory. The question of reversible and irreversible chemical reactions is beyond our present discussion as we have limited our discussion to the formation of a simple adsorbed layer rather than considering systems in which the surface reacts with the incident molecules to form a surface film with a new chemical composition such as an oxide layer or alloy. The question of memory is related to microscopic reversibility. The equivalence of the Maxwell–Boltzmann distributions for the incident and receding fluxes occurs because, when averaged over all molecules, both have exactly the same characteristics. This does not mean that when a molecule hits a surface, resides there for some time and subsequently desorbs it enters the gas phase with exactly the same energy that it had before it became adsorbed. The wonder of thermodynamics is that, even though no individual molecule returns to the gas phase with exactly the same energy and direction of propagation, when averaged over all molecules in the system, the energy and propagation directions do not vary in time for a system at equilibrium when averaged over a sufficiently long period of time. Fluctuations in equilibrium distributions do occur on short time scales [128] but this is beyond our current discussion.

On the basis of this new knowledge of surface processes, thermodynamic arguments and the recognition that several channels (desorption, inelastic and elastic scattering) return molecules to the gas phase, Clausing proposed three principles:

1. the second law of thermodynamics demands the validity of the cosine law for the angular distribution of the molecules that leave a surface

2. the principle of detailed balance demands that for every direction and every velocity the number of emitted molecules is equal to the number of incident molecules

3. every well-defined crystalline surface, that emits molecules has a sticking coefficient that is different from 1 and that is a function of the angle of incidence and velocity.

Clausing's principles are essentially correct, though in some circumstance s is unity to a good approximation. We now follow their implications, use them to understand the dynamics of adsorption and desorption and extend them to include the effects of molecular vibrations and rotations.

Consider, as in Figure 3.19, a gas at equilibrium with a solid covered with an adsorbed layer. For a system at equilibrium, the temperature of the gas, T_g, must be equal to the surface temperature, (i.e. $T_g = T_s$). Recall that, at equilibrium, chemistry does not stand still; rather, the rates of forward and reverse reactions exactly balance. In this case, the rates of adsorption and desorption are equal and therefore the coverage is constant.

Consider the case in which s is not a function of energy. Take $s = 1$, as did Knudsen and Gaede (a similar argument can be made as long as s is a constant, as assumed by Maxwell). Since the gas-phase molecules are at equilibrium, by definition all degrees of freedom of the gas-phase molecules are in equilibrium and these degrees of freedom – rotation, vibration and translation – can all be described by one temperature, T_g. For $s = 1$, an equivalent way of stating that the rates of adsorption and desorption are equal is to state that the flux of molecules incident upon the surface is precisely equal to the flux of molecules returned to the gas phase by desorption, $r_{ads} = r_{des}$. In order to maintain $T_g = T_s =$ a constant, as is demanded by the state of equilibrium, not only must the fluxes of adsorbing and desorbing molecules be equal but also the temperatures of the

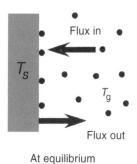

At equilibrium
$T_s = T_g$, $r_{ads} = r_{des}$
Flux in = flux out; θ is a constant

Figure 3.19 A gas in equilibrium with a solid. At equilibrium the fluxes of forward and reverse reactions are equal ($r_{ads} = r_{des}$), as are the temperatures of all phases ($T_s = T_g$). The balanced rates of adsorption and desorption lead to constant coverage. The balanced incident and exiting fluxes result in constant pressure and no net energy transfer between the gas and the solid. T_s, T_g, temperature of the surface and of the gas, respectively; r_{ads}, r_{des}, rate of adsorption and of desorption, respectively; θ, coverage

adsorbing and desorbing molecules must be equal. Furthermore, since the angular distribution of molecules incident upon the surface follows a cosine distribution, based on simple geometric arguments, the angular distribution of the desorbing molecules must also follow a cosine distribution.

These results at first seem trivial. It appears obvious that the molecules, which are at equilibrium with a surface at T_s, should desorb with a temperature T_s and that since the surface is in equilibrium with the gas, $T_s = T_g =$ the temperature of the desorbed molecules. Nonetheless, this state of affairs is only a special case. It is not generally valid, nor is it usually observed.

We have learned that the sticking coefficient depends on energy. Therefore we must consider the more general case of a sticking coefficient that is not constant. Consider the extreme case in which molecules at or below a certain energy, E_c, stick with unit probability but in which they do not stick at all above this energy, (i.e. $s = 1$ for $E \leq E_c$, and $s = 0$ for $E > E_c$), as shown in Figure 3.20.

At equilibrium we must have $T_g = T_s$ and $r_{ads} = r_{des}$. The adsorbed molecules are in equilibrium with the surface. When they desorb, we intuitively expect that the desorbates leave the surface with a temperature equal to T_s. Consider, however, the implications of Figure 3.20(b). From the form of the sticking coefficient as a function of energy, we know that only low-energy molecules stick to the surface. The high-energy molecules reflect and return to the gas phase. By comparing Figures 3.20(a) and 3.20(b) we see that a

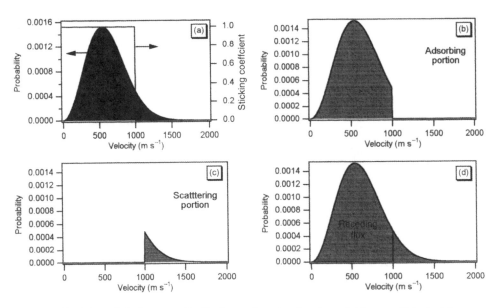

Figure 3.20   Maxwell–Boltzmann distribution and a step-function sticking coefficient: (a) energy distribution for incident flux, and the sticking function (as indicated by the arrows, the distribution relates to the left-hand axis, and the step function relates to the right-hand axis); (b) energy distribution of flux that adsorbs; (c) energy distribution of flux that does not adsorb; (d) energy distribution of total flux that leaves the surface

Maxwell–Boltzmann distribution (often simply called a Boltzmann distribution or a Maxwellian distribution) at T_s has a higher mean energy than the mean energy of the molecules that stick because of the presence of the high-energy tail. Therefore, if the desorbed molecules entered the gas phase with a Boltzmann distribution at T_s, they would on average have a higher energy than the molecules that stick. The gas phase would gradually warm up and the temperatures of the gas and the solid would no longer be equal. This violates our initial condition that the system is at equilibrium and because it spontaneously transports heat from a cool body to a warmer body it could be used to create a perpetual motion machine. Hence, something must be wrong in our analysis.

The error is in the assumption of a Boltzmann distribution at T_s for the desorbed molecules. Figure 3.20(b) displays the true energy distribution of the desorbed molecules. The energy distribution of the desorbed molecules must match *exactly* that of the adsorbed molecules. Consequently, whenever the sticking coefficient is a function of energy, the desorbed molecules will not leave the surface with a Boltzmann distribution at the surface temperature. Indeed, although the adsorbed layer is in equilibrium with the surface, it will not, in general, desorb from the surface with an energy distribution that is described by a Maxwell–Boltzmann distribution.

Figures 3.20(b)–3.20(d) describe how the system is able to maintain equilibrium. For a system in which the sticking coefficient is not unity, adsorption and desorption are not the only processes that occur. Scattering, which does not lead to sticking, must also be taken into account. Equilibration is not determined by adsorption and desorption alone: it is maintained by the sum of all dynamical processes. Therefore, the energy distribution of the incident flux must exactly equal the energy distribution of the sum of all fluxes that leave the surface. In other words, the energy distribution of the incident flux must equal the energy distribution of desorbed molecules plus scattered molecules, as is shown in Figure 3.20(d).

Clausing's conclusion that desorption alone need not exhibit cosine angular or Maxwellian velocity distributions was not tested for 58 years. Then, Comsa [55] reiterated and extended his arguments and within six months van Willigen measured noncosine distributions in the desorption of H_2 from iron, palladium and nickel surfaces. Soon after, molecular beam scattering was used to test these principles further [56, 129]. The adsorption and desorption experiments of Stickney, Cardillo and co-workers [57, 59, 60] led the way in demonstrating that adsorption and desorption can be linked with detailed balance.

Microscopic reversibility allows us to understand the nonintuitive result that an adsorbed layer *at equilibrium* does not desorb with an energetic or angular distribution characteristic of our expectations of an equilibrium distribution. The root cause of this is the energy and angle dependence of the sticking coefficient. In the next chapter, we again encounter a nonintuitive result arising from the energy dependence of the sticking coefficient. Namely, we will see that deviations from Arrhenius behaviour are expected for adsorption and desorption at high temperatures. This deviation can be understood again once the importance of microscopic reversibility is appreciated within the context of transition state theory.

3.9   The Influence of Individual Degrees of Freedom on Adsorption and Desorption

The applicability of microscopic reversibility (reciprocity) to adsorption and desorption means that we can discuss these two reactions in nearly equivalent terms. That is, what pertains to one must also pertain to the other simply by reversing the sense in which the particle traverses the potential energy curve. As soon as we consider more than one degree of freedom, we must recognize that the potential energy curves shown above are simplifications of the true multidimensional potentials that govern the dynamics. Because the potential is multidimensional, the potential energy is actually described by a potential energy hypersurface (the 'hyper' indicates that the 'surface' contains more than two dimensions). Although I use the acronym PES, I use the term hypersurface when referring to the potential energy to avoid confusion with the physical surface at an interface.

3.9.1   Energy exchange

What if a gas is not in equilibrium with the solid? How long does it take the system to attain equilibrium? This leads us to a discussion of what is know as accommodation. By accommodation, we mean the process by which a gas at one temperature and a solid at a different temperature come into equilibrium. If a cold gas is exposed to a hot surface, on average the collisions are inelastic, with the gas molecules gaining energy until the two phases attain the same temperature. In the older literature, and particularly in the applied and engineering literature, one encounters the concept of an accommodation coefficient introduced by Knudsen [117]. This is an attempt to quantify the efficiency of energy exchange between a solid and a gas. The thermal accommodation coefficient, α, is defined in terms of the energies of the incident, reflected and equilibrated molecules – E_i, E_r and E_s, respectively – as

$$\alpha = \frac{E_r - E_i}{E_s - E_i} \tag{3.20}$$

If the molecules transfer no energy at all in a single encounter with the surface, $E_r = E_i$ and $\alpha = 0$. At the other extreme, the molecules equilibrate in one bounce and therefore $E_r = E_s$ and $\alpha = 1$. Although the thermal accommodation coefficient is useful in engineering studies, it tells us little about the processes that lead to equilibration and is a poor quantity to use if we are to understand why certain gases equilibrate faster than do others.

We need to realize that not all degrees of freedom are created equal. In virtually all circumstances (for large characteristic dimensions of length and not too high temperatures) we can consider only three distinct degrees of freedom – translation, rotation and vibration – and can neglect electronic excitations and consider the translational degree of freedom to be a continuous variable. Rotations and vibrations are quantized and, in general, the spacing between rotational levels is significantly smaller than between vibrational levels. Energy exchange between collision partners depends on the relative spacing of the energy levels, the relative masses of the partners and the interaction potential. The closer together the spacing of energy levels and the more closely the energy levels of collision partners are to one another the easier it is to transfer energy between them. Thus, translational energy levels, which are separated by infinitesimal increments, are most readily available for energy exchange in a gas–surface collision. Rotational levels are generally separated by energies small compared with $k_B T$ and small compared with the Debye frequency. Therefore, they should exchange energy readily but possibly more slowly than translations. Vibrational levels, except for rather heavy small molecules such as I_2, or large molecules such as long-chain hydrocarbons, are generally separated by energies that are large compared with $k_B T$ and the Debye frequency. Hence, vibrations exchange energy at significantly slower rates. A full discussion of collisional energy transfer is found in chapter 6 of Levine and Bernstein [130]. As a rule of thumb, translations equilibrate in the order of a few (10^0) collisions, whereas rotations require roughly $10^0 – 10^1$ collisions; vibrations, unless they have a very low frequency, require $10^4 – 10^6$ collisions. Molecular hydrogen, on account of its unusually small mass and large rotational and vibrational energy spacings, requires at least one order of magnitude more collisions to equilibrate.

The properties of the surface affect the rate of energy exchange. The mass of surface atoms plays a role, as mentioned above, as do the vibrational properties of the surface (the phonon spectrum). For semiconductor and insulator surfaces, the only means of energy exchange (at normal temperatures) is via phonon excitations. For metals, however, low-energy electron–hole pairs can be excited. These low-energy excitations potentially play an important role in energy exchange at metal surfaces, especially for the excitation and de-excitation of the high-frequency vibrational modes of adsorbing and desorbing molecules [131–136].

The existence of an attractive potential well is also important for energy exchange. The more attractive the potential energy surface, the greater the likelihood of energy exchange, as demonstrated in the hard-cube model. In part, this results from the acceleration that the molecule experiences as it approaches the surface. In addition, a strongly attractive potential energy surface may be able to reorient and distort the incident molecule. This can lead to enhanced energy exchange between the molecule and the surface as well as exchange of energy between different degrees of freedom within the molecule.

3.9.2 Potential energy surface topography and the relative efficacy of energetic components

The shape of the PES defines the types of ad/desorption channels that are available: whether adsorption is activated or nonactivated, direct or precursor-mediated, molecular or dissociative. As can be seen in Table 3.2, the adsorption dynamics are determined by both the PES and the energy of the incident molecule. In the following sections we discuss each degree of freedom individually. First, a few introductory remarks about PES topology are in order.

Although it may seem obvious that vibrational excitation should facilitate dissociation, the degree to which vibrational energy can be converted into overcoming the activation barrier is determined by the exact shape of the PES. The topology of the PES also determines the relative efficacy of vibrational energy compared with translational and rotational energy [137]. To visualize the energetics of dissociative adsorption, we construct a two-dimensional PES such as those in Figure 3.18. Starting from the upper left, motion of the molecule toward the surface corresponds to motion along the z axis and the intramolecular vibration corresponds to motion along the R axis. In the lower right, motion along the z axis corresponds to the vibration of the chemisorbed atoms against the surface, and motion along the R axis corresponds to diffusion of the two atoms across the surface. Both the upper left and the lower right can be probed spectroscopically by measurements performed at equilibrium. The difficulty arises in that the transition state governs the reaction dynamics. Ever since the invention of the concept [138, 139], one of the great preoccupations of chemical dynamics has been the direct observation of the transition state [130]. Great progress has been made in probing the transitions states of gas-phase reactions [140], as witnessed by the Nobel Prizes of 1986 and 1999, awarded to Herschbach, Lee and Polanyi and to Zewail, respectively, but direct observation of transition states at surfaces still remains elusive and computationally challenging.

Three elbow potentials are represented in Figure 3.18. These are analogous to similar PESs drawn for gas-phase reactions and are similarly subject to the Polanyi rules developed from the study of gas-phase reactions [141–143]. For an early barrier, translational energy is most effective for overcoming the barrier, and rotational and vibrational energy are ineffectual. For a very late barrier, vibrational energy is most effective. Rotational energy can also facilitate reaction through the mechanism discussed for H_2/Cu below, but it is less effective than vibrational energy. Translational energy plays a subordinate role on such a PES. For the middle barrier of Figure 3.18(b), all of the molecular degrees of freedom play an important role in overcoming the barrier. Regardless of the barrier type, rotation always plays an important role because of its intimate relationship to the orientation of the molecule as it strikes the surface.

The extensive experimental results of Rettner, Auerbach and co-workers [43, 144–147] for both adsorption and desorption in the H_2/Cu system coupled with the trajectory calculations of Holloway, Darling and co-worker [42, 148–153] have demonstrated the ability of state-resolved experiments to probe the PES of a surface reaction. In addition, this body of work has shown that even with the high dimensionality of surface reactions, a relatively simple PES and microscopic reversibility can explain the most intimate details of the formation and dissociation of H_2 on a metal surface.

The PES is also dependent on the surface coordinates. Binding energies and activation barriers depend on the position within the unit cell. Vibrational excitation and surface relaxation may also be important in ad/desorption dynamics. As a first approximation, the PES is usually conceived of in terms of a rigid surface. However, corrugation or the lack thereof must always be considered.

3.10   Translations, Corrugation and Surface Atom Motions

3.10.1   *Effects on adsorption*

Potential energy hypersurfaces can be thought of very much like topographical maps, and we can use the intuition we have developed from our own experiences with geography and bobsledding to help us understand how molecules interact with surfaces. If you have ever taken a train from Berlin to Warsaw you have experienced an approximation of the flat potential in Figure 3.21(a). Once you have crossed the Oder River, the vast flat expanse of Wielkopolska gives one the impression that (if the tracks were perfectly frictionless) the engineer could turn off the engines and the train would simple roll ahead until it reached Warsaw. Apart from friction, the only things that could stop the train are occasional defects in the tracks and planar landscape. Figure 3.21(b) represents a corrugated landscape. The consequences of corrugation are that there are preferred sites for an adsorbate to come to rest, and some directions are easier to move in than are others. Corrugation also facilitates more extensive energy transfer between the adsorbate and the surface; for example, it leads to a coupling between normal and parallel momentum. Finally, as shown in Figure 3.21(c), not all regions are seismically inactive. Sometimes the surface must not be thought of as solely a rigid template but also its motions must be considered. A surface which is soft in comparison to the force with which an adsorbate strikes it is effective at dissipating energy, as energy can be transfer from the adsorbate to the phonons, and vice versa.

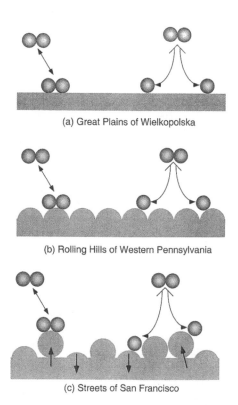

(a) Great Plains of Wielkopolska

(b) Rolling Hills of Western Pennsylvania

(c) Streets of San Francisco

Figure 3.21 The interaction of an ad/desorbing molecule with various models of solid surfaces: (a) a flat rigid surface; (b) a rigid corrugated surface; (c) a surface that is both corrugated and vibrationally active

Consider first the case of a rigid particle interacting with a flat, frictionless surface. This is represented in Figure 3.21(a) and is roughly equivalent to the experience of a xenon atom incident upon a Pt(111) surface. The rare-gas/Pt interaction potentials are extremely flat and nondirectional as they are dominated by van der Waals forces [154]. Therefore, xenon has a low probability of stopping on the terraces of the Pt(111) surface [155]. The xenon atoms tend to roam across the flat planes until they encounter defects in the landscape. These defects take the form either of steps or of previously adsorbed atoms (either xenon atoms or impurities).

Not all translational energy is the same. For Xe/Pt at low temperatures, the sticking coefficient is high; nevertheless, the atoms roam over large distances, as much as hundreds of angstroms, before coming to rest. Therefore, the important dynamical criterion for xenon sticking is not the loss of *all* translational energy upon the first encounter with the surface. Instead, *sufficient* energy must be lost from the normal component of translational energy, denoted $E_\perp$, or normal energy. The trapped but roaming xenon atom has a low value of $E_\perp$ but still has a significant parallel component of translational energy, $E_\parallel$. The potential energy surface is conducive to efficient accommodation of $E_\perp$, but $E_\parallel$ accommodates at a much slower rate. Xenon adsorption is nonactivated, so we expect that the sticking coefficient should decrease for sufficiently high kinetic energy. However, we can refine this statement and state that we expect the sticking coefficient of xenon to depend more strongly on $E_\perp$ than on $E_\parallel$ because it is the normal component that must be accommodated whereas the parallel component can relax on a much longer time scale.

Although Xe/Pt is an extreme case, it is a common feature of molecule–surface PESs that $E_\perp$ accommodates more rapidly than $E_\parallel$. As mentioned above, the occurrence of a mobile state that can roam the surface searching for an adsorption site is the basis of precursor-mediated adsorption. The concept of a mobile precursor was first proposed by Taylor and Langmuir [156]. A precursor is always bound above a well (even if this state does not appear in the $T_s = 0$ K PES [106, 107]) but the precursor need not be accommodated into the well. A precursor can be trapped in highly excited states in the well if a dynamical bottleneck prevents it from equilibrating rapidly.

Another important implication of transient mobility [157] is that even though the xenon atoms strike the surface at random positions they do not stop at completely random adsorption sites. The xenon atoms tend to form islands about the defects on the surface. Islanding, the process of forming patches of adsorbates, requires the adsorbing molecules to have mobility and a driving force which makes them stop at a preferred position. The driving force in this case is a change in the corrugation of the molecule–surface potential such as that presented by defects or adsorbate–adsorbate interactions.

Most PESs are not as flat as the Xe/Pt potential. Nevertheless, flatness is not the only consideration for the formation of a long-lived state prior to equilibration into the adsorption well [158]. Hydrogen is strongly bound on Cu(111), ($q_{ads} \approx 2.5$ eV) in a highly corrugated potential (binding at a hollow site approximately 0.5 eV stronger than an on-top site). Nonetheless, hydrogen atoms have a high probability of entering a mobile state that can live for the order of a picosecond or longer and roam more than 100 Å across the surface. The reason for this is that hydrogen is much lighter than copper and therefore exchanges momentum very slowly with the phonons. As hydrogen approaches the surface, the chemisorption well accelerates it to $E_K \approx 2.5$ eV. The corrugation in the potential is very effective at converting the normal momentum into parallel momentum on the first bounce. Afterward, however, momentum transfer is ineffective with the surface and it takes many collisions (in the order of tens of collisions) for the atom to equilibrate. As we shall see later in this chapter, such 'hot' hydrogen atoms are highly reactive.

A PES with significant corrugation is depicted in Figure 3.21(b). Corrugation leads to efficient exchange of energy between $E_\perp$ and $E_\parallel$. On a perfectly flat, frictionless potential, $E_\parallel$ never accommodates. Corrugation scrambles normal and parallel momentum and can lead to more efficient energy transfer with phonons. This conceptualization of a corrugated but rigid surface can go a long way to explaining many molecule–surface interactions. For instance, the dynamics of recombinative desorption of H_2 from Cu(1 0 0) can to a large extent be explained by visualizing a corrugated PES established by the symmetry of the Cu(1 0 0) surface and the motions of hydrogen atoms diffusing across it [42, 43, 47, 145, 148, 152, 159–161].

Figure 3.21(c) depicts a situation in which the surface plays a more active role in the adsorption and desorption dynamics. If a heavy molecule strikes a surface of comparatively light atoms it can easily push the surface atoms around. This is a purely mechanical instance in which the lack of rigidity is important. However, we have seen in Chapter 1 that the presence of adsorbates can lead to reconstruction of surfaces. Adsorbate-induced reconstruction implies that surface atom motion can be inherent to the adsorption/desorp-desorption dynamics. This is the case for the H_2/Si system, in which surface atom motion during the adsorption/desorption process must be considered if we are to achieve an understanding of the dynamics. Although hydrogen is far too light to distort the surface mechanically, the silicon atoms move under the influence of the forces accompanying the formation of chemical bonds. Surface atom excitations play a significant role in determining the fate of an incident hydrogen molecule because the height of the adsorption activation barrier is a function of the surface atom positions. Hence, it is not only the energy of the incident hydrogen molecules but also, more importantly, the energy (i.e. temperature) and positions of the surface silicon atoms that determine to a large extent the sticking coefficient of H_2 on Si [162–168].

3.10.2 Connecting adsorption and desorption with microscopic reversibility

We have mentioned previously that adsorption trajectories are reversible (e.g. see Section 3.8). That is, if we simply run a movie of adsorption backwards then we obtain a movie that describes the reverse process–namely, desorption. We can use this fact and the discussion of adsorption in the previous section to make predictions about desorption.

When xenon adsorbs on Pt(111) it roams the surface until it eventually finds its final adsorption site. Therefore, when xenon desorbs, we predict that it is kicked out onto the terrace, diffuses over a long distance then finally departs the surface from some random site on the terrace. Low-kinetic-energy atoms are the most likely to stick. Therefore, microscopic reversibility demands that low-energy atoms are also the most likely to be returned to the gas phase by desorption. The consequence is that, when averaged over the entire ensemble of desorbed atoms, the kinetic energy is lower than for an ensemble of molecules at the same temperature as the surface. If the energy distribution is Maxwellian, we can use a single number (the temperature T_{trans}) to describe the translational energy of the desorbed atoms. Thus, for xenon desorbing from Pt(1 1 0), we see that $T_{trans} < T_s$. This is known as translational cooling in desorption and is a direct consequence of the process described in Figure 3.20.

To understand better translational cooling, consider the following thought experiment for a molecule that experiences nonactivated adsorption. An adsorbed molecule has been excited all the way to the top of the chemisorption well, that is, to energy $E = 0$. We consider the desorption to be one-dimensional. All modes other than the normal component of translational energy do not contribute to desorption. These spectator modes all have an equilibrium energy distribution at T_s and, within the 1D model, they do not couple to the reaction coordinate. The molecule has one final inelastic collision with the surface. Consider the surface to be effectively an oscillator with an energy distribution characterized by an equilibrium distribution at T_s as in the hard-cube model. In the last encounter, the oscillator deposits all of its energy in the desorbing molecule. With no dynamical constraints on energy exchange, the surface oscillator transfers on average a distribution of energy characteristic of the equilibrium distribution at T_s into normal translation of the adsorbate. We call the amount of energy delivered $\langle kT \rangle$. The molecule then departs from the surface. The spectator modes all have an equilibrium distribution by definition in this model. The normal component of translational energy also has an equilibrium distribution. This description is what is demanded of a 1D model in which $s_0 = 1$ and is independent of kinetic energy.

Now consider a 1D model in which the oscillator finds it progressively more difficult to transfer larger amounts of energy. In this case, the surface oscillator delivers, on average, an amount of energy that is less than $\langle kT \rangle$. Whereas the spectator modes all have an equilibrium distribution of energy, the normal component of translational energy has less energy than expected for equilibrium at T_s. Translational cooling has been observed, for example, for desorption of NO/Pt(111) [65, 169], NO/Ge [170], Ar/Pt(111) [171] and Ar/H-covered W(100) [172]. It corresponds to the case in which s_0 decreases with increasing kinetic energy and is, therefore, a rather general phenomenon for systems that exhibit nonactivated adsorption. Just as it is difficult for the surface to take away sufficient energy to allow ever faster molecules to stick, so too it is difficult for the surface to produce ever faster molecules in desorption.

In a real desorption system, the molecules do not all desorb from the $E = 0$ level. Molecules from a range of levels near the top of the chemisorption well experience one last encounter with the surface and then desorb. The inability of the surface to deliver $\langle kT \rangle$ leads to a subthermal energy distribution in the desorption coordinate. The levels near the top of the well are rapidly depleted by desorption. However, the surface is unable to maintain an equilibrium population distribution in these levels because of the inefficiency of energy transfer at high energy. As we shall see in Chapter 4, this effect leads not only to translational cooling in desorption but also represents a dynamical factor that causes non-Arrhenius behaviour of the desorption rate constant at high temperature [173–175].

As we have seen in Section 3.5, H_2 adsorption on copper is activated. In a simple one-dimensional Lennard-Jones-type potential, we considered only the effect of energy along the reaction coordinate, which to a first approximation we take as the normal component of translational energy. Indeed, translational energy plays the most important role in determining the sticking coefficient of H_2 on copper: the sticking coefficient increases strongly with increasing H_2 normal energy [43, 176]. Therefore, microscopic reversibility demands that since fast H_2 molecules are the most likely to stick, they are also the most likely to desorb. This is the case, and it is found experimentally that the desorbed H_2 is very 'hot' in the translational degree of freedom. Furthermore, since molecules directed along the surface normal stick better than those incident at grazing angles, the desorbed molecules tend to be focused along the surface normal.

H_2 adsorption on silicon is also activated [163, 176]. However, the desorbed molecules are not translationally hot and therefore translational energy is not the primary means of promoting adsorption. The reason for this can be found in the role of surface excitations. Dissociative adsorption of H_2 on silicon is activated in the surface coordinates. Effectively, we need to excite the surface into preferential configurations for the H_2 to dissociate and stick as hydrogen atoms. Therefore, in desorption, the energy of the activation is not efficiently conveyed to the desorbing H_2 molecules. Instead, a large fraction of the activation energy remains in the silicon surface. The result is that molecules leave the surface not too hot, but a vibrationally excited surface is left behind.

3.10.3 Normal energy scaling

The Lennard-Jones model is a 1D model. Only energy directed along the reaction coordinate is effective in overcoming the barrier. Energy directed along other dimensions – so-called spectator coordinates – have no influence on the reaction dynamics. In the Lennard-Jones model, this dimension is taken as a barrier normal to the surface in which only the normal component of kinetic energy is capable of overcoming the barrier. Essentially, there was no evidence contradicting the 1D nature of the adsorption barrier until the work of Bernasek and co-workers [76] and Kubiak, Sitz and Zare [84, 85]. Their observations of superthermal population of the first vibrationally excited state could be accounted for only by including additional dimensions in the adsorption dynamics.

In the absence of corrugation, there is no coupling between the parallel and perpendicular components of momentum. Thus, in any flat surface system, if the dissociation barrier is directed along the surface normal, only the component of kinetic energy directed along the normal is effective in overcoming the barrier. In most systems, kinetic energy is the most efficient at overcoming an activation barrier. Thus, even in multidimensional systems, our first-order approximation is to expect normal energy scaling for the adsorption probability. Normal energy scaling is defined as follows. The normal component of the kinetic energy (the normal energy) varies with the angle of incidence, ϑ_i, as

$$E_n = E_i \cos^2 \vartheta \tag{3.21}$$

Since the amount of normal kinetic energy decreases as $\cos^2 \vartheta_i$ at constant E_i, the effective barrier height in a 1D system increases as $1/\cos^2 \vartheta_i$, as shown in Figure 3.22(c). If the adsorption probability scales with the magnitude of normal energy such that it

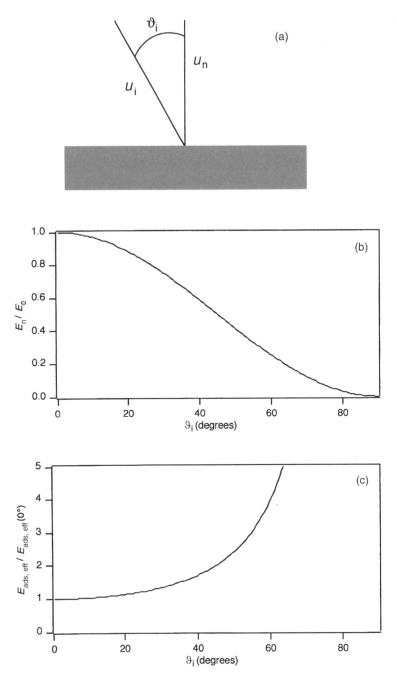

Figure 3.22 A description of normal kinetic energy and the effective height of a one-dimensional barrier: (a) the molecule is incident upon the surface at angle ϑ_i from the normal with velocity u_i and normal velocity u_n; (b) the normal energy falls with increasing angle of incidence according to Equation (3.21); (c) the decreasing normal energy of molecules incident at higher ϑ_i leads to an increasing effective barrier height if only the normal energy is capable of overcoming the barrier

decreases with angle as $\cos^2 \vartheta_i$, we say that adsorption exhibits normal energy scaling. This can be contrasted to total energy scaling, for which the exponent of $\cos^n \vartheta_i$ is $n = 0$; that is, the adsorption probability scales with total kinetic energy rather than the component along the normal.

Normal energy scaling is observed for the dissociative sticking of CH_4 on W(110) and Ni(111), and *n*-alkanes/Ni(1 0 0), whereas total energy scaling is observed for the dissociative sticking of N_2/W(1 0 0) [177]. For H_2 dissociation, normal energy scaling is commonly observed with the exception of Pt(111), Ni(1 0 0) and Fe(110) surfaces for which the parallel component inhibits dissociation. The presence of corrugation is expected to lead to a scaling characterized by $0 \le n \le 2$ [178]. Moreover, interaction potentials are corrugated, so how can we explain the prevalence of normal energy scaling?

Darling and Holloway [47] have shown how the combination of energetic and geometric corrugation combine counterbalancing effects that allow for normal energy scaling. Not only the energy of the activation barrier (energetic corrugation) but also the position of the barrier (geometric corrugation) varies with position across the surface. When they are combined in the right amounts, normal energy scaling results *because* the surface is corrugated rather than because it is flat. The reader is referred to the insightful discussion of Darling and Holloway for further details.

3.11 Rotations and Adsorption

3.11.1 *Nonactivated adsorption*

Our first impression might be that rotation should play a minor role in ad/desorption phenomena because it is, after all, the presence of energy in the normal component of kinetic energy that determines whether a molecule scatters or desorbs from a surface. Rotation, however, has two consequential effects upon chemical reactivity. The first is as a source of energy. If there is a way in which rotational energy can be channelled into the reaction coordinate, then rotational energy can be used to affect reactivity. One of the most important couplings is between rotational and kinetic energy. This channelling can occur either as rotation-to-translational energy transfer (R–T transfer) or in the opposite sense (T–R transfer). T–R transfer has been directly measured in, for example, the scattering of NO from silver and platinum surfaces [62, 64, 69, 70, 179]. Rotation also changes the orientation of the molecule with respect to the surface. From the discussion of the Blyholder model in Section 3.4.2, we have seen that the chemisorption potential is highly anisotropic. Molecules such as CO or NO bind in specific ways to a surface. A CO molecule on a transition metal surface chemisorbs only through the carbon end of the molecule and not through the oxygen end. Therefore, a CO that approaches the surface oxygen end first is not able to stick in its equilibrium chemisorption site unless it is able to rotate into the proper orientation. On the other hand, if a molecule is rotating rapidly, the surface might not be able to stop it and pull it into the proper orientation for sticking. Thus, a little rotation may be essential for sticking, but too much enhances scattering.

The adsorption of molecular NO on Pt(111) is direct and nonactivated with a sticking coefficient of c. 0.85 for room temperature NO [169, 180]. At low coverage NO occupies an on-top site and is bound perpendicularly to the surface through the nitrogen atom. NO sticks with a higher probability if it strikes the surface nitrogen end first [181, 182]. Just as for translations, not all rotations are the same. We can decompose rotational motion into two components: so-called helicopters, which rotate in the plane of the surface, and cartwheels, which rotate perpendicular to the surface. Think of the NO as an American or rugby football that is red on the nitrogen side and white on the oxygen side. It prefers to stick to the surface with the red side down and the white side up.

Several competing effects determine how rotation affects ad/desorption. These are related to the competing effects of R–T energy transfer and molecular orientation. A cartwheeling molecule can present the correct geometry for the chemisorbed state; however, as scattering measurements show [72, 179, 183–185], it can also efficiently couple its rotational energy into normal translational energy. This can give an added kick to the molecule and enhanced scattering. A helicoptering football never sticks red side down unless it is steered into an upright geometry. However, in-plane rotation does not couple at all to E_n on a perfectly flat surface (think of the analogy to an oval football) and therefore it allows a much softer landing at the surface with no prospect of enhanced desorption from R–T energy transfer.

When looked at from the perspective of the chemisorbed molecule, it appears that only cartwheeling motion can be excited in desorption. From the perspective of the surface, a platinum atom striking the normally bound NO from the bottom cannot excite in-plane rotations in an impulsive collision. Such a collision can excite only cartwheeling motion. Thus, we might predict that desorbed molecules should have a higher probability of being cartwheels. In adsorption, we might predict that helicopters would have a higher sticking coefficient because of the softer landing and the ability of the highly directional chemisorption well to steer the molecule along the incident trajectory. Our intuitive predictions appear to violate microscopic reversibility and can be resolved only by direct experimental observation.

First, we take the temperature of the desorbed molecules by using laser spectroscopy. We find rotational cooling in desorption [63, 64, 73, 175, 186]. This has been observed for NO/Ru(0 0 1) [73], NO/Pt(111) [63, 64], OH/Pt(111) [187] and CO/Ni(111) [188] and appears to be a general phenomenon for nonactivated adsorption systems. Analogously to the case of translational energy, this means that the sticking coefficient is a function of rotational energy and that it decreases for increasing rotational energy. Laser spectroscopy can also determine the degree of rotational alignment in a gas (i.e. the relative populations of the two components of rotational motion). Such an experiment has been performed for desorption of NO from Pt(111) [189]. For states with a low value of the rotational quantum number, J, there is no rotational alignment. However, as J increases there is a trend to a progressively greater number of in-plane rotors compared with out-of-plane rotors. In-plane rotation must, therefore, be more favourable for adsorption than is out-of-plane rotation. As J increases, cartwheeling motion becomes progressively more unfavourable for adsorption.

The observation of rotational alignment in desorption has an important consequence for our understanding of ad/desorption dynamics. As we have stated above, a normally bound NO molecule cannot be excited into in-plane rotation when it is struck from below by a platinum atom. Nevertheless, in-plane rotation is excited as efficiently as – and at high J even more efficiently than – out-of-plane rotation. This is direct proof that desorption (and therefore adsorption) is not a direct process in which a molecule leaves from the bottom of the chemisorption well and pops out into the gas phase. Instead, the molecule gradually percolates up to the top of the chemisorption well after many collisions with the surface. After this multistep process of climbing the ladder of vibrational states in the chemisorption well, the molecule leaves from the top of the chemisorption well. At the top of the well, the bonding interaction does not bear a strong resemblance to the well-oriented geometry at the bottom of the well. At the top of the well, where all of the several NO–surface vibrational modes are multiply excited, the adsorbate geometry is not well-defined and energy is easily transferred into both components of rotation. For the most highly excited rotational states, population is more readily built up in in-plane rotation. This is related to R–T energy transfer. Since in-plane rotation couples weakly to normal translation, in-plane rotation does not enhance desorption, and the desorption rate has little dependence on the extent of in-plane rotational excitation. Out-of-plane rotation does couple to normal translation, therefore, excitation of out-of-plane rotation enhances the desorption rate. The most highly excited out-of-plane rotation states therefore become progressively more depleted of population compared with the in-plane rotational states.

The implication for adsorption is that it, too, must be a multistep process. A slow molecule rotating in-plane in a low J state is the most likely to stick. The molecules that stick do not fall directly to the bottom of the chemisorption well. The adsorbing molecules require many collisions with the surface until they finally lose their energy, attain the correct bonding geometry and equilibrate with the surface. This may occur at the point of impact, but we can also expect that the molecule may exhibit some mobility before attaining its final chemisorption site. The extent to which the molecule roams about the surface is determined by the relative rates of accommodation and the corrugation of the molecule–surface potential.

3.11.2 Activated adsorption

The role of rotation in activated adsorption is best illustrated by the case of $H_2/Cu(1\,0\,0)$. Dissociative H_2 adsorption on copper is activated, as discussed in Section 3.5. The overlap of the $2\sigma*$ orbital with the metal orbitals is enhanced by a molecular orientation in which the internuclear axis is parallel to the surface [190]. We therefore predict rotational alignment in desorption with a preference for in-plane rotation, an expectation that is experimentally confirmed [147, 191] at low J.

The effect of rotational energy in H_2 dissociation has two distinct effects [145, 152]. For $J = 0$, the PES can reorient the molecule as it approaches the surface and bring it into the best possible orientation for entering the transition state. As J increases, steering becomes less effective. Therefore, increasing rotational excitation is a hindrance to dissociative adsorption at low J. At high J, a portion of the rotational energy can be channelled into overcoming the activation barrier. This occurs because of the shape of the PES. Bond extension is required to reach the transition state. The extension of the H–H distance in the transition state channels energy out of rotation and into the reaction coordinate. Hence, rotational excitation promotes dissociation at high J. Taking the two effects together, rotation hinders adsorption up to $J \approx 5$, whereupon the promotional effect of rotational energy becomes dominant. Similar effects should be observed in any system with an extended bond in the transition state.

3.12 Vibrations and Adsorption

For molecules that are not too large, the spacing between vibrational levels tends to be much greater than the energies of phonons and rotational states. Thus, vibrations tend not to couple efficiently to other degrees of freedom. Vibrations can couple to the electronic structure of a metal through electron–hole pair formation [132, 136, 192]. Because of the band gap found in semiconductors and insulators, electron–hole pair formation is not important at usual energies for these materials. Because of the weak coupling of vibrations to other degrees of freedom, molecular vibrations are not of utmost importance for nonactivated and, in particular, nondissociative sticking of small molecules. This is not true for large molecules or for the sticking of clusters at surfaces. Large molecules and clusters have a plethora of low-frequency vibrational modes that can effectively soak up translational energy. Their excitation upon collision with the surface can play an important role in the sticking dynamics.

For activated adsorption, vibrations can play a crucial role [42, 43]. For the dissociative adsorption of a diatomic molecule, the vibration is intimately related to the reaction coordinate. For H_2/Cu, H_2/Si and H_2/Pd, significant superthermal vibrational excitation in desorbed molecules has been measured [85, 193, 194]. These results demonstrate that vibrational excitation and the concomitant bond extension aid in reaching the transition state and overcoming the activation barrier. For any middle or late barrier PES, vibrational excitation is important in the ad/desorption dynamics.

3.13 Competitive Adsorption and Collision-Induced Processes

We now investigate what happens when a surface covered with one adsorbate is exposed to a different type of molecule. This is, of course, quite relevant to surface reactions as both types of adsorbates must usually be present on the surface for a reaction to ensue. The answer depends not only on the chemical identity of the surface and the adsorbing molecules but also sometimes on the order in which the molecules are exposed to the surface.

First consider the system $O_2 + CO$ on Pd(111) [36, 195]. When O_2 is exposed to a Pd(111) surface held at 100 K, three molecular chemisorbed states are formed. These three states have different binding energies to the surface. As the O_2–Pd bond strength increases the O–O bond is weakened and therefore the O–O stretch frequency is lowered. CO adsorbs in one molecular chemisorption state. Temperature-programmed desorption (TPD) studies (Section 4.7) have shown that CO is bound more strongly than O_2.

Systems tend to assume the lowest Gibbs energy. This is true at equilibrium; however, the road to equilibrium is not always a direct one. As a first approximation, we neglect entropic effects. This should be justified at the low temperatures considered and for the two similar diatomic molecules in our system. CO should be able to displace O_2 from the surface since it is more strongly bound. Furthermore, we might expect that the least tightly bound O_2 species would be the first to be displaced, followed by the intermediate and then the strongly bound O_2. That is, ω_3 before ω_2 before ω_1 (see Figure 3.8, page 94). The data in Figure 3.23 show that the first expectation is met but the second is not. That is, CO can displace O_2 from the surface; however, it is the most weakly bound species, ω_3, which is actually the most able to compete with CO for adsorption sites. Furthermore, we see that CO can also induce the conversion of one type of O_2 into another. The reasons for this lie in the dynamics of competitive adsorption. CO and O_2 must compete for sites on the palladium surface. In this competition, we must consider three factors: (1) the energetics of chemisorption bond formation, (2) what types of sites the adsorbates prefer and (3) the effect of the presence of one adsorbate on the energetics of a second adsorbate.

To understand the displacement of O_2 by CO on Pd(111), we have to consider not the relative energetics of the O_2 chemisorption states but the relative energies of the O_2 states bound on the sites onto which CO wants to adsorb. The results of Figure 3.23 suggest that

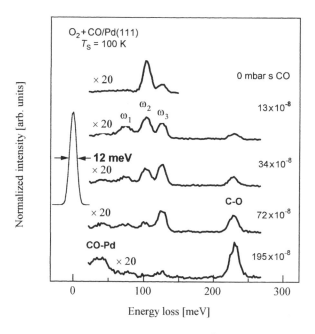

Figure 3.23 Competitive adsorption in the CO + O_2/Pd(1 1 1) system investigated by electron energy loss spectroscopy (EELS) measurements. The species associated with ω_1–ω_3 are illustrated in Figure 3.8, page 94. arb. units, arbitrary units; T_s, surface temperature. Redrawn from *Surf. Sci.* **334**, K. W Kolasinski, F. Cemič, A. de Meijere and E. Hasselbrink, "Interactions in co-adsorbed CO + O_2/ Pd(1 1 1) layers", 1995, page 19, with permission from Elsevier Science

the most tightly bound O_2 species must be bound to the same site onto which CO would like to adsorb. The ω_1 species is unable to compete with CO not only because the CO is more strongly bound, but also because the O_2 is not sufficiently strongly bound to establish a sizeable barrier to its displacement. The ω_1-O_2 has two choices when CO pushes it out of its site. It can either convert to the ω_2 or ω_3 state or it can desorb from the surface. Quantitative analysis of the EEL spectrum reveals that both of these fates await displaced ω_1-O_2. The reason why CO is more strongly bound than O_2 is that it more effectively withdraws electrons from the surface. Thus with increasing CO coverage the ability of palladium to donate electrons to adsorbates decreases. Since ω_3-O_2 requires the least amount of electron donation, and because it occupies a site different from that occupied by CO, it is able to survive to the highest coverages of CO coadsorption. The details of CO displacement of chemisorbed O_2 depend sensitively on the PES. Whereas CO displaces O_2 from Pd(111) with near unit efficiency, on Pt(111) the efficiency is only 0.09 per adsorbing CO [196].

The question remains as to the mechanism by which one adsorbate displaces another. Rettner and Lee [197] have shown that a strongly adsorbing species can effectively transfer a portion of its heat of adsorption to a preadsorbed molecule, leading to desorption. Specifically hydrogen, nitrogen or oxygen atoms incident upon O_2 chemisorbed on Pt(111) facilitate efficacious displacement of the O_2. A similar mechanism is also likely in the displacement of O_2 by CO.

Somorjai [198] and co-workers have investigated a number of instances in which an ordered array of adsorbates is pushed aside by a coadsorbate that is post-dosed to the surface. CO is able to compress preadsorbed sulfur layers on both Re(0 0 0 1) and Pt(111) surfaces, for example. The sulfur atoms move into regions of locally higher coverage. In the process, sites are freed up that become available for CO chemisorption. If CO desorbs from the surface, as caused by an increase in the temperature, the sulfur atoms move back into the ordered structure that they originally occupied. In other cases, adsorbing CO molecules can shepherd preadsorbed disordered molecules into an ordered phase. This has been observed for ethyne, ethylidyne, propylidyne, benzene, fluorobenzene, sodium, potassium and hydrogen on several metal surfaces. The common feature of all of these adsorbates is that they are relative electron donors compared with the electron-accepting CO. In all of these cases, the preadsorbed molecule has a dipole associated with it that is oriented opposite to that of the CO dipole. Conversely, in coadsorbed systems in which both adsorbates exhibit a dipole oriented in the same direction, disordering or segregation occurs. For instance in the CO + O_2/Pd(111) system no ordered structures are formed whereas in the CO + S/Pt(111) system segregation occurs.

In Chapter 6 we study another important example of competitive adsorption. This occurs during the formation of self-assembled monolayers (SAMs) of alkanethiols from solution. With regard to SAM formation, the important aspect is that relatively inert surfaces must be used. The surface interacts weakly with the solvent, but since it is present in a much greater concentration than the solute, the physisorbed solvent covers the surface initially. The alkanethiol slowly chemisorbs to the surface by displacing the solvent molecules. When it binds to the surface, the bond is quite strong and therefore the alkanethiol easily pushes aside the physisorbed species. Once bound, the alkanethiol cannot be removed from the surface by other adsorbates that bind less strongly.

In the above examples, the adsorption energy and lateral interactions were involved in displacement and shepherding of preadsorbed molecules. Another mechanism for displacement and dissociation of preadsorbed molecules utilizes the kinetic energy of an incident molecule. For moderately high incident kinetic energies, molecules that hit an adsorbed layer can impart sufficient momentum to cause chemical transformations [199, 200]. Two distinct processes need to be considered. The first is collision-induced desorption in which the incident molecule transfers sufficient momentum to the adsorbed molecule such that the adsorbate departs from the surface. Ceyer and co-workers [201, 202] demonstrated that argon incident on $CH_4/Ni(111)$ induces desorption of CH_4 via a direct impulsive collision. The second process is collision-induced dissociation [199, 201] in which the incident molecule fragments the adsorbate and the products remain, at least in part, on the surface. An impulsive collision transfers kinetic energy to the adsorbate, which then collides with the substrate. This mechanism leads to dissociation of CH_4 via a pathway that is equivalent to the dissociation of high-translational-energy CH_4 incident from the gas phase. The kinetic energy leads to a distortion of the molecule, which allows the molecule to overcome the dissociation barrier. For $CH_4/Ni(111)$, collision-induced desorption is always more probable than dissociation.

Advanced topic: High-energy collisions

Collisions of even higher energy are also important in surface science. When considering incident energies of the order of a kiloelectronvolt or larger, we usually are dealing with incident ions. At these translational energies, the differences between the dynamics of ions and molecules are largely inconsequential. At such high energies, the ions efficiently penetrate the surface; therefore, we must consider collisions not only with the adsorbed layer and the substrate surface but also with the bulk. Energy can be lost by direct collisions between the ion and the nuclei of the target. Electronic excitations and charge exchange occur as well. All three of these energy exchange mechanisms are dependent on the translational energy and make different contributions as the ion slows. Electronic losses dominate at the highest energies but cause only small changes in the scattering angle. Elastic nuclear collisions produce large-angle scattering, predominantly at low energies. Charge exchange is by far the least important relaxation pathway.

The nuclear scattering of ions from surfaces can be modelled quite successfully by classical hard-sphere kinematics. This allows for accurate predictions to be made about the scattering angles and has been exploited in the form of various ion-scattering spectroscopies that can be used to probe the structure of surfaces and the selvedge [203]. The depth of penetration of ions depends on the relative masses of the projectile and the target nuclei, the ion energy and the angle of incidence. By adjusting these parameters and the ion dose, the depth and concentration of the implanted ions can also be controlled. Ion implantation has a range of technical applications in materials science, including the formation of buried interfaces, semiconductor doping and the modification of surface chemical properties [204].

High-energy particles cause a variety of radiation damage phenomena. When the projectile collides with a target nucleus in a primary collision it transfers momentum, which leads to secondary collisions between the target and other substrate atoms. The atoms struck in the secondary collision go on to further generations of collisions. This process is known as a collision cascade. With the occurrence of so many collisions, it is unlikely that all of the atoms return to their equilibrium positions. Many defects are introduced into the lattice such as vacancies and the occupation of interstitial sites. Some atoms do not return to the substrate and are expelled into the gas phase. The process of removing substrate atoms by high-energy collisional processes is known as sputtering [205]. The number of atoms sputtered per incident ion (i.e. the sputter yield) is a function of the ion energy, angle of incidence, relative masses, the surface temperature and the ion flux. Sputter yields can range from 1–50 atoms per incident ion for typical mass combinations and collisions of the order of tens of kiloelectronvolts. Sputtering occurs not only because of collision cascades. The energy transfer processes that decelerate the ion lead to a variety of thermal and electronic excitations that also contribute to the ejection of substrate ions. Sputtering combined with postirradiation annealing to remove lattice defects is a commonly used method to prepare surfaces free of adsorbed impurities.

3.14   Classification of Reaction Mechanisms

Mass transport to, from and on the surface plays an essential role in surface reactions. When molecules are transferred between phases it takes a finite time for them to equilibrate. Furthermore, we know that not every collision between a molecule and a surface leads to sticking. These characteristics affect the course of reactions at surfaces.

3.14.1   Langmuir–Hinshelwood mechanism

The most common surface reaction mechanism is one in which both reactants are adsorbed on the surface, where they collide and form products. This is known as the Langmuir–Hinshelwood (L–H) mechanism [206]. Adsorption, desorption and surface diffusion play essential roles in the L–H mechanism. Although it might be expected that the reaction rate should depend on the surface coverage of both species, the rate law may be complex and depend on the reaction conditions. Ultimately, the rate law can be properly interpreted only when the complete reaction mechanism is understood. Nonetheless, determination of the rate law is an important component in determining the reaction mechanism.

The dynamics of an L–H reaction involves a convolution of the dynamics of adsorption, desorption and diffusion. Thus, in a sense, a convolution of all the dynamics we have studied up to this point is involved in L–H dynamics. The interplay between reaction dynamics and the rate of catalytic reactions are discussed thoroughly in Chapter 4.

In a multistep reaction mechanism, generally one reaction is the slowest and therefore determines the overall rate. This is known as the rate-determining step (RDS). The dynamics of the RDS is, therefore, the most important dynamics for any given L–H reaction system. The RDS can be any one of a number of different types of surface reactions – for example, adsorption, adsorbate decomposition, diffusion of an adsorbate to a reactive site or desorption of a product. Under normal conditions, the rate of ammonia synthesis in the Haber–Bosch process is determined by the rate of N_2 adsorption (see Section 5.2). The decomposition of Si_2H_6 on silicon or germanium depends on the presence of free sites. If performed at low temperature, the reaction is self-limiting. However, at high temperatures recombinative desorption of H_2 can liberate free sites. Thus the decomposition rate is limited by the desorption rate of H_2.

The identity of the RDS generally depends on the reaction conditions: the pressure of the reactants in the gas phase, surface temperature and coverage. Consider the reaction of O_2 and CO to form CO_2 on platinum-group metals [207]. The catalytic formation of CO_2 requires the reaction of adsorbed oxygen atoms with adsorbed CO molecules. However, in the section on competitive adsorption we have seen that whereas CO can adsorb on an O_2-covered surface (and similarly for O-covered surfaces), O_2 cannot adsorb on CO-covered surfaces. Thus at high CO coverage, the rate is limited by O_2 dissociation, and increases in the CO pressure are found to *inhibit* reaction. However, at low CO coverage, O_2 dissociation can occur rapidly. The reaction rate in this case depends on the coverage of both CO and O_2, and the rate increases with increases of O_2 and CO pressures. The extreme richness of CO oxidation kinetics is a result of changes in the reaction dynamics and can lead to rate oscillations and formation of standing and travelling concentration gradients on the surface. These phenomena are known as spatiotemporal pattern formation and are explored further in Chapter 5.

The energy distribution of the products of a L–H reaction informs us about the dissipation of energy during catalytic reactions. The oxidation of CO [75, 79, 81, 83, 208–211] and H_2 [212, 213] on platinum-group metals to form CO_2 and H_2O, respectively, are both highly exothermic reactions, and energy deposition into the reaction product has been studied in both cases. As can be seen in Figure 3.24, much of the exothermicity of the reaction is dissipated into the surface upon dissociative adsorption of O_2. In the case of CO_2, the product leaves the surface with high levels of rotational, vibrational and translational energy, carrying away approximately 80% of the energy it attains when it reaches the transition state. In contrast, H_2O leaves the surface with an energy that is much more characteristic of the surface temperature.

These energy distributions tell us something about the last interactions of the nascent product with the surface. H_2O essentially leaves the surface as if it were desorbed thermally. Thus, we can surmise that the H_2O formed in the reaction is able to accommodate and chemisorb before desorbing. This is possible if the H_2O formed in the reaction has a structure close to that of adsorbed water. CO_2, on the other hand, carries off a substantial fraction of the reaction exothermicity. In the reaction, a CO bound normal to the surface diffuses to an adsorbed oxygen atom and forms CO_2 in a bent configuration [214]. This, is in contrast to the normal adsorption geometry of CO_2 on a platinum surface in which CO_2 is very weakly bound in a linear configuration. The bent CO_2 is unable to accommodate into the physisorbed state and instead desorbs promptly into the gas phase without losing all of its energy to the surface. From these two case studies we surmise that whether or not some of the reaction exothermicity is returned to the gas phase depends on the formation and desorption dynamics of the product.

3.14.2 Eley–Rideal mechanism

A surface reaction need not involve two surface species. If a gas-phase molecule strikes an adsorbed molecule there is a possibility that the collision leads to reaction and that the

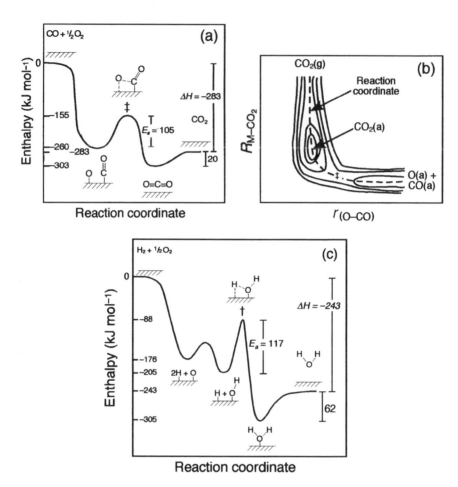

Figure 3.24 The oxidation of CO to form CO_2 and H_2 to form H_2O on Pt(1 1 1) follow Langmuir–Hinshelwood mechanisms. The changes in energy along the reaction pathway are as follows. (a) Enthalpy changes associated with $CO + O_2$ reaction. The transition state ‡ is a stretched and bent CO_2 entity. (b) A two-dimensional potential energy surface of the $CO + O_2$ reaction, portraying the energetic changes as a function of the CO_2–surface distance [$r(M-CO_2)$] and the forming OC bond length [$r(O-CO)$]. (c) Enthalpy changes associated with the $H_2 + O_2$ reaction. Two intermediates (2H + O, and H + OH) are formed during the reaction. The transition state reached prior to the formation of $H_2O_{(a)}$,† is also bent and stretched; however, † resembles $H_2O_{(a)}$ more closely than ‡ resembles $CO_{2(a)}$. (a) and (b) Reproduced from G. Ertl, *Ber. Bunsenges. Phys. Chem.* **86** (1982) 425. Values for (c) taken from M. P. D'Evelyn and R. J. Madix, *Surf. Sci. Rep.* **3** (1983) 413.

product escapes directly into the gas phase. This mechanism is known as the Eley–Rideal (E–R) mechanism [215]. The reaction rate is expected to depend on the coverage of the adsorbed species and the pressure of the other reactant. The products of such a reaction, in contrast to L–H products, should be highly energetic and have a memory of the initial conditions of the gas-phase reactant. This is because the incident reactant is not accommodated with the surface and does not give up part of the exothermicity of the reaction to the surface in the form of its heat of adsorption. Great efforts have been made to prove that this mechanism occurs. It is rather unlikely for molecules to undergo an E–R reaction. However, radicals are much more likely to react by this mechanism. For instance, hydrogen atoms incident on hydrogen covered copper, silicon and diamond surfaces have been shown to form H_2 through an E–R mechanism. Rettner provided the most convincing evidence for an E–R reaction by investigating the reaction of $H_{(g)} + D/Cu(111)$ and $D_{(g)} + H/Cu(111)$ [216]. Although kinetic measurements can be ambiguous, Rettner and Auerbach [217] subsequently measured the energy and angular distributions of the product HD to show that both atoms could not be chemisorbed on the surface. Yates and co-workers [218] have also shown that removal of halogen atoms from Si(100) by incident hydrogen atoms follows E–R kinetics. Lykke and Kay [219] and then Rettner and Auerbach [220, 221] observed a similar reaction for H + Cl/Au surfaces. Hydrogen-abstraction reactions play a role in the chemical vapour deposition (CVD; see Section 6.6.2) of diamond [222, 223] and may likewise play a role in the reactions of adsorbed hydrocarbons [224–226] and in ion pickup during the collision of large molecules with a hydrogen-covered surface [227, 228].

Why should molecules shun E–R dynamics whereas radicals may exhibit such dynamics? Reactions between molecules generally exhibit activation barriers. This is, after all, the reason why chemists search for catalysts that can lower the barriers to reactions. Weinberg [229] has applied transition state theory to argue quantitatively that whenever a barrier to reaction is present, L–H dynamics are preferred to E–R dynamics. E–R dynamics require the special case that a reaction is both barrierless and exothermic, and these are exactly the types of reactions that radical species are prone to partake in. Shalashilin, Jackson and Persson [230–235] have shown that even when these conditions are fulfilled, as in the case of hydrogen incident upon adsorbed hydrogen [H(a)], the probability of E–R dynamics is small. It is much more likely that the reaction in this case proceeds via an intermediate type of dynamics, which is explained in the next section.

3.14.3 Hot atom mechanism

The previous two mechanisms represent the extremes in equilibration or lack thereof of the reactants. Harris and Kasemo and co-worker [236, 237] considered what might happen if one of the reactants were adsorbed while the other was not yet fully accommodated to the surface. This is the hot-precursor or hot-atom (HA) mechanism. Such a mechanism would be quite interesting dynamically and would certainly lead to complex kinetics. Hints of this type of mechanism have been observed for oxygen atoms incident on CO/Pt(111) to form CO_2 [238]. Similarly, when O_2 and CO are adsorbed onto Pt(111) at 100 K or below and then heated to $c.$ 150 K, the adsorbed O_2 [$O_{2(a)}$] dissociates and it appears that some of the liberated oxygen atoms are able to react with

CO before they accommodate with the surface [239]. The dynamics of such a process are not general but are highly sensitive to the interaction potential, as a similar reaction is not observed on Pd(111) under the same conditions [195].

The clearest indication of HA reactions have been found for hydrogen atoms incident on hydrogen-covered copper, silver, nickel, aluminium and platinum surfaces. The classical trajectory studies of Shalashilin, Jackson and Persson on realistic PESs have elucidated the dynamics of the reaction mechanisms in great detail, in particular for the system H + D/Cu(111) and its isotopic analogues. In all of these systems, the reaction to form H_2 is barrierless and highly exothermic. These are necessary conditions for the occurrence of the E–R mechanism, nonetheless, the E–R reaction is quite improbable. For a surface covered with 0.5 monolayers of D, the probability of a direct E–R reaction with an incident hydrogen atom is only c. 0.04. Much more likely is that the incident atom is deflected and scoots along the surface. The hot hydrogen atom exchanges energy very inefficiently with the surface but scatters inelastically with much greater efficiency with adsorbed deuterium [$D_{(a)}$], losing about 0.1 eV per collision. A competitive process now sets in. The hot hydrogen atom can either play billiards with the $D_{(a)}$, eventually losing enough energy to stick (probability c. 0.5 of total incident atoms) or it can react with $D_{(a)}$ to form HD, which desorbs from the surface (probability c. 0.4). Occasionally (probability c. 0.02) one of the $D_{(a)}$ atoms that has suffered a collision with the hot hydrogen atom goes on to react with another adsorbed deuterium atom before it equilibrates with the surface (displacement reaction). Incident hydrogen atoms are particularly well suited to participate in HA reactions because they are highly reactive (and therefore can partake in barrierless exothermic reactions) and they exchange energy slowly with the surface (which increases the lifetime of the hot precursor state). The kinetics of the reactions induced by incident hydrogen atoms are rather complex because they involve a convolution of scattering, sticking, E–R reaction, HA reaction and displacement reaction. A model that successfully treats these kinetics has been developed by Küppers and co-workers [240, 241].

3.15   Measurement of Sticking Coefficients

Three types of sticking coefficients are encountered and measured. There is the initial sticking coefficient, s_0, which is the sticking coefficient at zero coverage. An integral sticking coefficient is obtained by dividing the total coverage by the total exposure. The differential or instantaneous sticking coefficient is the sticking coefficient at a specific coverage and is the quantity that is properly used in rate equations. The differential sticking coefficient (hereafter, as before, the sticking coefficient) is defined by

$$s(\sigma) = \frac{d\sigma}{d\varepsilon} = \lim_{\Delta \to 0} \frac{\Delta\sigma}{\Delta\varepsilon} \tag{3.22}$$

Thus, one method of determining $s(\sigma)$ is to take the derivative of an uptake curve (a plot of coverage against exposure). Some combination of TPD, X-ray photoelectron spectroscopy (XPS) or vibrational spectroscopy, perhaps supported by LEED measurements, is used to measure the coverage. For extremely low coverages, STM can be used to count adsorbates. The value of s_0 is then determined by extrapolation to $\sigma = 0$. Alternatively, s_0 can be calculated from $\Delta\sigma/\Delta\varepsilon$ in the limit of vanishing σ.

The method of King and Wells [242] is a particularly useful variant of sticking-coefficient measurement for reasonably large sticking coefficients. This utilizes a molecular beam in conjunction with the pressure changes that occur when the surface is exposed to the beam. The method is illustrated in Figure 3.25. The vacuum chamber has a base pressure of p_b when the beam is off. When the beam enters the chamber but is blocked from hitting the crystal by a movable shutter upon which no adsorption occurs the pressure rises to p_0. When the shutter is removed, the pressure, $p(t)$, changes as a function of time t as molecules adsorb onto and eventually saturate the surface. For a chamber with constant pumping speed, the sticking coefficient is

$$s(t) = \frac{p_0 - p(t)}{p_0 - p_b} \tag{3.23}$$

Assuming that molecules do not diffuse out of the beam-irradiated area, a subsequent measure of the total coverage at the end of the experiment can be used to calculate the absolute sticking coefficient. Thereby a measurement of $p(t)$, which leads directly to $s(t)$, can be transformed into the more physically meaningful quantity $s(\theta)$.

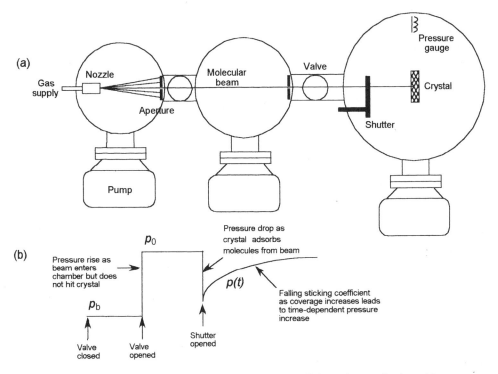

Figure 3.25 The King and Wells method of sticking-coefficient determmination: (a) apparatus, (b) pressure curve

3.16   Summary of Important Concepts

- Physisorption is a weak adsorption interaction in which polarization forces such as van der Waals interactions hold the adsorbate on the surface.

- Chemisorption is a strong adsorption interaction in which orbital overlap leads to chemical bond formation.

- Binding sites at surfaces are separated by energy barriers. Therefore, diffusion on surfaces is an activated process.

- The chemisorption bond is formed by hybridization of metal electronic states with molecular orbitals.

- As a first approximation, the interaction of frontier molecular orbitals with the substrate should be considered in order to understand chemisorption bonding and adsorbate structure.

- On transition metals, chemisorption bond formation can be considered as a two-step process. In step 1, the frontier orbitals are broadened and shifted by the interaction with the s band. In step 2, bonding and antibonding hybrids are formed by the interaction with the d band.

- The strength of the chemisorption bond depends on the position of the hybrid orbitals with respect to the Fermi energy, E_F.

- The strength of chemisorption correlates with the energy of the d-band centre. The lower the d band relative to E_F, the weaker the bond. Therefore, transition metals to the left of a row bind simple adsorbates more strongly than do those on the right.

- In general, a strengthening of adsorbate–surface bonding leads to a weakening of intramolecular bonds in the adsorbate.

- Sufficiently strong chemisorption can lead to the scission of intramolecular bonds in the adsorbate (dissociative chemisorption).

- Adsorption can either be a nonactivated or an activated process.

- Dissociative chemisorption is most commonly associated with activated adsorption. The height of the activation barrier depends on the molecular orientation and the impact position within the unit cell.

- For nonactivated adsorption, the sticking coefficient tends to one for low-energy molecules but decreases for very-high-energy molecules.

- For activated adsorption, sticking can occur only if the incident molecule has sufficient energy to overcome the adsorption barrier. Molecules with energy far in excess of the barrier height may have difficulty sticking as they cannot follow the minimum energy path.

- Adsorption occurs on a multidimensional potential energy hypersurface (PES) and the effect on the sticking coefficient of placing energy in any particular degree of freedom depends on the shape of the PES.

- Adsorption can either be direct or be precursor-mediated.

- Adsorption and desorption are connected by microscopic reversibility.

- In any system for which the sticking coefficient is a function of energy, the desorbed molecules do not have an energy distribution corresponding to an equilibrium distribution at the surface temperature.

- Corrugation is the variation of barrier heights across the surface.

Exercises

3.1 Given that the mean lifetime of an adsorbate is

$$\tau = \frac{1}{A} \exp\left(\frac{E_{\text{des}}}{RT_s}\right) \tag{3.24}$$

where A is the pre-exponential factor for desorption E_{des} is the desorption activation energy, R is the gas constant and T_s is the surface temperature show that the mean random walk distance travelled by an adsorbate is

$$\langle x^2 \rangle^{1/2} = \left[\frac{4D_0}{A} \exp\left(\frac{E_{\text{des}} - E_{\text{dif}}}{RT_s}\right)\right]^{1/2} \tag{3.25}$$

3.2 When pyridine adsorbs on various metal surfaces it changes its orientation as a function of coverage. Describe the bonding interactions that pyridine can experience and how this affects the orientation of adsorbed pyridine [243].

3.3 Cyclopentene, C_5H_8, is chemisorbed very weakly on Ag(111). Given that the double bond in C_5H_8 leads to a dipole that is oriented as shown in Figure 3.26, suggest a configuration for the molecule bound at low coverage on a stepped silver surface with (111) terraces [244].

Figure 3.26 Cyclopentene

3.4 When CO binds in sites of progressively higher coordination number (on-top to twofold bridge to threefold hollow to fourfold hollow) both the π contribution and the σ contribution to bonding increase in magnitude [245].

  (a) Predict the trends that are expected in the CO stretching frequency and chemisorption bond energy with change of site.

  (b) When the π bonding interaction with the surface is weak, which adsorption site is preferred?

3.5 The amount of energy, δE, transferred in the collision of a molecule with a chain of atoms in the limit of a fast, impulsive collision (i.e. a collision that is fast compared with the time that it takes the struck atom to recoil and transfer energy to the chain) is given by the Baule formula:

$$\delta E = \frac{4\mu}{(1+\mu^2)}(E_i + q_{ads}) \qquad (3.26)$$

where $\mu = M/m$, M is the mass of the molecule, m is the mass of one chain atom, E_i is the initial kinetic energy of the molecule before it is accellerated by q_{ads} (the depth of the attractive well; effectively the heat of adsorption). Estimate the energy transfer for H_2, CH_4 and O_2 incident upon copper or platinum chains. Take the incident energy to be

  (a) the mean kinetic energy at 300 K;

  (b) $E_K = 1.0$ eV.

Take the well depths to be 20 meV, 50 meV and 200 meV for H_2, CH_4 and O_2, respectively.

3.6 When a molecule strikes a surface it loses on average an amount of energy $\langle \Delta E \rangle$ given by [158]

$$\langle \Delta E \rangle = -\gamma \alpha_{dsp} E \qquad (3.27)$$

where γ is a constant characteristic of the potential energy surface (PES), α_{dsp} is a constant that depends on the collision partners, and E is the kinetic energy upon collision. For H/Cu(111), $\alpha_{dsp} = 0.0024$, $\gamma = 4.0$ and the binding energy of chemisorbed hydrogen is 2.5 eV.

  (a) For a hydrogen atom with an initial $E_K = 0.1$ eV 10 Å away from the surface, calculate the energy transfer on the first bounce.

  (b) Assuming the same amount of energy transfer on each subsequent collision, how many collisions are required for the hydrogen atom to reach the bottom of the well?

  (c) Given that α_{dsp} changes from one molecule to the next analogous to the Baule formula

$$\alpha_{dsp} = k \frac{4\mu}{(1+\mu)^2} \qquad (3.28)$$

where μ is calculated assuming one surface atom participates in the collision, calculate α_{dsp} for CO assuming the same proportionality factor as for hydrogen. Then make a rough estimate of the number of collisions CO requires to reach the bottom of a 1.2 eV chemisorption well, assuming $\gamma = 4.0$.

3.7 Classically, a chemical reaction cannot occur if the collision partners do not have sufficient energy to overcome the activation barrier. This and the thermal distribution of energy are the basis of the Arrhenius formulation of reaction rate constants. For an atom, the thermal energy is distributed among the translational degrees of freedom. The velocity distribution is governed by a Maxwell distribution:

$$f(v) = 4\pi \left(\frac{M}{2\pi RT} \right)^{3/2} v^2 \exp\left(\frac{-Mv^2}{2RT} \right) \tag{3.29}$$

where M is the molar mass, v is the speed, R is the gas constant and T is the temperature. Assuming that there is no steric requirement for sticking (i.e. that energy is the only determining factor) calculate the sticking coefficient of an atomic gas held at

(a) 300 K

(b) 1000 K

for adsorption activation barriers of $E_{ads} = 0$ eV, 0.1 eV, 0.5 eV and 1.0 eV.

3.8 A real molecule has quantized rotational and vibrational energy levels. The Maxwell–Boltzmann distribution law describes the occupation of these levels. The distribution among rotational levels is given by

$$N_{v,J} = N_v \frac{hc}{k_B T} (2J + 1) \exp\left(\frac{E_{rot}}{k_B T} \right) \tag{3.30}$$

where $N_{v,J}$ is the number of molecules in the rotational state with quantum numbers v and J, N_v is the total number of molecules in the vibrational state v, h is Planck's constant, c is the speed of light, k_B is the Boltzmann constant and T is the temperature. The energy of rigid rotor levels is given by

$$E_{rot} = hcB_v J(J + 1) \tag{3.31}$$

where B_v is the rotational constant of the appropriate vibrational state. The vibrational population is distributed according to

$$N_v = N \exp\left(\frac{-hcG_0(v)}{k_B T} \right) \tag{3.32}$$

where N is the total number of molecules and $G_0(v)$ is the wavenumber of the vibrational level v above the ground vibrational level. At thermal equilibrium the mean energy is distributed according to

$$\langle E \rangle = \langle E_{trans} \rangle + \langle E_{rot} \rangle + \langle E_{vib} \rangle \tag{3.33}$$

where, for a diatomic molecule,

$$\langle E_{trans} \rangle = 2k_B T \tag{3.34}$$
$$\langle E_{rot} \rangle = k_B T \tag{3.35}$$
$$\langle E_{vib} \rangle = \sum_{n>0} \frac{hv_n}{1 - \exp(hv_n/k_B T)} \tag{3.36}$$

Consider an extremely early barrier potential energy surface in which only translational energy is effective in overcoming the activation barrier.

(a) For the same temperatures and barrier heights as in Exercise 3.1, calculate the sticking coefficient.

(b) For the same barrier heights, calculate the sticking coefficient for molecules with mean total energies of 0.1 eV, 0.5 eV and 1.0 eV. Use NO as the molecule and assume that only $v = 1$ contributes to the vibrational energy for which $G_0 = 1904$ cm^{-1}.

3.9 Consider an extremely late barrier in which translational energy plays no role, vibrational energy is 100% effective at overcoming the barrier and rotational energy is 50% efficient. Calculate the classical sticking coefficient of H_2 and D_2 as a function of rovibrational state for the first three vibrational levels and an adsorption barrier of 0.5 eV. Assume that zero point energy plays no role and that molecules can be described as rigid rotors. The vibrational energy spacings and rotational constants of H_2 and D_2 are given in Table 3.3.

3.10 Consider the adsorption of D_2 (see Table 3.3). Assuming that normal translational energy is 100% effective and vibrational energy is 60% effective at overcoming the adsorption barrier, calculate the sticking coefficient of the first three vibrational levels as a function of normal translational energy. Neglect the effects of rotation.

3.11 The degeneracy of rotational states arises from the quantization of space. That is, both the magnitude and the direction (projection) of rotational angular momentum is quantized. This is reflected in the two quantum numbers J and m_J. Taking the surface normal as the quantization axis, $m_J = 0$ corresponds to out-of-plane rotation, and $m_J = J$ corresponds to in-plane rotation. Consider the case of a strong steric factor that influences the nonactivated adsorption of a diatomic molecule such that only in-plane rotating molecules stick, whereas all others scatter back into the gas phase.

(a) Assuming that there is no steering of the molecule along its approach to the surface, calculate the sticking coefficient as a function of rotational level.

(b) Now assume that steering occurs and that the probability of being steered into the proper orientation is a function of m_J such that it decreases from 1 for $m_J = 0$ in a Gaussian fashion. The width of the Gaussian is given by $\sigma = J/2$. Calculate s_0 as a function of J.

3.12 The flux of molecules striking a surface follows a cosine distribution, $\cos \vartheta$, where ϑ is the angle from surface normal. If the perpendicular component of translational energy is effective at overcoming the adsorption barrier and the parallel component is not, the angular distribution of the flux that sticks is tightly constrained about the surface normal. The desorbing flux is similarly peaked about the surface normal. It is often observed that the desorbing flux can be described by a $\cos^n \vartheta$ distribution in which $n > 1$; the greater the value of n, the more peaked the distribution. The angular distribution of D_2 desorbing from Cu(1 0 0) has been measured by Comsa and David [246]. They found a relationship between the normalized desorption intensity, $N(\vartheta)/N(0°)$, and the desorption angle measured from the surface normal, ϑ, as shown in Table 3.4. Determine n.

Table 3.3   Rotational constant B_v, and wavenumber $G_0(v)$ for vibrational state v for H_2 and D_2; see Exercises 3.9 and 3.10.

	H_2		D_2	
v	B_v/cm^{-1}	$G_0(v)/\text{cm}^{-1}$	B_v/cm^{-1}	$G_0(v)/\text{cm}^{-1}$
0	59.3	0.0	29.9	0.0
1	56.4	4161.1	28.8	2994.0
2	53.5	8087.1	28.0	5868.8

Table 3.4 The relationship between the normalized desorption intensity, $N(\vartheta)/N(0°)$, and the desorption angle measured from the surface normal, ϑ; see Exercise 3.12. Source: Comsa and David [246]

ϑ (°)	0	5	10	15	20	25	30	35	45
$\dfrac{N(\vartheta)}{N(0°)}$	1.00	0.99	0.98	0.77	0.63	0.48	0.38	0.21	0.06

3.13 The sticking of molecular hydrogen on silicon is highly activated in the surface coordinates. Bratu and Höfer [165] have determined the sticking coefficient of H_2 on Si(111)-(7×7) as a function of surface temperature. The data are recorded in Table 3.5. Using an Arrhenius formulation [Equation (3.37)], determine the pre-exponential factor and the activation barrier height.

3.14 If a chemical reaction proceeds with a single activation barrier, the rate constant should follow the Arrhenius expression. Accordingly, for the sticking coefficient S_d, we write

$$s_0 = A_s \exp\left(\frac{E_a}{RT}\right),\tag{3.37}$$

where A_s is a constant, E_a is the adsorption activation energy, R is the gas constant and T is the temperature. However, if a distribution of barriers rather than a single barrier participates in the reaction, a different form is followed. In the case of H_2 sticking on Cu(1 1 1), the sticking coefficient as a function of normal kinetic energy, E_n, is found to follow

$$s_0(E_n) = \frac{A_s}{2}\left[1 + \mathrm{erf}\left(\frac{E_n - E_0}{W}\right)\right]\tag{3.38}$$

where E_0 is the mean position of a distribution of barriers that has a width W. Given the data [247] in Table 3.6 for s_0 of $H_2(v = 0)$ and E_n, determine E_0 and W. Make plots of s_0 versus E_n with different values of E_0 and W to observe the effects these have on the shape of the sticking curve.

3.15 Classically we assign $\frac{1}{2}k_B T$ of energy to each active degree of freedom and, therefore, we assign a value of $\frac{3}{2}k_B T$ to the kinetic energy. This is true for a *volume* sample of a gas. For a *flux* of gas, such as that desorbing from a surface, the answer is different. Use the Maxwell velocity distribution to show that the equilibrium mean kinetic energy of a flux of gas emanating from (or passing through) a surface is

$$\langle E_{\mathrm{trans}}\rangle = 2k_B T_s\tag{3.39}$$

Table 3.5 Initial sticking coefficient, s_0, with surface temperature, T_s; see Exercise 3.13. Source: Bratu and Höfer [165]

T_s/K	587	613	637	667	719	766	826	891	946	1000	1058
s_0	2.8×10^{-9}	6.5×10^{-9}	1.3×10^{-8}	2.0×10^{-8}	5.3×10^{-8}	1.7×10^{-7}	5.4×10^{-7}	1.3×10^{-6}	2.3×10^{-6}	2.7×10^{-6}	5.0×10^{-6}

.able 3.6 Initial sticking coefficient, s_0, with normal kinetic energy E_n; see Exercise 3.14. Source: Michelsen *et al.* [247]

E_n/eV	0.1	0.2	0.3	0.4	0.5	0.6	0.7	0.8	0.9	1.0
s_0	4×10^{-6}	1×10^{-4}	0.0021	0.0167	0.0670	0.151	0.219	0.245	0.250	0.250

The mean kinetic energy is defined by the moments of the velocity distribution according to

$$\langle E_{\text{trans}} \rangle = \frac{1}{2} \frac{m M_3}{M_1} \tag{3.40}$$

where the moments are calculated according to

$$M_i = \int_0^\infty v_i f(v) dv. \tag{3.41}$$

3.16 For desorption from a rigid surface and in the absence of electron–hole pair formation or other electronic excitation, a desorbing molecule will not lose energy to the surface after it passes through the transition state. In the absence of a barrier the mean energy of the desorbed molecules is roughly equal to the mean thermal expectation value at the surface temperature, $\langle k_B T_s \rangle$. The excess energy above this value, as shown in Figure 3.27, is equal to the height of the adsorption activation energy. Therefore a measurement of the mean total energy of the desorbed molecules, $\langle E \rangle$, can be used to estimate the height of the adsorption

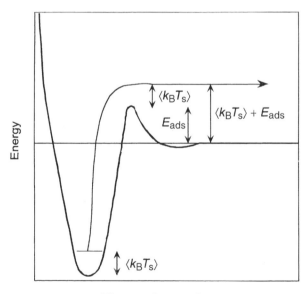

Figure 3.27 Energy as a function of molecular distance from surface, z; see Exercise 3.16. E_{ads}, Adsorption Activation Energy; T_s, surface temperature, k_B, Boltzmann constant

Table 3.7 Data for $D_2/Cu(1\,1\,1)$ and $D_2/Si(1\,0\,0)$-(2×1); see Exercise 3.16. Source: Rettner, Michelsen and Auerbach [248] and Kolasinski et al [162]

System	T_s (K)	$\langle E_{rot} \rangle$ (K)	T_{vib} (K)	$\langle E_{trans} \rangle$ (K)	E_{ads} (eV)
$D_2/Cu(1\,1\,1)$	925	330	1700	960	0.5
$D_2/Si(1\,0\,0)$	780	1020	1820	3360	0.8

Note: T_s, surface temperature; E_{rot}, E_{trans}, rotational and translational energies, respectively; E_{ads}, adsorption activation energy; T_{vib}, see text, Exercise 3.16.

barrier [162]. The mean thermal energy is given by

$$\langle E \rangle = \langle E_{trans} \rangle + \langle E_{rot} \rangle + \langle E_{vib} \rangle \tag{3.42}$$

where E_{trans}, E_{rot} and E_{vib} are the translational, rotational and vibrational energies, respectively.

(a) Given the data for D_2, $Cu(1\,1\,1)$ given in Table 3.7 [248], show that the above approximations hold and, therefore, to a first approximation we need not consider surface atom motions in the ad/desorption dynamics.

(b) Given the data for $D_2/Si(1\,0\,0)$–(2×1) in Table 3.7 [162] show that the above approximations do not hold and that therefore the static potential picture cannot be used to interpret the ad/desorption dynamics.

In both cases assume that the vibrational distribution is thermal and is described by a temperature T_{vib}.

Further Reading

M. R. Albert and J. T. Yates, Jr, *The Surface Scientist's Guide to Organometallic Chemistry* (American Chemical Society, Washington, DC, 1987).

J. A. Barker and D. J. Auerbach, "Gas-surface interaction and dynamics: thermal energy atomic and molecular beam studies", *Surf. Sci. Rep.* **4** (1985) 1.

G. P. Brivio and M. I. Trioni, "The adiabatic molecule–metal surface interaction: theoretical approaches", *Rev. Mod. Phys.* **71** (1999) 231.

M. J. Cardillo, "Concepts in gas–surface dynamics", *Langmuir* **1** (1985) 4.

S. T. Ceyer, "Dissociative chemisorption: dynamics and mechanisms", *Annu. Rev. Phys. Chem.* **39** (1988) 479.

G. Comsa and R. David, "Dynamical parameters of desorbing molecules", *Surf. Sci. Rep.* **5** (1985) 145.

M. C. Desjonquères and D. Spanjaard, *Concepts in Surface Physics* (Springer, Berlin, 1996).

M. P. D'Evelyn and R. J. Madix, "Reactive scattering from solid surfaces", *Surf. Sci. Rep.* **3** (1983) 413.

D. J. Doren, "Kinetics and dynamics of hydrogen adsorption and desorption on silicon surfaces", in *Advances in Chemical Physics, Volume 95*, eds I. Prigogine and S. A. Rice (John Wiley, New York, 1996), p. 1.

D. J. Doren and J. C. Tully, "Dynamics of precursor-mediated chemisorption", *J. Chem. Phys.* **94** (1991) 8428.

A. Gross, "Reactions at surfaces studied by ab initio dynamics calculations", *Surf. Sci. Rep.* **32** (1998) 291.

J. Harris, "On the adsorption and desorption of H_2 at metal surfaces", *Appl. Phys. A: Mater. Sci. Process.* **47** (1988) 63.

A. Hodgson, "State resolved desorption measurements as a probe of surface reactions", *Prog. Surf. Sci.* **63** (2000) 1.

U. Höfer, "Nonlinear optical investigations of the dynamics of hydrogen interactions with silicon surfaces", *Appl. Phys. A: Mater. Sci. Process.* **63** (1996) 533.

G. L. Kellogg, "Field ion microscope studies of single-atom surface diffusion and cluster nucleation on metal surfaces", *Surf. Sci. Rep.* **21** (1994) 1.

D. A. King and D. P. Woodruff (eds), *The Chemical Physics of Solid Surfaces and Heterogeneous Catalysis: Adsorption at Solid Surfaces*, Volume 2 (Elsevier, Amsterdam, 1983).

K. W. Kolasinski, "Dynamics of hydrogen interactions with Si(100) and Si(111) surfaces", *Int. J. Mod. Phys. B* **9** (1995) 2753.

R. D. Levine and R. B. Bernstein, *Molecular Reaction Dynamics and Chemical Reactivity* (Oxford University Press, New York, 1987).

M. C. Lin and G. Ertl, "Laser probing of molecules desorbing and scattering from solid surfaces", *Annu. Rev. Phys. Chem.* **37** (1986) 587.

C. T. Rettner and M. N. R. Ashfold (eds), *Dynamics of Gas–Surface Interactions* (The Royal Society of Chemistry, Cambridge, 1991).

C. T. Rettner, H. A. Michelsen, and D. J. Auerbach, "From quantum-state-specific dynamics to reaction rates: the dominant role of translational energy in promoting the dissociation of D_2 on Cu(111) under equilibrium conditions", *Faraday Discuss. Chem. Soc.* **96** (1993) 17.

M. W. Roberts and C. S. McKee, *Chemistry of the Metal–Gas Interface* (Clarendon Press, Oxford, 1978).

G. Scoles, *Atomic and Molecular Beam Methods*, Volumes 1 and 2, (Oxford University Press, New York, 1988 and 1992).

S. F. Bent, H. A. Michelsen, and R. N. Zare, "Hydrogen recombinative desorption dynamics", in *Laser Spectroscopy and Photochemistry on Metal Surfaces*, eds H. L. Dai and W. Ho (World Scientific, Singapore, 1995), p. 977.

G. A. Somorjai, *Introduction to Surface Chemistry and Catalysis* (John Wiley, New York, 1994).

J. C. Tully, "Theories of the dynamics of inelastic and reactive processes at surfaces", *Annu. Rev. Phys. Chem.* **31** (1980) 319.

J. C. Tully and M. J. Cardillo, "Dynamics of molecular motion at single-crystal surfaces", *Science* **223** (1984) 445.

J. L. Whitten and H. Yang, "Theory of chemisorption and reactions on metal surfaces", *Surf. Sci. Rep.* **24** (1996) 55.

H. Zacharias, "Laser spectroscopy of dynamical surface processes", *Int. J. Mod. Phys. B* **4** (1990) 45.

References

[1] R. M. Watwe, R. D. Cortright, M. Mavirakis, J. K. Nørskov and J. A. Dumesic, *J. Chem. Phys.* **114** (2001) 4663.

[2] K. Christmann, *Surf. Sci. Rep.* **9** (1988) 1.

[3] B. D. Kay, C. H. F. Peden and D. W. Goodman, *Phys. Rev. B* **34** (1986) 817.

[4] C. H. F. Peden, B. D. Kay and D. W. Goodman, *Surf. Sci.* **175** (1986) 215.

[5] R.Gomer, *Rep. Prog. Phys.* **53** (1990) 917.

[6] L. J. Lauhon and W. Ho, *Phys. Rev. Lett.* **85** (2000) 4566.

[7] B. S. Swartzentruber, *Phys. Rev. Lett.* **76** (1996) 459.

[8] G. Ehrlich and F. G. Hudda, *J. Chem. Phys.* **44** (1966) 1039.

[9] R. L. Schwoebel and E. J. Shipsey, *J. Appl. Phys.* **37** (1966) 3682.

[10] G. L. Kellogg, *Surf. Sci. Rep.* **21** (1994) 1.

[11] M. Bowker and D. A. King, *Surf. Sci.* **71** (1978) 583.

[12] M. Bowker and D. A. King, *Surf. Sci.* **72** (1978) 208.

[13] D. A. Reed and G. Ehrlich, *Surf. Sci.* **102** (1981) 588.

[14] G. P. Brivio and M. I. Trioni, *Rev. Mod. Phys.* **71** (1999) 231.

[15] E. Zaremba and W. Kohn, *Phys. Rev. B* **13** (1976) 2270.

[16] E. Zaremba and W. Kohn, *Phys. Rev. B* **15** (1977) 1769.

[17] F. Haber, *Z. Elektrochem.* **20** (1914) 521.

[18] I. Langmuir, *Phys. Rev.* **8** (1916) 149.

[19] I. Langmuir, *J. Am. Chem. Soc.* **40** (1918) 1361.

[20] M. C. Desjonquères and D. Spanjaard, *Concepts in Surface Physcis*, 2nd edn (Springer, Berlin, 1996).

[21] P. W. Anderson, *Phys. Rev.* **124** (1961) 41.

[22] T. B. Grimley, *Proc. R. Soc. London, Ser.* A **90** (1967) 751.

[23] T. B. Grimley, "Theory of chemisorption", in *The Chemical Physics of Solid Surfaces and Heterogeneous Catalysis*, Volume 2, eds D. A. King and D. P. Woodruff (Elsevier, Amsterdam, 1983) p. 333.

[24] D. M. Newns, *Phys. Rev.* **178** (1969) 1123.

[25] W. Brenig and K. Schönhammer, *Z. Phys.* **276** (1974) 201.

[26] I. Vasiliev, S. Ögüt and J. R. Chelikowsky, *Phys. Rev. Lett.* **82** (1999) 1919.

[27] D. J. Doren, "Kinetics and dynamics of hydrogen adsorption and desorption on silicon surfaces", in *Advances in Chemical Physics, Volume 95*, eds I. Prigogine and S. A. Rice (John Wiley, New York, 1996) p. 1.

[28] H. Haberland (ed.), *Clusters of Atoms and Molecules: Springer Series in Chemical Physics, Volume 52* (Springer, Berlin, 1993).

[29] H. Haberland (ed.), *Clusters of Atoms and Molecules II: Springer Series in Chemical Physics, Volume 56* (Springer, Berlin, 1994).

[30] J. C. Campuzano, "The adsorption of carbon monoxide by the transition metals", in *The Chemical Physics of Solid Surfaces and Heterogeneous Catalysis: Chemisorption Systems.* Volume 3A, eds D. A. King and D. P. Woodruff (Elsevier, Amsterdam, 1990) p. 389.

[31] G. Blyholder, *J. Phys. Chem.* **68** (1964) 2772.

[32] M. D. Alvey, M. J. Dresser and J. T. Yates Jr, *Surf. Sci.* **165** (1986) 447.

[33] M. R. Albert and J. T. Yates Jr, *The Surface Scientist's Guide to Organometallic Chemistry* (American Chemical Society, Washington, DC, 1987).

[34] A. Föhlisch, N. Nyberg, P. Bennich, L. Triguero, J. Hasselström, O. Karis, L. G. M. Pettersson and A. Nilsson, *J. Chem. Phys.* **112** (2000) 1946.

[35] P. Hu, D. A. King, M.-H. Lee and M. C. Payne, *Chem. Phys. Lett.* **246** (1995) 73.

[36] K. W. Kolasinski, F. Cemič and E. Hasselbrink, *Chem. Phys. Lett.* **219** (1994) 113.

[37] R. Imbihl and J. E. Demuth, *Surf. Sci.* **173** (1986) 395.

[38] H. Steininger, H. Ibach and S. Lehwald, *Surf. Sci.* **117** (1982) 685.

[39] G. A. Somorjai and K. R. McCrea, "Dynamics of reactions at surfaces", in *Advances in Catalysis, Volume 45*, ed. B. C. Gates and H. Knözinger (Academic Press, Boston, MA, 2000) p. 386.

[40] J. Harris, *Appl. Phys. A: Mater. Sci. Process.* **47** (1988) 63.

[41] A. E. DePristo, "Dynamics of dissociative chemisorption", in *Dynamics of Gas–Surface Interactions*, eds C. T. Rettner and M. N. R. Ashfold (The Royal Society of Chemistry, Cambridge, 1991) p. 47.

[42] G. R. Darling and S. Holloway, *Rep. Prog. Phys.* **58** (1995) 1595.

[43] H. A. Michelsen, C. T. Rettner and D. J. Auerbach, "The adsorption of hydrogen at copper surfaces: a model system for the study of activated adsorption", in *Surface Reactions: Springer Series in Surface Sciences*, Vol. 34, ed by R. J. Madix (Springer, Berlin, 1994) p. 185.

[44] B. I. Lundqvist, J. K. Nørskov and H. Hjelmberg, *Surf. Sci.* **80** (1979) 441.

[45] P. Nordlander, S. Holloway and J. K. Nørskov, *Surf. Sci.* **136** (1984) 59.

[46] J. K. Nørskov, A. Houmøller, P. K. Johansson and B. I. Lundqvist, *Phys. Rev. Lett.* **46** (1981) 257.

[47] G. R. Darling and S. Holloway, *Surf. Sci.* **304** (1994) L461.

[48] G. A. Somorjai, *Adv. Catal.* **26** (1977) 1.

[49] B. Hammer and J. K. Nørskov, *Surf. Sci.* **343** (1995) 211.

[50] B. Hammer and J. K. Nørskov, *Nature* **376** (1995) 238.

[51] B. Hammer and J. K. Nørskov, "Theoretical surface science and catalysis – calculations and concepts", in *Advances in Catalysis, Volume 45*, eds B. C. Gates and H. Knözinger (Academic Press, Boston, 2000) p. 71.

[52] G. Ertl, "Dynamics of reactions at surfaces", in *Advances in Catalysis, Volume 45*, eds. B. C. Gates and H. Knözinger (Academic Press, Boston, MA, 2000) p. 1.

[53] R. T. Brackmann and W. L. Fite, *J. Chem. Phys.* **34** (1961) 1572.

[54] J. N. Smith Jr and W. L. Fite, *J. Chem. Phys.* **37** (1962) 898.

[55] G. Comsa, *J. Chem. Phys.* **48** (1968) 3235.

[56] R. L. Palmer, J. N. Smith Jr, H. Saltsburg and D. R. O'Keefe, *J. Chem. Phys.* **53** (1970) 1666.

[57] A. E. Dabiri, T. J. Lee and R. E. Stickney, *Surf. Sci.* **26** (1971) 522.

[58] G. Marenco, A. Schutte, G. Scoles and F. Tommasine, *J. Vac. Sci. Technol.* **9** (1971) 824.

[59] M. Balooch and R. E. Stickney, *Surf. Sci.* **44** (1974) 310.

[60] M. J. Cardillo, M. Balooch and R. E. Stickney, *Surf. Sci.* **50** (1975) 263.

[61] F. Frenkel, J. Häger, W. Krieger, H. Walter, C. T. Campbell, G. Ertl, H. Kuipers and J. Segner, *Phys. Rev. Lett.* **46** (1981) 152.

[62] C. T. Campbell, G. Ertl and J. Segner, *Surf. Sci.* **115** (1982) 309.

[63] F. Frenkel, J. Häger, W. Krieger, H. Walter, G. Ertl, J. Segner and W. Vielhaber, *Chem. Phys. Lett.* **90** (1982) 225.

[64] M. Asscher, W. L. Guthrie, T.-H. Lin and G. A. Somorjai, *Phys. Rev. Lett.* **49** (1982) 76.

[65] W. L. Guthrie, T.-H. Lin, S. T. Ceyer and G. A. Somorjai, *J. Chem. Phys.* **76** (1982) 6398.

[66] M. Asscher, W. L. Guthrie, T.-H. Lin and G. A. Somorjai, *J. Chem. Phys.* **78** (1983) 6992.

[67] G. M. McClelland, G. D. Kubiak, H. G. Rennagel and R. N. Zare, *Phys. Rev. Lett.* **46** (1981) 831.

[68] J. E. Hurst Jr, G. D. Kubiak and R. N. Zare, *Chem. Phys. Lett.* **93** (1982) 235.

[69] G. D. Kubiak, J. E. Hurst, Jr., H. G. Rennagel, G. M. McClelland and R. N. Zare, *J. Chem. Phys.* **79** (1983) 5163.

[70] A. W. Kleyn, A. C. Luntz and D. J. Auerbach, *Phys. Rev. Lett.* **47** (1981) 1169.

[71] A. C. Luntz, A. W. Kleyn and D. J. Auerbach, *J. Chem. Phys.* **76** (1982) 737.

[72] A. C. Luntz, A. W. Kleyn and D. J. Auerbach, *Phys. Rev. B* **25** (1982) 4273.

[73] R. R. Cavanagh and D. S. King, *Phys. Rev. Lett.* **47** (1981) 1829.

[74] D. S. King and R. R. Cavanagh, *J. Chem. Phys.* **76** (1982) 5634.

[75] S. L. Bernasek and S. R. Leone, *Chem. Phys. Lett.* **84** (1981) 401.

[76] R. P. Thorman and S. L. Bernasek, *J. Chem. Phys.* **74** (1981) 6498.

[77] L. S. Brown and S. L. Bernasek, *J. Phys. Chem.* **82** (1985) 2110.

[78] D. A. Mantell, S. B. Ryali, B. L. Halpern, G. L. Haller and J. B. Fenn, *Chem. Phys. Lett.* **81** (1981) 185.

[79] D. A. Mantell, S. B. Ryali, G. L. Haller and J. B. Fenn, *J. Chem. Phys.* **78** (1983) 4250.

[80] D. A. Mantell, Y.-F. Maa, S. B. Ryali, G. L. Haller and J. B. Fenn, *J. Chem. Phys.* **78** (1983) 6338.

[81] D. A. Mantell, K. Kunimori, S. B. Ryali, G. L. Haller and J. B. Fenn, *Surf. Sci.* **172** (1986) 281.

[82] D. W. J. Kwong, N. DeLeon and G. L. Haller, *Chem. Phys. Lett.* **144** (1988) 533.

[83] G. W. Coulston and G. L. Haller, *J. Chem. Phys.* **95** (1991) 6932.

[84] G. D. Kubiak, G. O. Sitz and R. N. Zare, *J. Chem. Phys.* **81** (1984) 6397.

[85] G. D. Kubiak, G. O. Sitz and R. N. Zare, *J. Chem. Phys.* **83** (1985) 2538.

[86] U. Valbusa, "General principles and methods", in *Atomic and Molecular Beam Methods, Volume 2*, ed. G.Scoles (Oxford University Press, New York, 1992) p.327.

[87] G. Boato, "Elastic scattering of atoms", in *Atomic and Molecular Beam Methods, Volume 2*, ed. by G. Scoles (Oxford University Press, New York, 1992) p.340.

[88] O. Stern, *Naturwissenschaften* **21** (1929) 391.

[89] I. Estermann, R. Frisch and O. Stern, *Z. Phys.* **73** (1931) 348.

[90] R. Frisch and O. Stern, *Z. Phys.* **84** (1933) 430.

[91] H. Schlichting, D. Menzel, T. Brunner, W. Brenig and J. C. Tully, *Phys. Rev. Lett.* **60** (1988) 2515.

[92] G. Comsa and B. Poelsema, "Scattering from disordered surfaces", in *Atomic and Molecular Beam Methods, Volume 2*, ed. G. Scoles (Oxford University Press, New York, 1992) p. 463.

[93] R. B. Doak, R. E. Grisenti, S. Rehbein, G. Schmahl, J. P. Toennies and C. Woll, *Phys. Rev. Lett.* **83** (1999) 4229.

[94] R. B. Doak, "Single-phonon inelastic helium scattering", in *Atomic and Molecular Beam Methods, Volume 2*, ed. by G. Scoles (Oxford University Press, New York, 1992) p. 384.

[95] A. P. Graham and J. P. Toennies, *Surf. Sci.* **428** (1999) 1.

[96] J. A. Barker and D. J. Auerbach, *Surf. Sci. Rep.* **4** (1985) 1.

[97] M. Asscher and G. A. Somorjai, "Reactive scattering", in *Atomic and Molecular Beam Methods, Volume 2*, ed. G. Scoles (Oxford University Press, New York, 1992) p. 488.

[98] D. J. Auerbach, "Multiple-phonon inelastic scattering", in *Atomic and Molecular Beam Methods, Volume 2*, ed. by G. Scoles (Oxford University Press, New York, 1992) p. 444.

[99] P. Casavecchia, *Rep. Prog. Phys.* **63** (2000) 355.

[100] Y. T. Lee, *Science* **236** (1987) 793.

[101] J. B. Anderson, R. P. Andres and J. B. Fenn, *Adv. Chem. Phys.* **10** (1966) 275.

[102] G. Scoles (ed.) *Atomic and Molecular Beam Methods, Volume 1* (Oxford University Press, New York, 1988).

[103] G. Scoles (ed), *Atomic and Molecular Beam Methods, Volume 2* (Oxford University Press, New York, 1992).

[104] J. E. Lennard-Jones, *Trans.Faraday Society* **28** (1932) 333.

[105] F. Shimizu, *Phys. Rev. Lett.* **86** (2001) 987.

[106] D. J. Doren and J. C. Tully, *Langmuir* **4** (1988) 256.

[107] D. J. Doren and J. C. Tully, *J. Chem. Phys.* **94** (1991) 8428.

[108] C. R. Arumainayagam, M. C. McMaster, G. R. Schoofs and R. J. Madix, *Surf. Sci.* **222** (1989) 213.

[109] E. W. Kuipers, M. G. Tenner, M. E. M. Spruit and A. W. Kleyn, *Surf. Sci.* **205** (1988) 241.

[110] E. K. Grimmelmann, J. C. Tully and M. J. Cardillo, *J. Chem. Phys.* **72** (1980) 1039.

[111] R. M. Logan, J. C. Keck and R. E. Stickney, *J. Chem. Phys.* **44** (1966) 44.

[112] R. M. Logan and J. C. Keck, *J. Chem. Phys.* **49** (1968) 860.

[113] W. A. Dew and H. S. Taylor, *J. Phys. Chem.* **31** (1927) 281.

[114] O. Schmidt, *Z. Phys. Chem.* **133** (1928) 263.

[115] M. Head-Gordon, J. C. Tully, C. T. Rettner, C. B. Mullins and D. J. Auerbach, *J. Chem. Phys.* **94** (1991) 1516.

[116] J. C. Maxwell, *Philos. Trans. R. Soc. London* **170** (1879) 231.

[117] M. Knudsen, *Ann. Phys. (Leipzig)* **34** (1911) 593.

[118] M. v. Smoluchowski, *Ann. Phys. (Leipzig)* **33** (1910) 1559.

[119] W. Gaede, *Ann. Phys. (Leipzig)* **41** (1913) 289.

[120] R. A. Millikan, *Phys. Rev.* **21** (1923) 217.

[121] R. A. Millikan, *Phys. Rev.* **22** (1923) 1.

[122] P. Clausing, *Ann. Phys. (Leipzig)* **4** (1930) 36.

[123] I. Langmuir, *J. Am. Chem. Soc.* **38** (1916) 2221.

[124] R. C. Tolman, *Proc. Nat. Acad. Amer.* **11** (1925) 436.

[125] R. H. Fowler and E. A. Milne, *Proc. Nat. Acad. Amer.* **22** (1925) 400.

[126] E. P. Wenaas, *J. Chem. Phys.* **54** (1971) 376.

[127] I. Kuščer, *Surf. Sci.* **25** (1971) 225.

[128] D. A. McQuarrie, *Statistical Thermodynamics* (University Science Books, Mill Valley, CA, 1973).

[129] J. N. Smith Jr and R. L. Palmer, *J. Chem. Phys.* **56** (1972) 13.

[130] R. D. Levine and R. B. Bernstein, *Molecular Reaction Dynamics and Chemical Reactivity* (Oxford University Press, New York, 1987).

[131] S. Holloway, *Surf. Sci.* **299/300** (1994) 656.

[132] D. M. Newns, *Surf. Sci.* **171** (1986) 600.

[133] J. W. Gadzuk, *J. Vac. Sci. Technol. A* **5** (1987) 492.

[134] H. Metiu and J. W. Gadzuk, *J. Chem. Phys.* **74** (1981) 2641.

[135] J. C. Tully, *J. Electron Spectrosc. Rel. Phenom.* **45** (1987) 381.

[136] M. Persson and B. Hellsing, *Phys. Rev. Lett.* **49** (1982) 662.

[137] D. Halstead and S. Holloway, *J. Chem. Phys.* **93** (1990) 2859.

[138] H. Eyring and M. Polanyi, *Z. Phys. Chem. B* **12** (1931) 279.

[139] J. C. Polanyi, *Science* **236** (1987) 680.

[140] J. C. Polanyi and A. H. Zewail, *Accts. Chem. Res.* **28** (1995) 119.

[141] J. C. Polanyi and W. H. Wong, *J. Chem. Phys.* **51** (1969) 1439.

[142] M. H. Mok and J. C. Polanyi, *J. Chem. Phys.* **51** (1969) 1451.

[143] J. C. Polanyi, *Accts. Chem. Res.* **5** (1972) 161.

[144] C. T. Rettner, H. A. Michelsen, D. J. Auerbach and C. B. Mullins, *J. Chem. Phys.* **94** (1991) 7499.

[145] H. A. Michelsen, C. T. Rettner, D. J. Auerbach and R. N. Zare, *J. Chem. Phys.* **98** (1993) 8294.

[146] C. T. Rettner, H. A. Michelsen and D. J. Auerbach, *Faraday Discuss. Chem. Soc.* **96** (1993) 17.

[147] S. Gulding, A. Wodtke, H. Hou, C. Rettner, H. Michelsen and D. Auerbach, *J. Chem. Phys.* **105** (1996) 9702.

[148] G. R. Darling and S. Holloway, *Surf. Sci.* **268** (1992) L305.

[149] G. R. Darling and S. Holloway, *J. Chem. Phys.* **97** (1992) 5182.

[150] S. Holloway and G. R. Darling, *Surf. Rev. Lett.* **1** (1994) 115.

[151] G. R. Darling and S. Holloway, *J. Electron Spectrosc. Rel. Phenom.* **64/65** (1993) 571.

[152] G. R. Darling and S. Holloway, *J. Chem. Phys.* **101** (1994) 3268.

[153] G. R. Darling, Z. S. Wang and S. Holloway, *Phys. Chem. Chem. Phys.* **2** (2000) 911.

[154] D. Kulginov, M. Persson, C. T. Rettner and D. S. Bethune, *J. Phys. Chem.* **100** (1996) 7919.

[155] P. S. Weiss and D. M. Eigler, *Phys. Rev. Lett.* **69** (1992) 2240.

[156] J. B. Taylor and I. Langmuir, *Phys. Rev.* **44** (1933) 423.

[157] J. V. Barth, *Surf. Sci. Rep.* **40** (2000) 75.

[158] D. V. Shalashilin and B. Jackson, *J. Chem. Phys.* **109** (1998) 2856.

[159] C. T. Rettner, H. A. Michelsen and D. J. Auerbach, *J. Chem. Phys.* **102** (1995) 4625.

[160] C. T. Rettner, D. J. Auerbach and H. A. Michelsen, *Phys. Rev. Lett.* **68** (1992) 1164.

[161] H. A. Michelsen, C. T. Rettner and D. J. Auerbach, *Phys. Rev. Lett.* **69** (1992) 2678.

[162] K. W. Kolasinski, W. Nessler, A. de Meijere and E. Hasselbrink, *Phys. Rev. Lett.* **72** (1994) 1356.

[163] K. W. Kolasinski, *Int. J. Mod. Phys. B* **9** (1995) 2753.

[164] P. Bratu, W. Brenig, A. Groß, M. Hartmann, U. Höfer, P. Kratzer and R. Russ, *Phys. Rev. B* **54** (1995) 5978.

[165] P. Bratu and U. Höfer, *Phys. Rev. Lett.* **74** (1994) 1625.

[166] U. Höfer, *Appl. Phys. A: Mater. Sci. Process.* **63** (1996) 533.

[167] M. Dürr, M. B. Raschke, E. Pehlke and U. Höfer, *Phys. Rev. Lett.* **86** (2001) 123.

[168] E. J. Buehler and J. J. Boland, *Science* **290** (2000) 506.

[169] J. A. Serri, M. J. Cardillo and G. E. Becker, *J. Chem. Phys.* **77** (1982) 2175.

[170] A. Mödl, T. Gritsch, F. Budde, T. J. Chuang and G. Ertl, *Phys. Rev. Lett.* **57** (1986) 384.

[171] J. E. Hurst Jr, L. Wharton, K. C. Janda and D. J. Auerbach, *J. Chem. Phys.* **83** (1985) 1376.

[172] C. T. Rettner, E. K. Schweizer and C. B. Mullins, *J. Chem. Phys.* **90** (1989) 3800.

[173] E. K. Grimmelmann, J. C. Tully and E. Helfand, *J. Chem. Phys.* **74** (1981) 5300.

[174] J. C. Tully, *Surf. Sci.* **111** (1981) 461.

[175] C. W. Muhlhausen, L. R. Williams and J. C. Tully, *J. Chem. Phys.* **83** (1985) 2594.

[176] S. F. Bent, H. A. Michelsen and R. N. Zare, "Hydrogen recombinative desorption dynamics", in *Laser Spectroscopy and Photochemistry on Metal Surfaces*, edited by H. L. Dai and W. Ho (World Scientific, Singapore, 1995) p. 977.

[177] C. R. Arumainayagam and R. J. Madix, *Prog. Surf. Sci.* **38** (1991) 1.

[178] L.-Q. Xia, M. E. Jones, N. Maity and J. R. Engstrom, *J. Chem. Phys.* **103** (1995) 1691.

[179] D. C. Jacobs, K. W. Kolasinski, S. F. Shane and R. N. Zare, *J. Chem. Phys.* **91** (1989) 3182.

[180] M. J. Cardillo, *Annu. Rev. Phys. Chem.* **32** (1981) 331.

[181] E. W. Kuipers, M. G. Tenner, A. W. Kleyn and S. Stolte, *Phys. Rev. Lett.* **62** (1989) 2152.

[182] E. W. Kuipers, M. G. Tenner, A. W. Kleyn and S. Stolte, *Nature* **334** (1988) 420.

[183] C. Haug, W. Brenig and T. Brunner, *Surf. Sci.* **265** (1992) 56.

[184] A. W. Kleyn, A. C. Luntz and D. J. Auerbach, *Surf. Sci.* **152/153** (1985) 99.

[185] T. F. Hanisco, C. Yan and A. C. Kummel, *J. Chem. Phys.* **97** (1992) 1484.

[186] D. S. King, D. A. Mantell and R. R. Cavanagh, *J. Chem. Phys.* **82** (1985) 1046.

[187] D. S. Y. Hsu and M. C. Lin, *J. Chem. Phys.* **88** (1988) 432.

[188] M. A. Hines and R. N. Zare, *J. Chem. Phys.* **98** (1993) 9134.

[189] D. C. Jacobs, K. W. Kolasinski, R. J. Madix and R. N. Zare, *J. Chem. Phys.* **87** (1987) 5038.

[190] P. J. Feibelman, *Phys. Rev. Lett.* **67** (1991) 461.

[191] D. Wetzig, R. Dopheide, M. Rutkowski, R. David and H. Zacharias, *Phys. Rev. Lett.* **76** (1996) 463.

[192] H. Morawitz, *Phys. Rev. Lett.* **58** (1987) 2778.

[193] K. W. Kolasinski, S. F. Shane and R. N. Zare, *J. Chem. Phys.* **96** (1992) 3995.

[194] L. Schröter, R. David and H. Zacharias, *Surf. Sci.* **258** (1991) 259.

[195] K. W. Kolasinski, F. Cemič, A. de Meijere and E. Hasselbrink, *Surf. Sci.* **334** (1995) 19.

[196] C. Åkerlund, I. Zorić, J. Hall and B. Kasemo, *Surf. Sci.* **316** (1994) L1099.

[197] C. T. Rettner and J. Lee, *J. Chem. Phys.* **101** (1994) 10185.

[198] G. A. Somorjai, *Chem. Rev.* **96** (1996) 1223.

[199] J. D. Beckerle, A. D. Johnson, Q. Y. Yang and S. T. Ceyer, *J. Chem. Phys.* **91** (1989) 5756.

[200] J. D. Beckerle, Q. Y. Yang, A. D. Johnson and S. T. Ceyer, *J. Chem. Phys.* **86** (1987) 7236.

[201] J. D. Beckerle, A. D. Johnson, Q. Y. Yang and S. T. Ceyer, *J. Vac. Sci. Technol. A* **6** (1988) 903.

[202] J. D. Beckerle, A. D. Johnson and S. T. Ceyer, *Phys. Rev. Lett.* **62** (1989) 685.

[203] L. C. Feldman and J. W. Mayer, *Fundamentals of Surface and Thin Film Analysis* (North-Holland, Amsterdam, 1986).

[204] P. D. Townsend, J. C. Kelley and N. E. W. Hartley, *Ion Implantation, Sputtering and their Applications* (Academic Press, London, 1976).

[205] V. S. Smentkowski, *Prog. Surf. Sci.* **64** (2000) 1.

[206] C. N. Hinshelwood, *The Kinetics of Chemical Change* (Clarendon Press, Oxford, 1940).

[207] T. Engel and G. Ertl, *Adv. Catal.* **28** (1979) 1.

[208] C. A. Becker, J. P. Cowin, L. Wharton and D. J. Auerbach, *J. Chem. Phys.* **67** (1977) 3394.

[209] C. T. Campbell, G. Ertl, H. Kuipers and J. Segner, *J. Chem. Phys.* **73** (1980) 5862.

[210] T. Matsushima, K. Shobatake and Y. Ohno, *Surf. Sci.* **283** (1993) 101.

[211] K.-H. Allers, H. Pfnür, P. Feulner and D. Menzel, *J. Chem. Phys.* **100** (1994) 3985.

[212] S. T. Ceyer, W. L. Guthrie, T.-H. Lin and G. A. Somorjai, *J. Chem. Phys.* **78** (1983) 6982.

[213] A. de Meijere, K. W. Kolasinski and E. Hasselbrink, *Faraday Discuss. Chem. Soc.* **96** (1993) 265.

[214] A. Alavi, P. Hu, T. Deutsch, P. L. Silvestri and J. Hutter, *Phys. Rev. Lett.* **80** (1998) 3650.

[215] D. D. Eley and E. K. Rideal, *Nature* **146** (1946) 401.

[216] C. T. Rettner, *Phys. Rev. Lett.* **69** (1992) 383.

[217] C. T. Rettner and D. J. Auerbach, *Phys. Rev. Lett.* **74** (1995) 4551.

[218] C. C. Cheng, S. R. Lucas, H. Gutleben, W. J. Choyke and J. T. Yates Jr, *J. Am. Chem. Soc.* **114** (1992) 1249.

[219] K. R. Lykke and B. D. Kay, in *Laser Photoionization and Desorption Surface Analysis Techniques–SPIE Proceedings, Volume 1208*, ed. N. S. Nogar (SPIE, Bellingham, WA, 1990) p. 18.

[220] C. T. Rettner and D. J. Auerbach, *Science* **263** (1994) 365.

[221] C. T. Rettner, *J. Chem. Phys.* **101** (1994) 1529.

[222] C. Lutterloh, A. Schenk, J. Biener, B. Winter and J. Küppers, *Surf. Sci.* **316** (1994) L1039.

[223] D. D. Koleske, S. M. Gates, B. D. Thoms, J. N. Russell, Jr and J. E. Butler, *J. Chem. Phys.* **102** (1995) 992.

[224] M. Xi and B. E. Bent, *J. Vac. Sci. Technol. B* **10** (1992) 2440.

[225] M. Xi and B. E. Bent, *J. Phys. Chem.* **97** (1993) 4167.

[226] L. H. Chua, R. B. Jackman and J. S. Foord, *Surf. Sci.* **315** (1994) 69.

[227] E. W. Kuipers, A. Vardi, A. Danon and A. Amirav, *Phys. Rev. Lett.* **66** (1991) 116.

[228] E. R. Williams, G. C. Jones Jr, L. Fang, R. N. Zare, B. J. Garrison and D. W. Brenner, *J. Am. Chem. Soc.* **114** (1992) 3207.

[229] W. H. Weinberg, "Kinetics of surface reactions", in *Dynamics of Gas–Surface Interactions*, eds. C. T. Rettner and M. N. R. Ashfold (The Royal Society of Chemistry, Cambridge, 1991) p. 171.

[230] B. Jackson and M. Persson, *Surf. Sci.* **269/270** (1992) 195.

[231] M. Persson and B. Jackson, *J. Chem. Phys.* **102** (1995) 1078.

[232] M. Persson and B. Jackson, *Chem. Phys. Lett.* **237** (1995) 468.

[233] D. V. Shalashilin, B. Jackson and M. Persson, *Faraday Discuss. Chem. Soc.* **110** (1998) 287.

[234] D. V. Shalashilin, B. Jackson and M. Persson, *J. Chem. Phys.* **110** (1999) 11038.

[235] M. Persson, J. Strömquist, L. Bengtsson, B. Jackson, D. V. Shalashilin and B. Hammer, *J. Chem. Phys.* **110** (1999) 2240.

[236] J. Harris, B. Kasemo and E. Törnqvist, *Surf. Sci.* **105** (1981) L288.

[237] J. Harris and B. Kasemo, *Surf. Sci.* **105** (1981) L281.

[238] C. B. Mullins, C. T. Rettner and D. J. Auerbach, *J. Chem. Phys.* **95** (1991) 8649.

[239] T. Matsushima, *Surf. Sci.* **123** (1982) L663.

[240] T. Kammler, D. Kolovos-Vellianitis and J. Küppers, *Surf. Sci.* **460** (2000) 91.

[241] A. Dinger, C. Lutterloh and J. Küppers, *J. Chem. Phys.* **114** (2001) 5304.

[242] D. A. King and M. G. Wells, *Surf. Sci.* **29** (1972) 454.

[243] C. M. Mate, G. A. Somorjai, H. W. K. Tom, X. D. Zhu and Y. R. Shen, *J. Chem. Phys.* **88** (1988) 441.

[244] M. D. Alvey, K. W. Kolasinski, J. T. Yates Jr and M. Head-Gordon, *J. Chem. Phys.* **85** (1986) 6093.

[245] A. Föhlisch, M. Nyberg, J. Hasselström, O. Karis, L. G. M. Pettersson and A. Nilsson, *Phys. Rev. Lett.* **85** (2000) 3309.

[246] G. Comsa and R. David, *Surf. Sci.* **117** (1982) 77.

[247] H. A. Michelsen, C. T. Rettner, D. J. Auerbach and R. N. Zare, *J. Chem. Phys.* **98** (1993) 8294.

[248] C. T. Rettner, H. A. Michelsen and D. J. Auerbach, *J. Chem. Phys.* **102** (1995) 4625.

4

Thermodynamics and Kinetics of Surface Processes

In this chapter we begin with a discussion of the thermodynamics of surface processes and move on to kinetics. Here we treat adsorption and desorption and the influence of lateral interactions. By concentrating on a statistical mechanical approach to kinetics, we see the importance of dynamics in surface processes. We also broaden our previous discussions by treating explicitly surface reactions at finite coverages. This is, of course, a necessary extension to be able to handle the kinetics of surface chemical reactions.

4.1 Thermodynamics of Ad/Desorption

4.1.1 Binding energies and activation barriers

The energetics of the potential energy hypersurface are important for dynamics, thermodynamics and kinetics. We can use Lennard–Jones diagrams to define the relationships between a number of quantities. The simplest case is that of an atom approaching a surface along the z coordinate, though, again, we can generalize this by considering the reaction coordinate. Figure 4.1(a) depicts the case of nonactivated adsorption, and activated adsorption is shown in Figure 4.1(b). As drawn, both potentials include a physisorption well. Note that the physisorption well is located further from the surface than is the chemisorption well. This is consistent with the usual trend in chemistry that shorter bonds correspond to stronger bonds. At low temperatures, a species can be trapped in a physisorbed stated even though a more strongly bound chemisorption state exists. Once an adsorbate settles into a physisorption well, it must overcome a small barrier to pass into the chemisorbed state.

At absolute zero for a classical system, there is no ambiguity in defining the heat released by adsorption, q_{ads}, the desorption activation energy, E_{des}, the adsorption activation energy, E_{ads}, and the adsorption bond binding energy (bond strength), $\varepsilon(M-A)$. In the case of nondissociative, nonactivated adsorption, $E_{ads} = 0$, and these relations are almost trivial

$$\varepsilon_{non}(M-A) = E_{des} \tag{4.1}$$

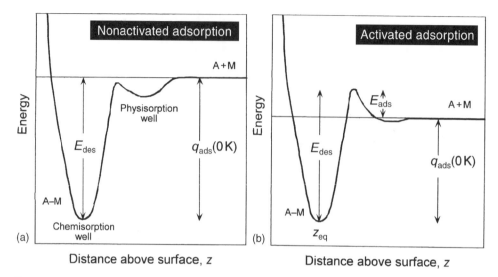

Figure 4.1 One-dimensional potential energy curves for molecular adsorption: (a) nonactivated adsorption; (b) activated adsorption. E_{ads}, E_{des}, adsorption activation energy and desorption activation energy, respectively; q_{ads}, heat released by adsorption; z_{eq}, adsorbate–surface bond length

and

$$\varepsilon_{non}(M-A) = q_{ads} \qquad (4.2)$$

For a quantum mechanical system, the energy differences must be calculated from the appropriate zero-point energy levels (see Fig. 4.5, page 174); $\varepsilon(M-A)$ is temperature independent, whereas, as we shall see in the next section, q_{ads} depends on temperature. For the moment these caveats need not concern us.

For activated adsorption, $E_{ads} > 0$ and the following relationships hold. Here, q_{ads} is the difference between the bottom of the chemisorption well and the zero of energy, taken as the energy of the system when the adsorbate is infinitely far from the surface. E_{des} is the difference between the bottom of the chemisorption well and the top of the adsorption barrier. E_{ads} is the height of the activation barrier when approaching the surface from $z = \infty$. The defining relationships, for activated adsorption, are now written as:

$$E_{des} = E_{ads} + \varepsilon(M-A) \qquad (4.3)$$

and

$$\varepsilon(M-A) = q_{ads} = E_{des} - E_{ads} \qquad (4.4)$$

In dissociative adsorption, the intramolecular adsorbate bond with dissociation energy $\varepsilon(A-A)$ is also broken. Figure 4.2 depicts activated dissociative adsorption of a diatomic molecule A_2. The dissociation energy of the atomic fragments and heat of adsorption are then given by

$$\varepsilon(M-A) = \tfrac{1}{2}[E_{des} - E_{ads} + \varepsilon(A-A)] \qquad (4.5)$$

and

$$q_{ads} = 2\varepsilon(M-A) - \varepsilon(A-A) \qquad (4.6)$$

4.1.2 Thermodynamic quantities

Here we give a brief account of surface thermodynamics. More extensive discussions can be found elsewhere [1–4]. The most fundamental quantity in thermodynamics is the Gibbs energy, G. Any system will relax to the state of lowest Gibbs energy *as long as no dynamical or kinetic constraints exist* that block it from reaching global equilibrium. Therefore, a spontaneous change is always accompanied by a decrease in Gibbs energy, that is, $\Delta G < 0$ for all spontaneous processes. The relationship between the change in Gibbs energy, entropy, S, and enthalpy, H, and temperature, T, is

$$\Delta G = \Delta H - T\Delta S \qquad (4.7)$$

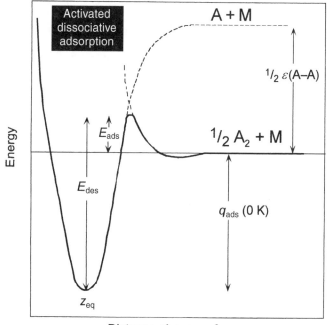

Figure 4.2 Activated dissociative adsorption. E_{ads}, E_{des}, adsorption activation energy and desorption activation energy, respectively; q_{ads}, released by adsorption; z_{eq}, adsorbate–surface bond length

Adsorption confines a gas to a surface, which results in an unfavourable entropy change, $\Delta_{ads}S < 0$, unless by some unusual process the substrate experiences an extraordinarily large positive entropy change that can compensate for this. Therefore, $\Delta_{ads}H$ must be negative (exothermic) for spontaneous adsorption.

4.1.3 Some definitions

As always in thermodynamics, it is essential to make clear and consistent definitions of the symbols used in the mathematical treatment. Table 4.1 defines several of the symbols used here.

Two different definitions of σ_0 appear in the literature. The reader must always be attentive as to whether coverage is defined with respect to the number of sites or to the number of surface atoms. These two definitions are equivalent in some instances. However, there are instances in which the saturation coverage is, say, one adsorbate for every two surface atoms. A saturated layer of adsorbates then has a coverage of 0.5 monolayers (ML) when defined with respect to the number of surface atoms but a coverage of 1.0 ML when defined with respect to sites. The definition with respect to

Table 4.1 Definition of symbols

Symbol	Definition and units of measurement
N_{ads}	Number of adsorbates (molecules or atoms, as appropriate)
N_0	Number of surface sites or atoms (as defined by context)
N_{exp}	Number of atoms or molecules exposed to (incident upon) the surface
A_s	Surface area (in m^2 or, more commonly, cm^2)
σ	Areal density of adsorbates (adsorbates cm^{-2}); $\sigma = N_{ads}/A_s$
σ_0	Areal density of sites or surface atoms (sites cm^{-2} or atoms cm^{-2})[a]
$\sigma*$	Areal density of empty sites (cm^{-2})
σ_{sat}	Areal density that completes a monolayer (cm^{-2})
θ	Coverage; fractional number of adsorbates (monolayers, ML); also sometimes called fractional coverage; $\theta \equiv \sigma/\sigma_0$
θ_{sat}	Saturation coverage; $\theta_{sat} \equiv \sigma_{sat}/\sigma_0$, where σ_0 is the density of surface atoms
δ	Relative coverage defined with respect to saturation; $\delta \equiv \sigma/\sigma_{sat} \equiv \theta/\theta_{sat}$
ε	Exposure; amount of gas incident on the surface (cm^{-2} or Langmuir)
$\varepsilon(M-A)$	Binding energy of the M−A bond
q_{ads}	Heat released when a single particle adsorbs (positive for exothermic change)
L	Langmuir; unit of exposure; $1\,L = 1 \times 10^{-6}$ torr $\times 1$ s
s	Sticking coefficient; $s = \sigma/\varepsilon$
	Integral sticking coefficient: total coverage divided by exposure; meaningful only if s is constant or as $\theta \to 0\,ML$
	Instantaneous or differential sticking coefficient at coverage θ: $s(\theta) = dN_{ads}/dN_{exp} = d\sigma/d\varepsilon$, evaluated at a specific value of θ
s_0	Initial sticking coefficient; sticking coefficient at $\theta \to 0\,ML$

[a] σ_0 may also be defined as the density of surface atoms; in some circumstances the two are equivalent (see Section 4.1.3).

atoms is more absolute, but the definition with respect to sites can lead to useful simplifications in kinetics calculations and also has the operational simplification of not requiring any knowledge about the surface structure: a monolayer is simply defined as the number of molecules in the saturated layer. So as not to confuse these two measures of coverage, θ is defined in this text with respect to the number of surface atoms, and δ is used when fractional coverage is defined with respect to the number of sites.

A clean surface is also something of a matter of definition. The normal sensitivity of many surface analytical techniques, apart from scanning tunnelling microscopy (STM) and a few others, is roughly 0.01 ML. Thus, we often consider a surface 'clean' if it has $\leq$0.01 ML of impurities. Detection limits are a source of major experimental difficulty in surface science and these difficulties must be kept in mind when results are analysed.

4.1.4 Heat of adsorption

Returning now to thermodynamics, we have two goals. First, we seek to define q_{ads} more precisely and understand its behaviour. Second, we would like to be able to relate the heat release measured in calorimetry [5] to thermodynamic parameters. The enthalpy is defined by

$$H = U + pV \tag{4.8}$$

where U is the internal energy, p is the pressure and V is the volume. For an ideal gas in molar units

$$H_g = U_g + p_g V_g = U_g + RT \tag{4.9}$$

where R is the gas constant. For the adsorbed gas, the pV term is negligible, thus

$$H_a = U_a \tag{4.10}$$

The enthalpy change in going from the gas to the adsorbed phase is, therefore

$$\Delta_{ads}H = H_a - H_g = U_a - U_g - RT \tag{4.11}$$

In Figures 4.1 and 4.2 we have chosen the origin of energy such that the internal energy of the system is zero at infinite separation and 0 K. The internal energy depends on the sum of translational, rotational and vibrational energies of the gas (or adsorbate) – a quantity with an obvious temperature dependence. We now make the following two identifications:

$$-q_{ads} = U_a - U_g \tag{4.12}$$

and

$$q_c = RT \tag{4.13}$$

From Equation (4.12) we can appreciate that although q_{ads} is related to $\varepsilon(M-A)$, the equality of the two is valid only at 0 K. The quantity q_c is the heat of compression arising from the transformation of a gas of finite volume into an adsorbed layer of essentially zero volume. By convention, the heat of adsorption (a positive quantity for exothermic adsorption) is quoted in surface science rather than the adsorption enthalpy (a negative quantity for exothermic adsorption).

In general, the heat of adsorption is a coverage-dependent quantity, hence

$$-\Delta_{ads}H(\theta) = q_{ads}(\theta) + q_c(\theta) = q_{st}(\theta) \tag{4.14}$$

where $q_{st}(\theta)$ is the isosteric heat of adsorption, $\Delta_{ads}H(\theta)$ is the differential adsorption enthalpy, and $q_{ads}(\theta)$ is the differential heat of adsorption. At room temperature q_c is only $2.5\,kJ\,mol^{-1}$; hence, in practice, it is usually negligible.

The isosteric heat of adsorption is defined through the Clausius–Clapeyron equation:

$$q_{st}(\theta) = RT^2 \left(\frac{\delta \ln p}{\delta T}\right)_\theta = -R\left(\frac{\delta \ln p}{\delta (1/T)}\right)_\theta \tag{4.15}$$

where p is the equilibrium pressure that maintains a coverage θ at temperature T. It can be shown [5] that the heat measured in a single-crystal adsorption calorimetry experiment is the isosteric heat of adsorption.

One final quantity of interest is the integral adsorption enthalpy. This represents the total enthalpy change (generally in molar units) recorded when the coverage changes from zero to some final value θ_f. The integral adsorption enthalpy is related to the heat of adsorption by

$$\Delta_{ads}H_{int} = \left[\int_0^{\theta_f} -q_{ads}(\theta)d\theta\right]\left[\int_0^{\theta_f} d\theta\right]^{-1} \tag{4.16}$$

Figure 4.3 displays several possible scenarios for the dependence of q_{ads} on θ. If the surface has only one type of site and if all of these sites adsorb particles independently, then q_{ads} will be a constant, as shown in curve (1) of Figure 4.3, and $\Delta_{ads}H_{int} = -q_{ads}$ (in molar units). For a surface that has two independent adsorption sites with different characteristic adsorption energies that fill sequentially, a step-like behaviour such as that found in curve (3) of Figure 4.3 will be observed. Chemisorption involves charge transfer and the capacity of a surface to accept or donate charge is limited. Consequently, as more and more particles adsorb, the ability of the surface to bind additional adsorbates will likely drop. Thus, q_{ads} drops with increasing θ. This type of behaviour is mimicked in curve (2) of Figure 4.3. In addition, as the coverage increases, the distance between adsorbates decreases. Lateral interactions will become increasingly likely, and these can influence q_{ads}, leading to changes as a function of θ (see Section 4.3).

4.2   Adsorption Isotherms: Thermodynamic Approach

Consider adsorption onto a solid substrate in which adsorption occurs from an ideal monatomic gas in equilibrium with the solid. There are σ_0 equivalent sites on the surface and not more than one adsorbate can bind on each. The adsorbates are noninteracting, thus the binding energy, ε, is independent of coverage.

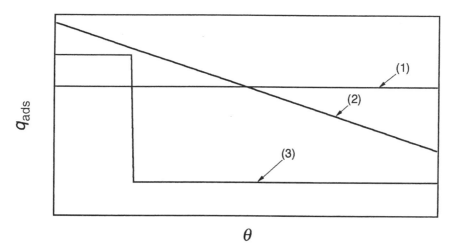

Figure 4.3   Three different behaviours of the heat of adsorption, q_{ads}, as a function of coverage, θ. Case (1): the surface is composed of one and only one type of noninteracting site. Case (2): q_{ads} decreases linearly with θ. Case (3): the surface is composed of two types of sites with different binding energies that fill sequentially. As shown in Section 4.3 of text, case (3) can also arise from strong lateral interactions

The probability of having N_{ads} adsorbed atoms is given by the ratio

$$P(N_{ads}) = \frac{1}{\Xi} Q_{ads} \exp\left(\frac{N_{ads}\mu_{ads}}{k_B T}\right) \tag{4.17}$$

where Q_{ads} is the canonical partition function, Ξ is the grand canonical partition function, μ_{ads} is the chemical potential of the adsorbed atoms (see Exercise 4.1), and k_B is the Boltzmann constant. Equation (4.17) can be used to calculate the equilibrium fractional coverage of adsorbates with respect to the number of sites.

$$\theta = \left[\exp\left(\frac{\mu_{ads} + \varepsilon}{k_B T}\right)\right]\left[1 + \exp\left(\frac{\mu_{ads} + \varepsilon}{k_B T}\right)\right]^{-1} \tag{4.18}$$

At equilibrium, the chemical potential of all phases present must be equal. Therefore, we can calculate μ_{ads} by equating it to the chemical potential of monatomic ideal gas, which can be readily found [2]. Thus,

$$\mu_{ads} = k_B T \ln\left[\frac{p}{k_B T}\left(\frac{h^2}{2\pi m k_B T}\right)^{3/2}\right] \tag{4.19}$$

where p is the equilibrium pressure of the gas, m is the mass of the atom, and h is the Planck constant. Substitution into Equation (4.18) leads to

$$\theta = \frac{p}{p + p_0(T)} \tag{4.20}$$

where

$$p_0(T) = \left(\frac{2\pi m k_{\mathrm{B}} T}{h^2}\right)^{3/2} k_{\mathrm{B}} T \exp\left(\frac{-\varepsilon}{k_{\mathrm{B}} T}\right) \tag{4.21}$$

is the pressure required to obtain an equilibrium coverage of $\theta = 0.5\,\mathrm{ML}$ at temperature T.

Equation (4.21) describes the equilibrium coverage found on a surface as a function of the adsorbate binding energy, pressure and temperature. This equation is known as the Langmuir isotherm. In Section 4.6 the Langmuir isotherm is derived from kinetics. The physical interpretation of this isotherm is that the equilibrium coverage is determined by the balance between the rate of adsorption and desorption. The shape of the isotherm is given in Figure 4.4. This general isotherm holds in many real systems, at least at low coverage. Note that no *a priori* knowledge of the ad/desorption dynamics is required, nor are any details of the adsorption kinetics apart from the assumption of independent sites that adsorb no more than one adsorbate. Accordingly, no dynamical information can be obtained from isotherms. Kinetics affect the time required to attain equilibrium but they do not actually determine the shape of the isotherms.

There are numerous isotherms bearing names of many catalytic chemists [6]. An exposition of all of these is of little more than academic interest. The occurrence of other isotherms arises because of the breakdown in the assumptions of the Langmuir model. The two most suspicious assumptions are that adsorption stops when the N_0 sites on the surface are filled; that is, that $\theta_{\mathrm{sat}} = 1\,\mathrm{ML}$, and that the adsorbates are noninteracting.

Relaxation of the assumption of saturation at 1 ML leads to the Brunauer–Emmett–Teller (BET) isotherm. This isotherm is used in countless measurements every year to determine the surface area of powders and porous solids, such as high-surface-area supported catalysts. The BET isotherm is obtained as above but with the addition that occupied sites can be filled with a second layer, of binding energy ε'. This leads to the isotherm equation [2]

$$\theta = p p_0(T) \left\{ \left[p_0(T) + p - p \exp\left(\frac{\varepsilon' - \varepsilon}{k_{\mathrm{B}} T}\right) \right] \left[p_0(T) - p \exp\left(\frac{\varepsilon' - \varepsilon}{k_{\mathrm{B}} T}\right) \right] \right\}^{-1} \tag{4.22}$$

4.3   Lateral Interactions

Relaxation of the noninteracting-adsorbate assumption leads to isotherms that are much more complex. The interactions are built into the isotherm equation with modified partition functions. Two of the most widely used approximations to treat effects of lateral interactions are the Bragg–Williams approximation and the quasi-chemical approximation. An in-depth discussion of these approximations can be found in Desjonquères and Spanjaard [2]. The inclusion of lateral interactions is necessitated by the observation not only of deviations from the Langmuir isotherm but also of a range of two-dimensional phase transitions [7].

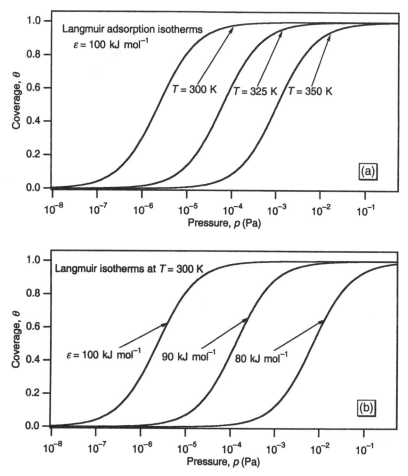

Figure 4.4 Langmuir isotherms: (a) constant heat of adsorption for various temperatures, T; (b) constant temperature for various adsorption energies, ε. Note: Langmuir isotherms exhibit a dependence on temperature and binding energy

Adsorbate interactions occur in four classes [8, 9]:

- Direct interactions due to wavefunction overlap: the formation of chemical bonds between adsorbates is an extreme case of attractive interaction. Brown and King, for example, have observed the formation of NO dimers on metal surfaces [10]. Usually, however, such interactions are repulsive because of Pauli repulsion.

- Indirect (substrate-mediated) interactions: the binding of an adsorbate often shifts the d states of neighbouring transition metal atoms downwards. This causes a weaker metal-atom–adsorbate interaction for subsequent adsorbates.

- Elastic interactions: a local distortion of the lattice in the vicinity of an adsorbate leads to a repulsive interaction with neighbouring adsorbates.

- Nonlocal electrostatic effects: dipole–dipole (or, higher, multipole–multipole) and van der Waals forces can exist between adsorbates. Dipole interactions can be either repulsive or attractive depending on the relative orientation of the interacting dipoles.

Indirect interactions can arise either from electronic or from structural changes induced by the presence of an adsorbate. Chemisorption is associated with charge transfer between the adsorbate and the surface. This means that the binding of an adsorbate to one site can lead to changes in the binding affinity of neighbouring sites. The changes can lead either to repulsive or to attractive interactions; indeed, the sign of the interaction may change with distance from the adsorbate. For example, the surface phase diagram of H/Pd(100) is best fit by assuming a nearest-neighbour repulsion of $-1.6 \, \text{kJ mol}^{-1}$, a next-nearest neighbour attraction of $+1.2 \, \text{kJ mol}^{-1}$ and a nonpairwise 'trio' interaction of $+0.81 \, \text{kJ mol}^{-1}$ [11]. The formation of a chemisorption bond is also accompanied by changes in the substrate atom positions. This may amount to a simple relaxation or a well-developed reconstruction of the substrate. In Chapter 6 we explore some of the consequences of the strain fields that arise during chemisorption, and adsorbate-induced reconstructions were explored in Section 1.2.2. A restructuring of the surface is often a local event that affects only those adsorbates that attempt to bind in the restructured region.

All mechanisms of lateral interactions display some type of distance dependence. In general, the importance of lateral interactions grows with decreasing distance between adsorbates. Hence, lateral interactions are important at high coverage. Nevertheless, even at low coverages lateral interactions cannot always be neglected. In many adsorption systems the formation of islands is observed even at low temperatures. The formation of islands is influenced both by lateral interactions and by dynamics. Once an island is formed the local coverage within the island is high compared with the globally averaged coverage. This circumstance can lead to invalidation of the assumption of noninteracting isolated particles even for very low global coverages.

A straightforward model to treat the effects of lateral interactions can be formulated with the following assumption [12–14]:

- adsorption occurs on a fixed number of equivalent sites;

- interactions are limited to a nearest-neighbour interaction of energy ω, which is independent of coverage;

- desorption is an equilibrium process;

- adsorption is confined to just one layer on top of the surface.

With this set of approximations, the heat of adsorption is found to be

$$q_{\text{ads}}(\theta) = q_0 + \frac{1}{2}z\omega \left[1 - \frac{1 - 2\theta}{\left\{ 1 - 4\theta(1 - \theta)[1 - \exp(\omega/k_{\text{B}}T)] \right\}^{1/2}} \right] \tag{4.23}$$

where q_0 is the heat of adsorption at zero coverage, and z is the maximum possible number of nearest neighbours. In this convention, repulsive interactions correspond to negative values of ω. Equation (4.23) has two interesting limits. For $\omega/T \to 0$, that is, for temperatures sufficiently large to ensure a completely disordered layer, Equation (4.23) reduces to

$$q_{ads}(\theta) = q_0 + zw\theta \qquad (4.24)$$

In the limit of $w/T \rightarrow -\infty$, the heat of absorption resembles a step function switching between the values q_0 and $(q_0 + zw)$ at $\theta = 0.5\,ML$. These two extremes correspond to the behaviour depicted in curves (2) and (3), respectively, of Figure 4.3. Such step-function behaviour approximates what is observed for $CO/Ru(0\,0\,1)$ [15]. Therefore, the behaviour of curve (3) of Figure 4.3 is ambiguous as it can be caused either by sequential filling of two sites or by strong repulsive interactions.

4.4   Rate of Desorption

The kinetics of surface reactions has many similarities with kinetics in other areas of chemistry [16, 17]. However, several differences also exist, the most essential of which is that *reactions can occur on only a limited number of sites*. The possibility of sites with different reactivities is another complicating factor. We begin by treating the simplest of surface reactions – adsorption and desorption – and move on from there to consider chemical reactivity at surfaces.

The rate of a chemical process is described by the Polanyi–Wigner equation. In this case we are interested in the rate of desorption, r_{des}, which is the rate of change in the coverage as a function of time. In its most general form, the Polanyi–Wigner equation is written

$$r_{des} = -\frac{\partial \theta}{\partial t} = v_n \theta^n \exp\left(-\frac{E_{des}}{RT_s}\right) \qquad (4.25)$$

where v_n is the pre-exponential factor of the chemical process of order n, E_{des} is the desorption activation energy, and T_s is the surface temperature. As written in Equation (4.25), the rate is in terms of monolayers per second. It could equally well be formulated in terms of molecules per unit area per second by substituting the absolute coverage $\sigma = \sigma_0 \theta$, where θ is referred to the number of surface atoms.

4.4.1   First-order desorption

Consider a first-order process (i.e. atomic or nonassociative molecular desorption). The absolute rate can be written in terms of the concentration term and the rate constant, k_{des},

$$r_{des} = \sigma_0 \theta k_{des} \qquad (4.26)$$

for which the rate constant can be written in an Arrhenius form:

$$k_{des} = A \exp\left(-\frac{E_{des}}{RT_s}\right) \qquad (4.27)$$

Conventionally, the pre-exponential factor of a first-order process is denoted A.

Adsorption and desorption are often reversible processes. For a simple first-order process, we can express the lifetime of the adsorbate on the surface, τ, through the Frenkel Equation:

$$\tau = \frac{1}{k_{des}} = \frac{1}{A}\exp\left(\frac{E_{des}}{RT_s}\right) = \tau_0\exp\left(\frac{E_{des}}{RT_s}\right) \tag{4.28}$$

Equation (4.28) is an important result to keep in mind. It states that an adsorbate has a finite lifetime at the surface that depends on the temperature and on the desorption activation energy. As the temperature increases, the lifetime drops exponentially, but even at high temperatures there is a finite lifetime. Obviously, in catalytic chemistry it is important for the surface lifetime of reactants to be long compared with the time it takes for adsorbates to diffuse across the surface and react.

Equation (4.25) can easily be extended to higher-order processes. It would appear that the kinetics of desorption are completely equivalent to the kinetics of chemical reactions in other phases. There are, however, important complications that arise for desorption. Even for noninteracting adsorbates, we have seen that more than one binding state can be populated at the surface. Each one of these binding states has associated with it unique values of A and E_{des}. In this way the desorption rate depends not only on the identity of the adsorbate but also on the binding state that it occupies. Surface diffusion between sites of different binding energy can lead to further complications, as is the case for NO/Pt(111) for which binding at the steps is significantly stronger than on the terraces [18].

Work, for instance by King [19, 20], Menzel [15], Schmidt [21] and co-workers has shown that lateral interactions can play an important role in the kinetics of desorption, particularly at high coverage. Lateral interactions affect the rate of desorption by making both A and E_{des} coverage dependent. Consequently, Equation (4.27) is more generally written

$$k_{des} = A(\theta)\exp\left(-\frac{E_{des}(\theta)}{RT_s}\right) \tag{4.29}$$

such that the coverage dependence of the kinetic parameters is accounted for. The Bragg-Williams and quasi-chemical approximations [2, 3] can be invoked to formulate expressions for the changes in A and E_{des} with coverage. Temperature-programmed desorption (Section 4.7) and adsorption isotherms (Section 4.2) provide methods for determining the magnitude of such interactions.

Frequently, A and E_{des} vary in concert. The relationship

$$\ln A(\theta) = \frac{E_{des}(\theta)}{RT_\theta} + c \tag{4.30}$$

where T_θ is the isokinetic temperature [16], is followed more or less in these systems. Such a relationship between A and E_{des} is known as the compensation effect. A compensation effect can result from a number of sources such as

- a heterogeneous surface that contains adsorption sites with a range of binding energies [22];

- lateral interactions [23], in particular if they are strong enough to give rise to coverage-dependent phase changes in the adsorbed layer [24];

- adsorbate-induced changes in the substrate structure [25].

More discussion on the compensation effect follows in Section 4.4.3.

4.4.2 Transition-state-theory treatment of first-order desorption

We want to develop a deeper understanding of A and E_{des}. For the moment, we implicitly neglect the coverage dependence of the kinetic parameters. Direct atomic and simple molecular desorption are examples of one-step reactions, known as elementary reactions, of the type

$$A* \rightleftharpoons A + * \qquad\qquad [4.1]$$

in which it is implicit that desorption occurs directly from one adsorbed phase into the gas phase. The reaction is in some ways equivalent to unimolecular dissociation. The rate of an elementary reaction is simply the product of the concentration and the rate constant, as in Equation (4.26). The rate constant can be interpreted via thermodynamic or statistical mechanical routes. Transition-state theory is the foundation for these formulations. In transition-state theory, the reactants follow a multidimensional potential energy hypersurface, as discussed in Sections 3.7–3.12. The reactants and products are separated by a transition state, which is at some position along the potential energy surface (PES). The chemical entity at the transition state is called the activated complex. If an activation barrier separates the reactants and products, the transition state is located at the top of the barrier. If there is no maximum in PES, the definition of the transition state is somewhat arbitrary.

The main assumptions of conventional transition-state theory (CTST) are [17]:

- once the transition state is reached, the system carries on to produce the products;

- the energy distributions of the reactants follow Maxwell–Boltzmann distributions;

- the whole system need not be at equilibrium, but the concentration of the activated complex can be calculated based on equilibrium theory;

- the motion along the reaction coordinate is separable from other motions of the activated complex;

- motion is treated classically.

Extensions to CTST can be formulated that improve upon each of these assumptions, but most of these extensions need not concern us here. As shown by Tully and co-workers [26], however, it is essential to relax the first assumption. This is done by introducing the transmission coefficient [17], κ, which defines the probability with which an activated complex proceeds into the product channel. This is a number that is strictly less than or equal to 1. The transmission coefficient is a dynamical correction to CTST. We shall soon see the fundamental importance of κ.

First, we note that there are three equivalent ways to write an equilibrium constant. For the general reaction

$$aA + bB \rightleftharpoons cC + dD \qquad [4.2]$$

the equilibrium constant can be written in terms of concentrations (more accurately, activities), rate constants or molecular partition functions.

$$K = \frac{[C]^c [D]^d}{[A]^a [B]^b} = \frac{\vec{k}}{\overleftarrow{k}} = \frac{q_C^c q_D^d}{q_A^a q_B^b} \exp\left(\frac{-E_0}{RT}\right) \qquad (4.31)$$

In Equation (4.31), the square brackets indicate concentrations, the $\vec{k}$ and $\overleftarrow{k}$ represent the rate constant of the forward and reverse reaction, respectively, and the q_Xs are molecular partition functions per unit volume ($X = A$, B, C and D). The energy E_0 is the molar energy change accompanying the conversion of reactants to products at $0\,K$, everything being in the appropriate standard state. The final expression follows from the CTST formulation of the rate constant, which for the forward reaction is

$$\vec{k}_{CTST} = \frac{k_B T}{h} \frac{q_{\ddagger}}{q_A q_B} \exp\left(\frac{-E_0}{RT}\right) \qquad (4.32)$$

The molecular partition function is the product of the partition functions for all degrees of freedom:

$$q_X = q_{trans} q_{rot} q_{vib} q_{elec} \qquad (4.33)$$

The electronic partition function,

$$q_{elec} = \sum_i g_{ei} \exp\left(\frac{-\varepsilon_{elec,\,i}}{k_B T}\right) \qquad (4.34)$$

where g_{ei} is the degeneracy of the electronic state of energy $\varepsilon_{elec,\,i}$, is usually unity because excited electronic states tend to be high in energy and the ground state is often a singlet. The translational partition function is

$$q_{trans} = \prod_i \frac{(2\pi m k_B T)^{1/2}}{h} \qquad (4.35)$$

where m is the mass, and i is the dimensionality. For instance, for a system confined to two dimensions, $i = 2$ and $q_{trans} = (2\pi m k_B T)/h^2$. The rotational partition function depends on whether the molecule is linear

$$q_{rot} = \frac{8\pi^2 I k_B T}{\sigma h^2} \qquad (4.36)$$

or nonlinear

$$q_{rot} = \frac{8\pi^2(8\pi^3 I_A I_B I_C)^{1/2}(k_B T)^{3/2}}{\sigma h^3} \tag{4.37}$$

where I (or I_x, $X = $ A, B or C is the moment of inertia, and σ is the symmetry number, which is an integer determined by the symmetry of the reactants, the transition state and the products. The rules for determining σ are given by Laidler [17]. The vibrational partition function is written as the product of the partition functions of all the $(3N - 5)$ or $(3N - 6)$ normal modes of the molecule,

$$q_{vib} = \prod_i \frac{1}{1 - \exp(h\nu_i/k_B T)} \tag{4.38}$$

where ν_i is the fundamental frequency of the ith oscillator. In the partition function of the transition state, the reaction coordinate is not bound, and this motion is not included in the partition function.

Laidler, Gladstone and Eyring [17, 27] used absolute rate theory to express the rate of desorption in terms of the partition functions. From the assumption of equilibrium between the transition state and the adsorbed phase, we write

$$K = \frac{\sigma_{\ddagger}}{\sigma_a} = \frac{q^{\ddagger}}{q_a} \exp\left(-\frac{E_{des}}{RT}\right) \tag{4.39}$$

where σ_a and $\sigma_{\ddagger}$ are the surface concentrations (per unit area) of the adsorbate and transition state, respectively. The precise definition of E_{des} is the activation energy at 0 K; that is, the energy required to elevate an adsorbed molecule in the lowest vibrational state to the lowest vibrational state of the activated complex. Note that in Equation (4.39) the partition function of the surface has been omitted. The partition function of the surface is difficult to calculate and is generally assumed to change little upon adsorption. Changes in the electronic and vibrational degrees of freedom of the substrate can occur upon adsorption. It has been argued that the electronic changes should not result in a significant change in the partition function [28]; therefore, their neglect is justified. If the adsorbate does not significantly affect the phonon spectrum of the substrate, then the vibrational changes are also insignificant. However, in instances where adsorption leads to significant substrate rearrangement and phonon spectrum changes, such as with H/Si or H/W(100), the neglect of the surface partition function becomes tenuous and must be tested experimentally. One further caveat is that, as discussed by Menzel [29], a configurational partition function for the adsorbed layer may need to be included. This is required to explain experimental results in the CO/Ru(001) system [15, 30] and is, in general, important for adsorbed layers that can undergo structural phase transitions. This term can be neglected at low coverage and for noninteracting adsorbates.

The concentration of the activated complex is calculated from Equation (4.39):

$$\sigma_{\ddagger} = \sigma_a \frac{q^{\ddagger}}{q_a} \exp\left(-\frac{E_{des}}{RT}\right) \tag{4.40}$$

The activated complex can be thought of as having one loose vibrational mode that corresponds to the motion leading to desorption. This is expressed as $k_B T/h\nu$ and, when factored out of $q^{\ddagger}$ to give $q_{\ddagger}$, we have

$$\sigma_{\ddagger} = \sigma_a \frac{k_B T}{h\nu} \frac{q_{\ddagger}}{q_a} \exp\left(-\frac{E_{des}}{RT}\right) \qquad (4.41)$$

This rearranges to

$$\nu\sigma_{\ddagger} = \sigma_a \frac{k_B T}{h} \frac{q_{\ddagger}}{q_a} \exp\left(-\frac{E_{des}}{RT}\right) \qquad (4.42)$$

The left-hand term is the concentration of the activated complex multiplied by the frequency with which it leaves the transition state. This corresponds to the reaction rate. Recall, however, that we must correct the CTST result by introducing the transmission coefficient,

$$k_{des} = \kappa k_{CTST} \qquad (4.43)$$

thus

$$r_{des} = \sigma_a \kappa \frac{k_B T}{h} \frac{q_{\ddagger}}{q_a} \exp\left(-\frac{E_{des}}{RT}\right) \qquad (4.44)$$

According to Equation (4.44) then, the pre-exponential factor in desorption is

$$A = \kappa \frac{k_B T}{h} \frac{q_{\ddagger}}{q_a} \qquad (4.45)$$

Note that although A has the units of frequency, it should not be confused with an 'attempt' frequency. The frequency along the desorption coordinate was already subsumed into the rate in Equation (4.44). The value of A tells us about the relative floppiness of the adsorbed phase compared with the transition state through the ratio of the partition functions as well as the dynamical corrections to CTST through κ.

If we assume that $\kappa \approx 1$ and that there is no change in the partition function between the adsorbed phase and the transition state ($q_{\ddagger}/q_a \approx 1$), we obtain $k_B T/h = 6.3 \times 10^{12}$ s^{-1} at room temperature whence the assumption that $A \approx 10^{13}$ s^{-1}. This is no more than a rough guess. For example, Ibach, Erley and Wagner [31] showed that for CO/Ni(100), $A \approx 6 \times 10^{16}$ s^{-1} and that it *increases* with coverage. Values of c. 10^{13} s^{-1} can be expected only if no dynamical corrections to CTST occur, if the adsorbates are noninteracting and if no change in the vibrational, rotational and translational degrees of freedom occur. This seems highly unlikely and, indeed, for CO adsorption on metal surfaces at low coverage, A varies from 10^{13}–10^{17} s^{-1} [25]. Reasonable values of A can range from 10^{11}–10^{19} s^{-1}. The value of 10^{13} s^{-1}, however, does serve as a useful benchmark. Assuming for the moment that $\kappa \approx 1$, which is often true for simple atomic and nondissociative molecular adsorption, values of A greater than 10^{13} s^{-1} indicate that $q_{\ddagger}/q_a > 1$. This means that the transition state is much 'looser' than the adsorbed phase. A loose transition state in this sense means that it has degrees of freedom that are more easily excited by thermal energy than the adsorbed phase. An example of a loose transition state is a localized adsorbate that desorbs through an activated complex that is a two-dimensional gas. If A is less than 10^{13} s^{-1}, the transition state is constrained. A constrained transition state occurs if the molecule must take on a highly specific configuration in the activated complex. As a rule, pre-exponential values in excess of 10^{13} s^{-1} are found for nonactivated, simple adsorption. Constrained transition states and low values of κ are generally associated with activated adsorption, as we will see shortly.

4.4.3   Thermodynamic treatment of first-order desorption

Comparing Equations (4.31), (4.32) and (4.43) we write the rate constant for desorption as

$$k_{\mathrm{des}} = \kappa \frac{k_{\mathrm{B}}T}{h} K^{\ddagger}$$

(4.46)

where $K^{\ddagger}$ is the equilibrium constant for formation of the activated complex. Recalling that

$$\Delta G^0 = -RT \ln K$$

(4.47)

we write

$$k_{\mathrm{des}} = \kappa \frac{k_{\mathrm{B}}T}{h} \exp\left(\frac{-\Delta^{\ddagger}G^0}{RT}\right)$$

(4.48)

where $\Delta^{\ddagger}G^0$ is the standard Gibbs energy of activation; as usual, it can be split into the standard enthalpy and entropy of activation as $\Delta^{\ddagger}H^0 - T\Delta^{\ddagger}S^0$:

$$k_{\mathrm{des}} = \kappa \frac{k_{\mathrm{B}}T}{h} \exp\left(\frac{\Delta^{\ddagger}S^0}{R}\right) \exp\left(\frac{-\Delta^{\ddagger}H^0}{RT}\right)$$

(4.49)

The standard enthalpy of activation is related to a general activation energy E_{a} by [17]

$$E_{\mathrm{a}} = \Delta^{\ddagger}H^0 + RT$$

(4.50)

Thus for desorption we write

$$k_{\mathrm{des}} = \kappa \frac{k_{\mathrm{B}}T}{h} \exp\left(\frac{\Delta^{\ddagger}S^0}{R}\right) \exp\left(\frac{-(E_{\mathrm{a}} - RT)}{RT}\right) = e\kappa \frac{k_{\mathrm{B}}T}{h} \exp\left(\frac{\Delta^{\ddagger}S^0}{R}\right) \exp\left(\frac{-E_{\mathrm{des}}}{RT}\right)$$

(4.51)

In Equation (4.51) E_{des}, the energetic difference between the zero-point levels of the reactant and transition state, is equated with E_{a} of desorption. As can be seen in Figure 4.5, the two energies are identical at 0 K. At any other temperature, E_{des} and E_{a} are defined between two different sets of energy levels. Nonetheless, to a close approximation, they are equal.

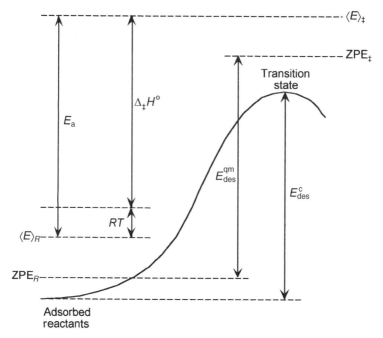

Figure 4.5   The classical barrier height, $E_{\text{des}}^{\text{c}}$, is the energy difference from the bottom of the well to the top of the activation barrier. The quantum mechanical barrier, $E_{\text{des}}^{\text{qm}}$, is a similar difference defined between reactant and transition-state zero-point energy levels ZPE_{R} and $\text{ZPE}_{\ddagger}$, respectively. E_{a} is defined as the difference between the mean energy of the reactants, $\langle E \rangle_{\text{R}}$ and the mean energy of the molecules in the transition state, $\langle E \rangle_{\ddagger}$. The standard enthalpy of activation, $\Delta^{\ddagger}H^0$ is also shown

Returning now to the Arrhenius formulation of the rate constant, we see that

$$A = e\kappa \frac{k_{\text{B}}T}{h} \exp\left(\frac{\Delta^{\ddagger}S^0}{R}\right) \tag{4.52}$$

Within this framework, and assuming $\kappa \approx 1$, we interpret pre-exponential factors of $c.~10^{13}~\text{s}^{-1}$ as being associated with $\Delta^{\ddagger}S^0 = 0$. Larger pre-exponentials have $\Delta^{\ddagger}S^0 > 0$, and smaller factors have $\Delta^{\ddagger}S^0 < 0$. A greater entropy is associated with a greater number of accessible configurations for the system. The direct relationship to the partition functions should now be clear.

These results can also be used to gain further insight into compensation effects. Equation (4.48) shows us that the rate of desorption depends fundamentally on the standard Gibbs energy of activation. Equations (4.50) and (4.52) articulate how the activation energy and the pre-exponential factor depend on the standard enthalpy and entropy of activation, respectively. Therefore, A and E_{des} are fundamentally linked to $\Delta^{\ddagger}G^0$. The compensation effect arises because $\Delta^{\ddagger}H^0$ and $\Delta^{\ddagger}S^0$ vary strongly with coverage but in such a way that $\Delta^{\ddagger}G^0$ is relatively constant in accord with

$$\Delta^{\ddagger}G^0(\theta) = \Delta^{\ddagger}H^0(\theta) - T\Delta^{\ddagger}S^0(\theta) \approx \text{constant} \tag{4.53}$$

Those systems that exhibit a compensation effect are those in which either (1) an increase in $\Delta^{\ddagger}H^0$ is accompanied by an increase in $\Delta^{\ddagger}S^0$, or (2) a decrease in $\Delta^{\ddagger}H^0$ is accompanied by a decrease in $\Delta^{\ddagger}S^0$. Compensation effects are well known for reactions performed in a series of different solvents or for homologous reactions carried out with a series of different substituents introduced into one of the reactants; see Laidler [17]. Laidler has explained the solvent effect by suggesting that stronger binding between a solute molecule and the solvent decreases the enthalpy. This also simultaneously lowers the entropy by restricting the rotational and vibrational freedom of the solvent molecules. Transposing this argument to surfaces, higher adsorption energies tend to decrease the freedom of diffusion and of frustrated translational and rotational modes. Regardless of whether the range of adsorption enthalpies is caused by surface heterogeneity or lateral interactions, the result is the same: $\Delta^{\ddagger}H^0$ and $\Delta^{\ddagger}S^0$ rise and fall in unison. In analogy to reactions in solution, we can think of an adsorbate at high coverage as being 'solvated' by its neighbouring adsorbates. Attractive or repulsive lateral interactions increase with the number of neighbours, and the compensation effect observed in desorption can be considered in the same terms as those observed in solution.

4.4.4 *Non-first-order desorption*

In the absence of lateral interactions and for well-mixed adlayers there is no ambiguity in using the Polanyi–Wigner equation to describe desorption, and it is trivial to write down the rate law expected for simple desorption reactions. For evaporation, or desorption from any phase that has a constant coverage because it is being replenished by another state, zero-order kinetics governs desorption; that is, for

$$A_{(l)} \rightarrow A_{(g)} \qquad [4.3]$$

we may write

$$r_{\mathrm{des}} = k_{\mathrm{des}} \qquad (4.54)$$

For simple atomic and nonassociative molecular desorption, first-order kinetics is expected; that is, for

$$A_{(a)} \rightarrow A_{(g)} \qquad [4.4]$$

we may write

$$r_{\mathrm{des}} = \theta k_{\mathrm{des}} \qquad (4.55)$$

In the case of recombinative desorption such as $H + H \rightarrow H_2$ the bimolecular character of the reaction suggests second-order kinetics; that is, for

$$A_{(a)} + A_{(a)} \rightarrow A_{2(g)} \qquad [4.5]$$

we may write

$$r_{\mathrm{des}} = \theta^2 k_{\mathrm{des}} \qquad (4.56)$$

With the same caveats as for first-order desorption, a 'normal' pre-exponential factor for second-order desorption can be calculated. This is of the order of 10^{-2} cm^2 s^{-1} but can range from 10^{-4}–10^{-1} cm^2 s^{-1}.

The presence of lateral interactions not only make A and E_{des} coverage dependent, but also can change the effective reaction order. On Si(1 0 0) surfaces, the formation of dimers leads to a (2×1) reconstruction. These dimers are stabilized by what may be considered either a π bond or a Peierls interaction [32]. When a hydrogen atom binds to the dimer, this added stabilization is lost, leaving a dangling bond on the other end of the dimer to which the hydrogen atom has bonded. When a second hydrogen atom attempts to bind to the surface, it can bind either on the same dimer or on a neighbouring dimer that still enjoys its full stabilization. The former option is the lower-energy proposition; therefore, hydrogen preferentially pairs up on surface dimers [33]. In a series of investigations, D'Evelyn and co-workers have shown that this pairing effect can lead to first-order desorption of H$_2$ from Si(1 0 0) surfaces, and similarly for H$_2$ and HBr from Ge(1 0 0) [34–37]. First-order desorption occurs because the H$_2$ desorbs preferentially from doubly occupied dimers. The desorption rate is linearly dependent on the coverage of doubly occupied dimers. Since the concentration of doubly occupied dimers varies close to linearly with θ(H), desorption is effectively first-order in θ(H). Deviations from linearity are predicted at low θ(H), and Höfer, Li and Heinz [38] have observed such deviations. Consequently, the effective reaction order with respect to θ(H) of H$_2$ desorption from Si(1 0 0) is coverage dependent.

Reider, Höfer and Heinz [39] have shown that the desorption of H$_2$ from Si(1 1 1) follows noninteger desorption kinetics. For coverages below 0.2 ML, the effective reaction order is 1.4–1.7. This number has no absolute meaning but it demonstrates that desorption involves a more complex mechanism than implied by Equation (4.56). The kinetics of H$_2$ desorption from Si(1 1 1) can be explained by the presence of two types of binding sites, A and B, possibly the rest atoms and adatoms, with binding energies differing by c. 0.15 eV. Under the assumption that desorption is a bimolecular reaction that occurs only from one of the sites, the desorption kinetics can be accurately modelled. A difference in the binding energies between the rest atom and adatom sites has been confirmed theoretically, which lends more support to this model [40].

The desorption of H$_2$ from Ag(1 1 1) is another illustrative example. Hodgson and co-workers [41] have shown that hydrogen can occupy surface or subsurface sites. Zero-order and half-order H$_2$ desorption kinetics are observed. Zero-order kinetics reigns when the desorbing phase has a constant concentration. In this case, the desorbing phase maintains a constant coverage because hydrogen diffusing out of the subsurface sites continuously replenishes it. Half-order kinetics is rather unusual. However, if it is assumed that hydrogen forms islands on the surface and that desorption occurs only from the perimeter of the islands, half-order kinetics is a direct consequence. This is consistent with low-energy electron diffraction (LEED) measurements on the system. Island formation arises from lateral interactions, which again highlights how lateral interactions can strongly affect the kinetics of surface reactions. This example also demonstrates that the reaction order only explicitly contains information regarding how the rate of reaction is affected by the supply of reactants rather than containing detailed dynamical information. A reaction mechanism must be consistent with the measured kinetics. The kinetics deliver insight into the mechanism; however, it cannot unambiguously determine the mechanism. For more information on the kinetic effects of lateral interactions, in particular their effect on surface reactions, see the review of Lombardo and Bell [42].

4.5 Kinetics of Adsorption

4.5.1 The conventional transition-state theory approach to adsorption kinetics

The mechanics used to determine the rate of adsorption are much the same as for desorption. First, we recognize that as a result of microscopic reversibility there is one and only one transition state for adsorption and desorption. That is, the transition state is the same regardless of the direction from which it is approached. One subtlety in the kinetics of adsorption is that only a limited number of sites on the surface can react. In surface kinetics we must always keep track not only of the number of sites but also of whether these sites are indistinguishable and whether the number of sites occupied by an adsorbate is greater than, equal to, or less than one.

We start with a clean surface and again assume that an equilibrium distribution between the gas phase and the activated complex prevails. Thus

$$\frac{\sigma_{\ddagger}}{c_g \sigma_*} = \frac{q^{\ddagger}}{q_g q_*} \exp\left(\frac{-E_{ads}}{RT}\right) \tag{4.57}$$

where $\sigma^{\ddagger}$ and σ_* are the areal densities of the activated complex and empty sites, respectively, c_g is the number density of gas-phase molecules and the qs are partition functions labelled accordingly. The activation energy is again strictly referenced to absolute zero.

Using the same procedure as for desorption, we identify one vibrational mode to extract from $q^{\ddagger}$ and equate $vc^{\ddagger}$ with the rate of adsorption, which yields

$$r_{ads} = c_g \sigma_* \frac{k_B T}{h} \frac{q_{\ddagger}}{q_g q_*} \exp\left(\frac{-E_{ads}}{RT}\right). \tag{4.58}$$

Equation (4.58) is valid only on the clean surface. In order to proceed and to characterize the rate of adsorption for increasing coverage, we need to know something about the adsorption dynamics.

4.5.2 Langmuirian adsorption: nondissociative adsorption

The name of Irving Langmuir is indelibly linked with surface science so it should come as no surprise that the most useful starting point for the understanding of adsorption kinetics is the Langmuir model of adsorption [43, 44]. First, let us get an idea of some orders of magnitude. There are roughly 10^{15} surface atoms cm^{-2}. The flux of molecules attempting to stick on these atoms is given by the Hertz–Knudsen equation

$$Z_w = \frac{N_a p}{(2\pi MRT)^{1/2}} = \frac{p}{(2\pi m k_B T)^{1/2}} \tag{4.59}$$

where N_a is Avogadro's number, M is the molar mass (in $kg\,mol^{-1}$), m the mass of a particle (kg) and p the pressure (Pa). For $M = 28$ g mol^{-1} at standard ambient temperature and pressure (SATP; at SATP, $p = 1$ atm, $T = 298$ K), $Z_w = 2.92 \times 10^{23}$ cm^{-2} s^{-1}. That is, about a mole per second of molecules of molar mass 28 (e.g. N_2 and CO) hits the surface per square centimetre at SATP. This value of the impingement rate does not have much of an intuitive feel to it. If, on the other hand, we calculate the impingement rate for $M = 28$ g mol^{-1}, at 1×10^{-6} torr, we find $Z_w(T = 298$ K, $p = 1 \times 10^{-6}$ torr, $M = 28$ g mol$^{-1}) = 3.84 \times 10^{14}$ cm^{-2} s^{-1}. Since surfaces have roughly 10^{15} atoms cm^{-2}, this exposure is roughly the equivalent of exposing each surface atom to one molecule from the gas phase. If all of these molecules had stuck, the coverage would be roughly 1 ML. Hence a convenient unit of exposure is the Langmuir: $1\,L \equiv 1 \times 10^{-6}$ torr $\times 1$ s $= 1 \times 10^{-6}$ torr s. Langmuirs are defined in terms of torr, rather than the SI unit Pa, because, historically, they were the units of choice. Thus, a rule of thumb is that 1 L exposure leads to *c*. 1 ML coverage *if the sticking coefficient is unity and independent of coverage.*

The next step is to define the behaviour of the sticking coefficient, s, as a function of coverage. In the Langmuir model, we assume $s = 1$ on empty sites and $s = 0$ on filled sites. Therefore, for nondissociative adsorption, the sticking coefficient is equal to the probability of striking an empty site

$$s = 1 - \frac{\sigma}{\sigma_0} = 1 - \theta \qquad (4.60)$$

and varies with coverage as shown in Figure 4.6. This behaviour indicates that the sticking coefficient decreases because of simple site blocking rather than because of any chemical or electronic effects. Note that this model does not mean that the adsorbing molecule sticks where it hits. This statement is too restrictive. It merely means that the adsorbing molecule '*makes the decision*' of whether it sticks or not on the first bounce.

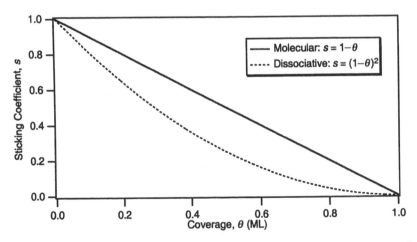

Figure 4.6   Langmuir models (molecular and dissociative) of sticking coefficient, s, as a function of coverage, θ

The observation of Langmuirian adsorption kinetics does not rule out the possibility of transient mobility after the first collision with the surface.

Further assumptions are (1) adsorption stops when all of the sites are full, that is saturation occurs at 1 ML, (2) the surface is homogeneous, thus containing only one type of site, and (3) the adsorbates are noninteracting. Thus, the Langmuir model can be thought of as describing ideal adsorption – a basis from which we can model all adsorption and the deviations from ideal adsorption. We now have sufficient information to calculate the rate of adsorption and its dependence on pressure, temperature and surface coverage. The rate of change of surface coverage is simply the impingement rate multiplied by the sticking coefficient:

$$\frac{\partial \sigma}{\partial t} = r_{ads}$$

$$= Z_w s$$

$$= \frac{p}{(2\pi m k_B T)^{1/2}}(1 - \theta)$$

$$= c_A k_{ads}(1 - \theta) \tag{4.61}$$

where c_A is the number density in the gas phase.

Several extensions to Langmuirian adsorption, which do not change its essential features, are quite useful. One is that we might want to define the saturation coverage to be θ_{max} rather than strictly unity. This allows for absolute comparison between molecules that have different absolute coverages. Second, the initial sticking coefficient need not be exactly unity; thus, we give it the more general symbol s_0. Finally, we allow for the possibility that adsorption may be activated. The generalized adsorption rate is then

$$r_{ads} = Z_w s_0 (1 - \theta) \exp\left(-\frac{E_{ads}}{RT}\right)$$

$$= \frac{p}{(2\pi m k_B T)^{1/2}} s_0 (1 - \theta) \exp\left(\frac{-E_{ads}}{RT}\right) \tag{4.62}$$

The coverage at time t is given by

$$\sigma(t) = s\varepsilon = sZ_w t \tag{4.63}$$

where ε is the exposure. Exposure, as we have mentioned above, is often expressed in the experimentally convenient units of pressure multiplied by time, or Langmuirs. Adsorption within this model would lead to a growth in coverage with exposure, as shown in Figure 4.7.

Extensions to the Langmuir model are essentially provisions for nonideal behaviour. Values of $s_0 < 1$ indicate the importance of dynamical corrections to the sticking coefficient as discussed in Sections 3.7–3.12. For instance, not every empty site on the surface might be equally capable of adsorbing molecules. It could also be indicative of steric constraints related to the orientation of the molecule when it strikes the surface. Lateral interactions could lead to island formation. Filled sites might not be totally inert

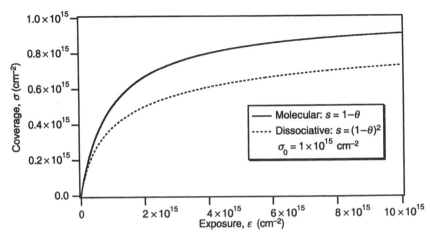

Figure 4.7   Langmuir models (molecular and dissociative) of coverage, σ, as a function of exposure, ε

toward sticking. Below we discuss the phenomenon of precursor-mediated adsorption, in which sticking occurs even when a filled site is encountered by the incident molecule.

Comparing Equation (4.62) with our CTST result in Equation (4.58) seems to bring little joy. First, remember that the CTST result is valid only at $\theta = 0$. Next, assume for the moment that neither the adsorbate nor the activated complex is localized. The activated complex is a two-dimensional (atomic) gas, whereas the gas phase is three dimensional. The ratio of the partition functions is $h/(2\pi m k_B T)^{1/2}$, hence

$$r_{ads}^{CTST} = c_g \frac{k_B T}{h} \frac{h}{(2\pi m k_B T)^{1/2}} \exp\left(\frac{-E_{ads}}{RT}\right) \tag{4.64}$$

Identifying $c_g k_B T$ as the pressure (where c_g is the density of the gas), the final CTST result is

$$r_{ads}^{CTST} = \frac{p}{(2\pi m k_B T)^{1/2}} \exp\left(\frac{-E_{ads}}{RT}\right) \tag{4.65}$$

At $\theta = 0$, Equation (4.62) reduces to

$$r_{ads} = \frac{p}{(2\pi m k_B T)^{1/2}} s_0 \exp\left(\frac{-E_{ads}}{RT}\right) \tag{4.66}$$

These results agree exactly except for the factor of s_0. Recalling that the results of CTST need to be corrected for barrier re-crossings by the transmission coefficient, we make the identification

$$s_0 = \kappa \tag{4.67}$$

Thus, the study of sticking coefficients and their dependence on various experimental parameters is itself a study of the validity of CTST and it corrections. The implication of

Equation (4.67) has been further investigated by Tully and co-workers [26, 45, 46]. As we have seen in Sections 3.7–3.12, s_0 is a function of energy (temperature) and not a constant. When we recall that κ and, therefore, s_0 are related to the Arrhenius pre-exponential factor A by Equation (4.52), we see that A has an additional dependence on energy that was not envisioned in the original conception of A. At high temperatures, deviation from Arrhenius behaviour is expected as a consequence. That is, a plot of $\ln k_{des}$ against $1/T$ will yield straight lines at low and moderate temperatures but will exhibit curvature to lower values for sufficiently high temperatures. This is essentially a consequence of microscopic reversibility. The dynamical correction to adsorption kinetics (s_0) equivalently applies to desorption kinetics (κ).

4.5.3 Langmuirian adsorption: dissociative adsorption

With the same set of assumptions as defined in Section 4.5.2, we can also treat dissociative adsorption. Assuming that dissociation requires two adjacent empty sites, the sticking probability varies with coverage as does the probability of finding two adjacent empty sites. This probability is given by $(1 - \theta)^2$ and, therefore,

$$s = s_0(1 - \theta)^2 \exp\left(-\frac{E_{ads}}{RT}\right) \tag{4.68}$$

Figures 4.6 and 4.7 show how s changes with θ, and how σ increases with ε for Langmuirian dissociative adsorption.

Deviations from ideal behaviour are more common and pronounced for dissociative adsorption than for simple adsorption. As discussed in Sections 3.5 and 3.11 regarding H_2 dissociation, the orientation of the molecule is often crucial in determining the sticking coefficient. The position in the unit cell where the molecule strikes is also important; that is, sticking is not uniform with regard to the point of impact. Often, defects exhibit a much higher sticking coefficient than do terrace sites. This can lead to a sticking coefficient that changes suddenly with coverage if the defect sites become decorated with immobile adsorbates. Adsorption can be self-poisoning. In other words, the adsorbate may lead to changes in the electronic structure that decrease the sticking coefficient at a more rapid rate than predicted by simple site blocking. This observation is sometimes modelled phenomenologically by assuming that an adsorbed molecule blocks more than one site or that the number of sites blocked is coverage dependent. Such models are of limited mechanistic value because the root cause of such effects is generally electronic. Deviations from ideal behaviour are all the more likely for adsorption that is highly activated.

4.5.4 Dissociative Langmuirian adsorption with lateral interactions

Consider now a model in which all of the assumptions of Langmuirian adsorption are kept except that we allow for $s_0 \neq 1$ and pairwise lateral interactions characterized by the interaction energy ω. Dissociative adsorption requires two adjacent empty sites. King and Wells [47] have shown that the number of adjacent empty sites is given by

$$\theta_{OO} = 1 - \theta - \frac{2\theta(1 - \theta)}{[1 - 4\theta(1 - \theta)(1 - \exp(\omega/k_B T_s))]^{1/2} + 1} \tag{4.69}$$

The strength of the lateral interactions can be characterized by the term

$$B = 1 - \exp\left(\frac{\omega}{k_B T_s}\right) \tag{4.70}$$

Introducing

$$s = s_0 \theta_{req} \tag{4.71}$$

we define θ_{req} as the coverage dependence of the sites required for adsorption. For Langmuirian dissociative adsorption, $\theta_{req} = (\theta_*^2) = (1 - \theta)^2$, whereas in the present model, $\theta_{req} = \theta_{OO}$.

In the limit of weak interactions or sufficiently high temperature to ensure no short-range order in the overlayer, $B = 0$. Substitution into Equations (4.69) and (4.71) yields

$$s = s_0(1 - \theta)^2 \tag{4.72}$$

as expected. For large repulsive interactions $\omega \ll 0$ and $B = 1$. Therefore

$$s = \begin{cases} s_0(1 - 2\theta) & \text{for } \theta \leq 0.5 \\ 0 & \text{for } \theta \geq 0.5 \end{cases} \tag{4.73}$$

This dependence on θ arises from the adsorbates spreading out across the surface in an ordered array in which every other site is occupied. For large attractive interactions $\omega \gg 1$, $B \rightarrow \infty$ and, therefore,

$$s = s_0(1 - \theta) \tag{4.74}$$

This is the same result as for nondissociative adsorption because the adsorbates coalesce into close-packed islands leaving the remainder of the surface completely bare. Thus, the functional form of s against θ depends on the magnitude of ω, as shown in Figure 4.8.

4.5.5 Precursor mediated adsorption

Taylor and Langmuir [48] observed that the sticking coefficient of caesium on tungsten did not follow Equation (4.60) but rather it followed the form found in Figure 4.9. They suggested that adsorption is mediated by a precursor state through which the adsorbing atom passes on its way to the chemisorbed state. Precursor-mediated adsorption has subsequently been observed in numerous systems (see Table 3.2, page 112). Precursor states can be classified as either intrinsic or extrinsic. An intrinsic precursor state is one associated with the clean surface whereas an extrinsic precursor state is associated with the presence of adsorbates. The explanation of how a precursor state manages to keep the

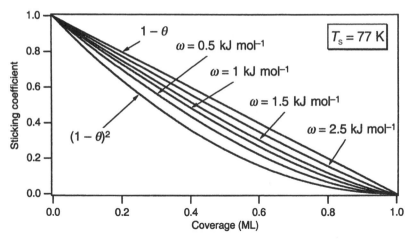

Figure 4.8 The effect of lateral interactions on the dissociative sticking coefficient as a function of interaction strength, ω, and coverage, θ. T_s, surface temperature

sticking coefficient above the value predicted by the Langmuir model is that the incident molecule enters the precursor state and is mobile. Therefore, the adsorbing molecule can roam around the surface and hunt for an empty site. This greatly enhances the sticking probability because an adsorbed molecule has a certain probability of sticking even if it collides with a filled site.

Kisliuk [49] was the first to provide a useful kinetic model of precursor-mediated adsorption. The basis of the Kisliuk model is that the total rate of adsorption is viewed as a competitive process, as outlined in Figure 4.10. King and co-workers [20, 47] extended this by including the effects of lateral interactions and considering the effects on

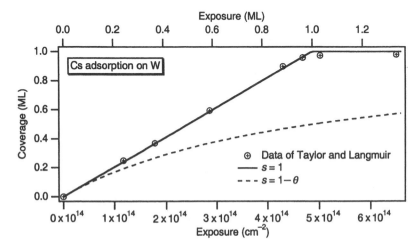

Figure 4.9 Sticking of caesium on tungsten. Replotted from data in J. B. Taylor and I. Langmuir, *Phys. Rev.* **44** (1933) 423

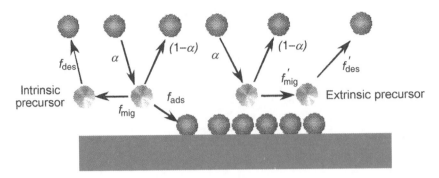

Figure 4.10 The Kisliuk model of precursor-mediated adsorption. Incident molecules trap into intrinsic or extrinsic precursors. Thereafter, sticking becomes a competitive process between desorption out of the precursor and transfer into the stable chemisorbed state. α, probability of entering the precursor state

adsorption and desorption kinetics. Madix [50], Weinberg [51] and co-workers have presented modified Kisliuk models that account for a combination of direct and precursor-mediated adsorption as well as an initial direct sticking coefficient that differs from the trapping probability into the precursor state.

The Kisliuk model has an intuitive formulation. Consider nondissociative molecular adsorption on a finite number of equivalent sites. When a molecule strikes an empty site, it has a probability f_{ads} of becoming adsorbed. Otherwise it has a probability f_{mig} of migrating to a neighbouring site or a probability f_{des} of desorbing. If a molecule strikes an occupied site, it cannot chemisorb but it can migrate or desorb with probabilities, f'_{mig} and f'_{des}, respectively. King introduced a trapping coefficient α that describes the probability of entering the precursor state. In our treatment α is independent of coverage and is the same for intrinsic and extrinsic precursors.

The incident molecule hops across the surface and makes a decision at each hop of whether to chemisorb, migrate or desorb. By summing up the probability over all possible hops, the sticking coefficient is calculated:

$$s = \alpha \left(1 + \frac{f_{des}}{f_{ads}} \right)^{-1} \left[1 + K \left(\frac{1}{\theta_{req}} - 1 \right) \right]^{-1} \tag{4.75}$$

where

$$K = \frac{f'_{des}}{f_{ads} + f_{des}} \tag{4.76}$$

The initial sticking coefficient is

$$s_0 = \alpha \left(1 + \frac{f_{des}}{f_{ads}} \right)^{-1} \tag{4.77}$$

and

$$\frac{f_{\text{des}}}{f_{\text{ads}}} = \frac{r_{\text{des}}}{r_{\text{ads}}} \tag{4.78}$$

where r_{des} and r_{ads} are the desorption rate and adsorption rate, respectively, from the precursor state. In other words, s_0 is determined by α and the competition between adsorption and desorption from the precursor state. As both adsorption and desorption are activated processes, we expect that sticking through a precursor state should have a temperature dependence. If $E_{\text{des}} > E_{\text{ads}}$, increasing the surface temperature decreases the sticking coefficient. However if $E_{\text{des}} < E_{\text{ads}}$, increasing T_s favours sticking.

The ratio of the s to s_0 is

$$\frac{s(\theta)}{s_0} = \left[1 + K\left(\frac{1}{\theta_{\text{req}}} - 1\right)\right]^{-1} \tag{4.79}$$

As written, Equation (4.79) can be used to describe either dissociative or nondissociative adsorption as long as the correct form of θ_{req} is inserted. If a plot of $s(\theta)/s_0$ against coverage can be fit by a unique value of K, the precursor model describes the sticking behaviour. Determination of K can then provide some idea of the relative rates of sticking onto bare sites compared with the desorption rates from the intrinsic and extrinsic precursor states. It can also be used to measure the affects of lateral interactions in dissociative adsorption. Figure 4.11 shows how s depends on K. In addition, it demonstrates that a precursor can actually decrease sticking compared with Langmuirian adsorption if desorption out of the precursor is rapid.

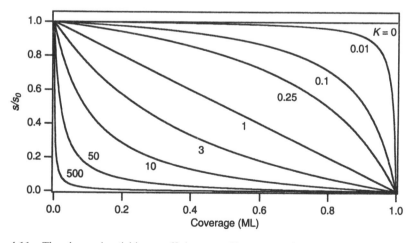

Figure 4.11 The change in sticking coefficient, s, with coverage for precursor-mediated adsorption. The change is characterized by the parameter K. For $K = 0$ the sticking coefficient is constant, whereas for $K = 1$ it drops linearly with coverage as in Langmuirian adsorption. Large values of K decrease s relative to Langmuirian adsorption

4.6   Adsorption Isotherms: Kinetics Approach

In Section 4.2 we introduced the adsorption isotherm, that is, an equation that describes the equilibrium coverage at constant temperature. There are various named isotherms that depend on the assumptions made to describe the rates of adsorption and desorption [6]. The basis of all isotherms is that we consider a system at equilibrium. Therefore, $T_S = T_g = T$, $r_{des} = r_{ads}$, and the coverage is constant, $d\theta/dt = 0$.

4.6.1   Langmuir isotherm

Using the set of assumptions posited in the Langmuir model of adsorption, we can derive the Langmuir isotherm. Equating the rates of desorption and adsorption yields

$$pk_{ads}(1 - \theta) = k_{des}\theta \tag{4.80}$$

A note on units is called for. Both sides of Equation (4.80) are written such that the rate is expressed in molecules per unit area per second. Thus, k_{des} has a factor of σ_0 included in it, and $k_{ads} = (2\pi mk_BT)^{-1/2}$. Rearranging yields

$$\frac{\theta}{1 - \theta} = p\frac{k_{ads}}{k_{des}} \tag{4.81}$$

and, since the ratio of forward and reverse reactions is equal to the equilibrium constant, we can write

$$\frac{\theta}{1 - \theta} = pK \tag{4.82}$$

Rearranging, we obtain an equation describing the equilibrium coverage as a function of pressure:

$$\theta = \frac{pK}{1 + pK} \tag{4.83}$$

4.6.2   Thermodynamic measurements via isotherms

The van't Hoff equation relates the equilibrium constant to the isosteric heat of adsorption:

$$\left(\frac{\partial \ln K}{\partial T}\right)_\theta = \frac{-q_{st}}{RT^2} \tag{4.84}$$

Substituting from Equation (4.82) for K and using $\partial(1/T)\partial T = -1/T^2$ yields

$$\left(\frac{\partial \ln p}{\partial(1/T)}\right)_\theta = -\frac{q_{st}}{R} \tag{4.85}$$

which is the same result as Equation (4.15), which was derived from the Clausius–Clapeyron equation. Thus, a plot of $\ln p$ at constant coverage against $1/T$ yields a line of slope $-q_{st}/R$. A series of such measurements as a function of θ leads to the functional form of the dependence of q_{st} on θ. Should the plot of $\ln p$ against $1/T$ not yield a straight line, then one of the assumptions of the Langmuir model must be in error. One can try, for instance, to fit the measured curves by introducing modified sticking coefficients. Plots of q_{st} versus θ can be used to investigate lateral interactions by determining whether models such as the Bragg–Williams or quasi-chemical approximation can accurately predict the plots.

4.7 Temperature-Programmed Desorption

4.7.1 The basis of temperature-programmed desorption

Temperature-programmed desorption (TPD) [14, 52–55] is a conceptually straightforward technique, as illustrated in Figure 4.12. A surface is exposed to a gas. Exposure can be performed by several means. The most uniform method is to backfill the chamber with the desired gas to some predetermined pressure for a measured length of time. Alternatively, a tube may be brought close to the sample and the gas be allowed to flow through it. This has the advantage of exposing the sample to relatively more gas than the rest of the chamber but has the disadvantage of supplying a highly nonuniform gas flux, which can lead to nonuniform coverage across the surface. An array doser [56, 57] attached to the end of the tube greatly improves the uniformity. In special cases, a molecular beam may be used to dose the surface. This method is particularly relevant to

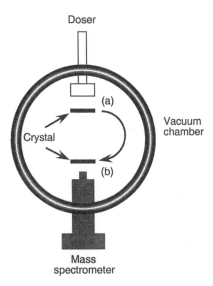

Figure 4.12 Temperature-programmed desorption. The process begins by dosing the crystal at position (a). The crystal is then moved to position (b), where mass-selective detection is performed with a mass spectrometer

dynamical studies in which a molecular beam is used so that the energetic characteristics of the molecules in the molecular beam can be systematically varied. The compendium of Yates [58] should be consulted for an explanation of a wide range of experimental techniques that can be used to improve the performance of TPD.

After exposure, the sample is rotated to face the detector. Early flash desorption experiments, based on the method of Taylor and Langmuir [48] and revived by Ehrlich [59], merely detected the pressure rise caused by rapid increase in the sample temperature. Redhead [60] slowed down the heating rate and showed how a thermal desorption spectrum could be used to determine surface kinetics. A quadrupole mass spectrometer (QMS) is employed in the modern technique. A QMS has the advantage of mass-selectively measuring the pressure rise. Therefore, it determines whether multiple products are formed and discriminates against background gases.

Consider the situation in Figure 4.12. Molecules desorb from the walls at a rate L and are removed from the chamber of volume V by a vacuum pump operating at a pumping speed S. The vacuum chamber contains a gas of density c_g and molecules desorb from the sample of surface area A_s with a time-dependent rate $A_s r_{ads}(t)$. The time dependence is important because a temperature ramp, usually linear in time, is applied to the sample according to

$$T_s = T_0 + \beta t \tag{4.86}$$

where T_s is the surface temperature, T_0 is the base temperature (the temperature at $t = 0$) and β is the heating rate. The change in the number of gas-phase molecules is given by

$$V \frac{dc_g}{dt} = A_s r_{des}(t) + L - c_g S \tag{4.87}$$

Assume that L and S are constant, that is, that the sample holder does not warm up leading to spurious desorption and that the pressure rise that accompanies desorption is not large enough to cause displacement of molecules from the walls or to affect the pumping speed. At the base temperature, which is low enough to ensure no desorption from the sample, the gas phase attains a steady-state composition given by

$$V \frac{dc_g}{dt} = L - c_g S = 0 \tag{4.88}$$

The steady-state solution is

$$c_{gs} = \frac{L}{S} \tag{4.89}$$

which corresponds to a steady-state pressure of

$$p_s = k_B T_g c_g = \frac{k_B T_g L}{S} \tag{4.90}$$

The pressure change caused by desorption is

$$\Delta p = p - p_s \tag{4.91}$$

The relationship between the pressure change and the rate of desorption can be written

$$V\frac{d\Delta p}{dt} + S\Delta p = kT_g A_s r_{des}(t) \tag{4.92}$$

In the limit of high pumping speed, the first term on the left-hand side is negligible and

$$\Delta p = \frac{k_B T_g A_s}{S} r_{des}(t) \tag{4.93}$$

In other words, the measured pressure change is directly proportional to the desorption rate, and the position and shape of the desorption peak contains information about the kinetics parameters, including lateral interactions, that affect the rate.

4.7.2 Qualitative analysis of temperature-programmed desorption spectra

Figure 4.13 displays thermal desorption spectra measured for cyclopentene adsorbed on Ag(2 2 1) [61]. These desorption spectra contain several features. Four different peaks, α_1–α_4, appear in the spectra with relative sizes that depend on the amount of cyclopentene exposed to the surface. The α_1 and α_2 peaks have a constant peak temperature, T_p, whereas the α_3 and α_4 peaks shift with increasing coverage. The lowest-temperature peaks do not exhibit saturation whereas the two highest-temperature peaks do. Two obvious parameters that can be used to characterize the peaks are the peak area, A_p, and the temperature measured at the peak maximum, T_p (the peak temperature).

Integrating a desorption spectrum over all time,

$$A_p \propto \frac{S}{A_s k_B T_g} \int_0^\infty \Delta p\, dt = \int_0^\infty r_{des}(t)\, dt = \sigma \tag{4.94}$$

we see that the peak area (or sum of peak areas for a multiple-peak spectrum) is directly proportional to the adsorbate coverage. Furthermore, if we integrate from $t = 0$ up to some intermediate time t_i, the included peak area is proportional to the amount desorbed. This allows us to determine the coverage relative to the total initial coverage at every point along the desorption curve. If an absolute coverage measurement is available for the initial coverage, for instance by X-ray photoelectron spectroscopy (XPS), the desorption curve can be used to calculate the absolute coverage at every point along the curve. Note, however, that for a multiple-peak spectrum resulting from multiple adsorption sites, the area under each peak is not necessarily equal to the initial coverage of the different adsorption sites. Interconversion between the sites can occur during the acquisition of the spectrum, as occurs, for example, in the H/Si system when dihydride and monohydride sites are both occupied.

Temperature-programmed desorption is the beauty and the beast of surface kinetics. Its beauty lies in its simplicity. TPD rapidly leads to an approximate picture of what is going on in surface kinetics. The beast lies in the accurate and unambiguous interpretation of the data. The sensitivity of TPD is generally of the order of 0.01 ML unless special measures are taken that can take this limit down by up to two orders of magnitude. As long as the pumping speed of the chamber in which desorption measurements are

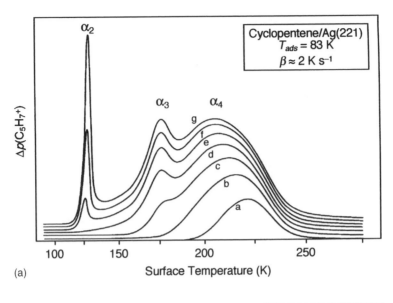

(a)

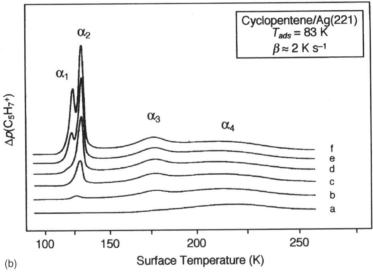

(b)

Figure 4.13 Temperature-programmed desorption (TPD) spectra for cyclopentene/Ag(2 2 1) at (a) low coverage and (b) high coverage, in parts (a) and (b), adsorption was carried out at a surface temperature of $T_{ads} = 83$ K, and the heating rate was 2 K s^{-1}. (T_{ads} is used to denote the surface temperature at the time of adsorption as distinguished from the surface temperature during the TPD experiment) The exposures to cyclopentene in part (a) are: spectrum a, 1.2×10^{14} cm^{-2}; spectrum b, 2.5×10^{14} cm^{-2}; spectrum c, 3.1×10^{14} cm^{-2}; spectrum d, 3.7×10^{14} cm^{-2}; spectrum e, 4.9×10^{14} cm^{-2}; spectrum f, 6.2×10^{14} cm^{-2}; spectrum g, 6.3×10^{14} cm^{-2}. The exposures to cyclopentene in part (b) are: spectrum a, 7.4×10^{14} cm^{-2}; spectrum b, 8.6×10^{14} cm^{-2}; spectrum c, 9.3×10^{14} cm^{-2}; spectrum d, 12.5×10^{14} cm^{-2}; spectrum e, 15.6×10^{14} cm^{-2}; spectrum f, 18.8×10^{14} cm^{-2}. α_1–α_4 refer to desorption states. Reproduced with permission from M. D. Alvey, K. W. Kolasinski, J. T. Yates Jr and M. Head-Gordon, *J. Chem. Phys.* **85** (1986) 6093. ©1986 by the American Institute of Physics

performed is sufficiently high, the desorption rate is directly proportional to the pressure rise in the chamber. In most modern surface science chambers this condition is easily met. However, under special situations it may still be of concern. For instance, the pumping speed of H_2 is comparatively low in turbomolecular pumped systems. Additionally, if the conductance between the sample and the pressure measurement device is constricted in some way or if spurious desorption from the walls or sample holder occur then interferences arise. Since these experimental concerns can be solved, TPD or its cousin – temperature-programmed reaction spectrometry (TPRS) – remains the technique of choice for initial investigations of surface kinetics. In one quick experiment, the identity of reaction products and the temperatures at which they appear can be determined.

Returning to the Polanyi–Wigner formulation of reaction kinetics [Equation (4.25)], we see that a direct measurement of reaction rate can lead to information on the kinetic parameters governing the reaction. Substituting a linear temperature ramp into Equation (4.25) and assuming that v_n and E_{des} are *independent of coverage*, yields

$$\frac{E_{des}}{RT_p^2} = \frac{nv^n}{\beta} \theta_p^{n-1} \exp\left(\frac{-E_{des}}{RT_p}\right).$$ (4.95)

For *first-order* desorption

$$\frac{E_{des}}{RT_p^2} = \frac{v}{\beta} \exp\left(\frac{-E_{des}}{RT_p}\right)$$ (4.96)

and for *second-order* desorption

$$\frac{E_{des}}{RT_p^2} = \frac{2v^2}{\beta} \theta_p \exp\left(\frac{-E_{des}}{RT_p}\right)$$ (4.97)

where θ_p is the coverage at T_p. Equations (4.96) and (4.97) show that T_p is independent of initial coverage, θ_i, for first-order desorption, and T_p shifts to lower temperatures as θ_i increases for second-order desorption. An increasing T_p with increasing coverage is indicative of an order $0 < n < 1$. Zero-order desorption strictly does not lead to a peak. Integration of the Polanyi–Wigner equation shows that for $n = 1$

$$\theta_p = \frac{\theta_i}{e}$$ (4.98)

and that for $n > 1$

$$\theta_p^{n-1} = n^{-1}\theta_0^{n-1}$$ (4.99)

The peak shape can also be used to analyse desorption. Zero-order desorption from finite coverages leads to a series of peaks that all share the same leading edge. Eventually, as the coverage drops during the desorption experiment, the desorption order changes to $n > 0$ and therefore a peak is observed. First-order desorption leads to symmetric peaks.

Second-order desorption leads to asymmetric peaks. All of these generalizations apply only if A and E_{des} are coverage independent.

Returning to the desorption spectra in Figure 4.13, we can make several qualitative conclusions about the adsorption of cyclopentene on Ag(2 2 1). The large peaks at low temperature do not saturate and the leading edges of the peaks as a function of increasing coverage overlap. These are clear indications of evaporation from a physisorbed layer. The occurrence of two peaks in the low-temperature range, well-separated from the higher-temperature peaks attributable to chemisorption, are indicative of a change in binding energy between the first (few) physisorbed layer(s) on top of the chemisorbed layer compared with the physisorbed layers on top of the first physisorbed layer. This frequently occurs. Menzel and co-workers [62] have made high-resolution TPD studies of the desorption of rare gases. They have shown that the desorption peaks from the first through fourth layers for argon physisorbed on Ru(0 0 1) can be resolved. The binding energy eventually converges on the sublimation energy of the bulk material and indicates a decreasing influence of the metal substrate on the binding of the physisorbate as the separation from the surface increases.

The two higher-temperature peaks in Figure 4.13 are attributable to the chemisorbed layer. Since the adsorption of cyclopentene is molecular, desorption should be first order. However, the α_4 peak shifts to lower energy even before the α_3 peak appears. This is an indication that strong lateral interactions are present because a first-order desorption peak should not shift with increasing coverage. As shown by Adams [13] sufficiently strong lateral interactions can lead not only to shifting and broadening of TPD peaks but also to the formation of separate peaks although only one binding state is occupied. Proof of the influence of lateral interactions for cyclopentene on Ag(2 2 1) would require an in-depth analysis [14] that is beyond the scope of this discussion, but it provides a likely explanation for the appearance of two peaks in the chemisorption region even though only one binding site is occupied. The review of Lombardo and Bell [42] should be consulted for more details on techniques for simulating TPD spectra and the effects of lateral interactions.

4.7.3   Quantitative analysis of temperature-programmed desorption spectra

Figure 4.14 exhibits a variety of simulated TPD spectra which demonstrate how the shape and size of the desorption peak varies with reaction order, coverage, E_{des} and A. Note that, independent of reaction order, the desorption rate at T_p from a full monolayer is of the order of 0.1 ML s^{-1}. This rule of thumb works for a variety of adsorbates. Strongly bound adsorbates have large A factors and weakly bound ones have small A factors. At the end of the day, nature likes to desorb c. 0.1 ML s^{-1} from a full layer for normal heating rates.

Redhead [60] showed that the activation energy for desorption for first-order desorption is related to the peak temperature by

$$E_{des} = RT_p \left[\ln\left(\frac{AT_p}{\beta}\right) - 3.46 \right] \tag{4.100}$$

Equation (4.100) is often used with an assumed value of A. However, it should be used as no more than a rule of thumb. It is a good first approximation to obtain a feeling for the binding energies, but a final value can be arrived at only by a full quantitative analysis.

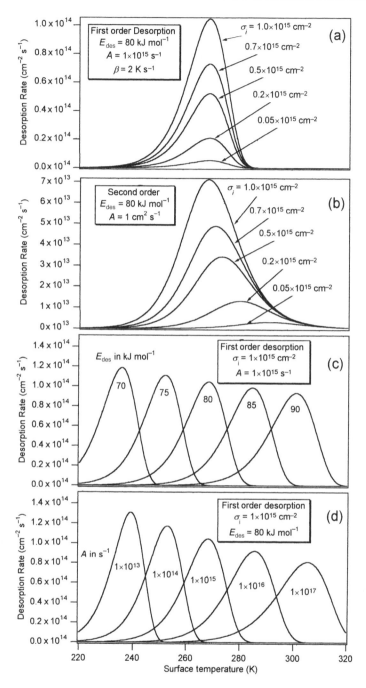

Figure 4.14 Simulated thermal desorption spectra. In all cases the desorption activation energy, E_{des}, and the pre-exponential factor of the rate equation, A, are independent of coverage. (a) First-order desorption for various coverages σ (note the asymmetric peak with constant T_p). (b) Second-order desorption for various coverages σ (note the symmetric peak with T_p that decreases with increasing coverage). (c) Effect of change in E_{des} of c. 5 kJ mol^{-1}. (d) Effect of change in A of c. 10^1. The changes illustrated in parts (c) and (d) lead to similar shifts in peak position for first-order desorption in this temperature and coverage range. β, heating rate; σ_i initial coverage

The only circumstances under which Equation (4.100) should be applied is when A is known from experiment and for peaks that are well resolved showing no indication for lateral interactions.

Numerous methods have been developed to analyse TPD spectra and these have been critically reviewed by de Jong and Niemantsverdriet [63] and by Yates and co-workers [64]. The relatively quick technique of Falconer and Madix [65] provides a useful approximation to the kinetic parameters. Consider a desorption rate written as a general function of the (non-negative order) expression of the rate dependence on coverage, $g(\theta)$. This yields

$$r_{\text{des}} = vg(\theta) \exp\left(\frac{-E_{\text{des}}}{RT}\right) \tag{4.101}$$

The rate reaches a maximum at T_p, thus by substituting from Equation (4.86) for T and setting the derivative with respect to t equal to zero, we find

$$\ln\left(\frac{\beta}{T_p^2}\right) = \ln\left[\frac{Rv}{E_{\text{des}}}\left(\frac{dg(\theta)}{dc}\right)_{T_p}\right] - \left(\frac{E_{\text{des}}}{R}\right)\left(\frac{1}{T_p}\right) \tag{4.102}$$

Therefore, a plot of $\ln(\beta/T_p^2)$ against $(1/T_p)$ results in a straight line with a slope of $-E_{\text{des}}/R$ provided that E_{des} is independent of coverage. This method provides a moderately accurate value as long as the heating rate is varied by at least two orders of magnitude. It is also a convenient method of determining whether lateral interactions are significant, as deviations from linearity in the plots are a clear indication of their presence. The pre-exponential factor can then be determined by returning to the Polanyi–Wigner equation and fitting the measured desorption curves.

The only fireproof method of determining the kinetics parameters is the complete analysis of TPD spectra, as illustrated in Figure 5.15. This method is involved but accurate. A family of desorption curves is measured as a function of initial coverage $-\theta_i$ or σ_i, depending on whether an absolute coverage is known–as shown in Figure 4.14(a). These are used to construct a family of $\sigma(t)$ curves via Equation (4.94). Because of the known (linear) heating rate, this corresponds to a knowledge of $\sigma(T_s)$ as a function of σ_i. Then an arbitrary value of coverage, σ_1, is chosen that is contained in each of the desorption curves. The desorption rate at this coverage, $r_{\text{ads}}(\sigma_1)$, and the temperature at which this rate was obtained, T_1, are then read off each of these desorption curves. A plot of $\ln[(r_{\text{ads}}(\sigma_1)]$ against $1/T_1$ is used to determine $E_{\text{des}}(\sigma_1)$ and $A(\sigma_1)$. This follows directly from the Polanyi–Wigner equation, restated as

$$\ln[r_{\text{des}}(\sigma)] = \ln(v_n(\sigma)\sigma^n) - \frac{E_{\text{des}}(\sigma)}{R}\frac{1}{T_S} \tag{4.103}$$

Equation (4.103) shows that the slope unambiguously determines $E_{\text{des}}(\sigma)$. By choosing a set of coverage values, the functional form of $E_{\text{des}}(\sigma)$ can be determined. The pre-exponential factor and the reaction order are determined from the intercept. This can lead to ambiguity in the values of $A(\sigma)$ and n if only a relative rather than absolute coverage is known. Likewise, the functional form of $A(\sigma)$ is determined by a series of intercepts determined for different coverages.

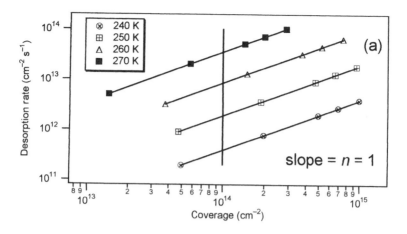

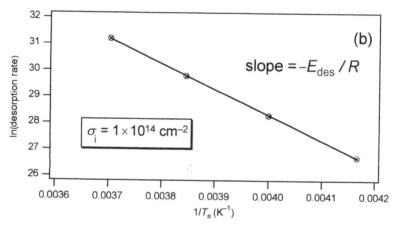

Figure 4.15   The complete analysis of temperature-programmed desorption curves. The analysis in part (b) is performed for the spectra in Figure 4.14(a) at the arbitrarily chosen coverage of $\sigma = 1 \times 10^{14}\,\text{cm}^{-2}$. (a) Desorption rate plotted against coverage, for various surface temperatures, T_s; (b) ln (desorption rate) plotted against the reciprocal of T_s at coverage $1 \times 10^{14}\,\text{cm}^{-2}$. The straight lines in (a) and (b) confirm that desorption is first order and that the kinetic parameters are coverage independent, respectively.

4.8   Summary of Important Concepts

- Whereas adsorption can be activated or nonactivated, desorption is always activated. Correspondingly, adsorption is always exothermic.

- The heat of adsorption per molecule generally depends on the number of adsorbates. Changes in the heat of adsorption are associated with adsorption at different types of sites, interaction between adsorbates (lateral interactions) and/or changes in electronic structure associated with adsorption.

- The equilibrium coverage is determined by the heat of adsorption, the temperature and the pressure in the gas phase (adsorption isotherms).

- The compensation effect yields desorption rate constants that change little with coverage because of counterbalancing changes in the enthalpy and entropy of activation.

- The kinetics of adsorption and desorption share many aspects (Polanyi–Wigner and Arrhenius formulations) in common with kinetics in other phases. However, on surfaces there are a limited number of adsorption sites and, therefore, the concentration terms that are used in surface kinetics differ fundamentally from those used in other phases.

- Noninteger orders in desorption kinetics demonstrate that desorption is more complicated than an elementary reaction. More complex reaction orders can arise from strong lateral interactions, multiple binding sites and/or nonrandom adsorbate distributions.

- The kinetics of adsorption can be calculated only if a model is assumed to describe how the sticking coefficient changes with coverage.

- Within the Langmuir model of adsorption, adsorption occurs only if the incident molecule strikes an empty site, adsorption saturates at 1 ML and adsorption is random. The sticking coefficient then drops as $(1 - \theta)$ for molecular adsorption and $(1 - \theta)^2$ for dissociative adsorption requiring two adjacent empty sites.

- In precursor-mediated adsorption, the sticking coefficient does not follow $(1 - \theta)$. Adsorption occurs via a competitive process. Molecules enter a mobile state from which desorption competes with the search for an empty site.

- Adsorption isotherms can be derived by equating the rates of adsorption and desorption at equilibrium.

- Kinetic and thermodynamic parameters can be determined by the measurement of isotherms and temperature-programmed desorption spectra.

Exercises

4.1   The canonical partition function appropriate for the Langmuir model is [2]

$$Q_{ads} = C_{N_0}^{N_{ads}} \exp\left(\frac{N_{ads}\varepsilon}{k_B T}\right) \tag{4.104}$$

and the grand canonical partition function is

$$\Xi = \sum_{N_{ads}}^{N_0} Q_{ads} \exp\left(\frac{N_{ads}\mu_{ads}}{k_B T}\right) \tag{4.105}$$

where $C_{N_0}^{N_{ads}}$ is the number of configurations and μ_{ads} is the chemical potential of the adsorbate phase. Derive Equation (4.17) by calculating

$$\theta = \frac{\langle N_{ads} \rangle}{N_0} \tag{4.106}$$

where $\langle N_{ads} \rangle$ is the average number of adsorbed atoms.

4.2 Consider nondissociative adsorption. Assuming both adsorption and desorption are first-order processes, write down an expression for the coverage of an adsorbate as a function of time. The system is at a temperature T, there is only one component in the gas phase above the surface at pressure p, and the saturation surface coverage is given by θ_{max}. Assume adsorption is nonactivated and that the sticking is direct.

4.3 (a) What is the gas flux striking a surface in air at 1 atm and 300 K?
(b) Calculate the pressure necessary to keep a $1\,cm^2$ Pt(100) surface clean for 1 h at 300 K, assuming a sticking coefficient of 1, no dissociation of the gas upon adsorption, and that 'clean' means less than 0.01 ML of adsorbed impurities.

4.4 Given that the kinetic parameters for diffusion are $D_0 = 5 \times 10^{-6}\,cm^2\,s^{-1}$ and $E_{dif} = 20.5\,kJ\,mol^{-1}$ and those for desorption are $A = 10^{12}\,s^{-1}$ and $E_{des} = 110\,kJ\,mol^{-1}$ (first order) for CO on Ni(100), how far does the average CO molecule roam across the surface at $T_S = 480$ K (the top of the temperature-programmed desorption peak).

4.5 Show that for precursor-mediated adsorption described by the Kisliuk model, a plot of $\ln[(\alpha/s_0) - 1]$ against $1/T_S$ is linear with a slope of $-(E_{des} - E_{ads})/R$, where E_{des} and E_{ads} are the activation energies for desorption and adsorption out of the precursor state.

4.6 Prove mathematically that, in precursor-mediated adsorption, if $E_{des} > E_{ads}$ then increasing the surface temperature decreases the sticking coefficient and that if $E_{des} < E_{ads}$ then increasing T_S favours sticking.

4.7 Consider the absorption of two molecules A and B that adsorb nondissociatively on the same sites on the surface. Assuming that both follow Langmuirian adsorption kinetics and that each site on the surface can bind only either one A or one B molecule, show that the equilibrium coverages of A and B are given by

$$\theta_A = \frac{K_A[A]}{1 + K_A[A] + K_B[B]} \tag{4.107}$$

$$\theta_B = \frac{K_B[B]}{1 + K_A[A] + K_B[B]} \tag{4.108}$$

4.8 Many different pressure units are encountered in surface science and conversions are inevitable. Show that

$$Z_w = 3.51 \times 10^{22}\,cm^{-2}\,s^{-1} \frac{p}{(MT)^{1/2}} \tag{4.109}$$

where p is given in torr, M in $g\,mol^{-1}$, and T in Kelvin. Derive a similar expression that relates the SI units of pressure (Pa) and molar mass ($kg\,mol^{-1}$) to the flux in $m^{-2}\,s^{-1}$.

4.9   Show that for dissociative adsorption within the Langmuir model the isotherm equation is given by

$$\theta = \frac{(pK)^{1/2}}{1 + (pK)^{1/2}} \tag{4.110}$$

4.10   Use the simple pairwise model of lateral interactions to investigate the effects of lateral interactions on a temperature-programmed desorption spectrum. Assume that the overlayer is completely disordered at the desorption temperature, that adsorption is nonactivated and therefore $E_{des} = q_{ads}$ and that the pre-exponential factor is independent of coverage. Use kinetic parameters typical of CO on a bcc($1\,0\,0$) lattice (bcc, body-centred cubic): $A = 1 \times 10^{16}$ s^{-1}, $E_{des}^0 = 150$ kJ mol^{-1}. Calculate and plot spectra for $\omega = 0$, ± 5 and ± 10 kJ mol^{-1}.

4.11   Given that A adsorbs in the threefold hollow site on a fcc(111) lattice (fcc, face-centred cubic) with an energy $q_{ads}^0 = 100$ kJ mol^{-1} and using all of the assumptions of the Langmuir model of adsorption apart from noninteracting adsorbates, calculate the isotherms expected for repulsive lateral interactions of 0, 2, 5 and 10 kJ mol^{-1}. Make a series of plots to demonstrate the effects of repulsive interactions on the isotherms.

4.12   Derive Equation (4.102) and show that if the Polanyi–Wigner equation describes the coverage dependence of the rate then

$$\ln\left(\frac{\beta}{T_p^2}\right) = \ln\left[\frac{Rv_n n\theta_0^{n-1}}{E_{des}}\right] - \left(\frac{E}{R}\right)\left(\frac{1}{T_p}\right) \tag{4.111}$$

4.13   Derive Equations (4.72), (4.73) and (4.74).

4.14   Write out expressions of Equation (4.80) in the limits of
(a) large desorption rate from the precursor,
(b) large adsorption rate into the chemisorbed state,
(c) large desorption rate from the chemisorbed state.
Explain the answers for (b) and (c) with recourse to the value of s_0.

4.15   In recombinative desorption, the pre-exponential factor can be written [16]

$$A = D\bar{v} \tag{4.112}$$

where D is the mean molecular diameter and $\bar{v}$ is the velocity of surface diffusion,

$$\bar{v} = \frac{\lambda}{\tau} \tag{4.113}$$

where λ is the mean hopping distance (the distance between sites) and τ is the mean time between hops,

$$\tau = \tau_0 \exp\left(\frac{E_{dif}}{RT}\right) \tag{4.114}$$

where τ_0 is a constant. Use this information and the expression for the desorption rate constant to show that a compensation effect could be observed in a system that exhibits a coverage-dependent diffusion activation energy, E_{dif}.

Further Reading

Michel Boudart and G. Djéga-Mariadassou, *Kinetics of Heterogeneous Catalytic Reactions* (Princeton University Press, Princeton, NJ, 1984).

S. Černý, "Adsorption microcalorimetry in surface science studies. Sixty years of its development into a modern powerful method", *Surf. Sci. Rep.* **26** (1996) 1.

M. C. Desjonquères and D. Spanjaard, *Concepts in Surface Physics* (Springer, Berlin, 1996).

David A. King, "Thermal desorption from metal surface: a review", *Surf. Sci.* **47** (1975) 384.

D. A. King and D. P. Woodruff (eds), *The Chemical Physics of Solid Surfaces and Heterogeneous Catalysis: Adsorption at Solid Surfaces*, Volume 2 (Elsevier, Amsterdam, 1983).

Keith James Laidler, *Chemical Kinetics* (HarperCollins, New York, 1987).

Theodore E. Madey and J. T. Yates Jr, "Desorption methods as probes of kinetics and bonding at surfaces", *Surf. Sci.* **63** (1977) 203.

D. Menzel, "Thermal Desorption", in *Chemistry and Physics of Solid Surfaces IV*, Volume 20, eds R. Vanselow and R. Howe (Springer, New York, 1982) p. 389.

M.W. Roberts and C.S. McKee, *Chemistry of the Metal–Gas Interface* (Clarendon Press, Oxford, 1978).

J.T. Yates Jr, "Thermal desorption of adsorbed species", in *Solid State Physics: Surfaces. Methods of Experimental Physics*, Volume 22, eds Robert L. Park and Max G. Lagally (Academic Press, New York, 1985) p. 425.

References

[1] S. Černý, "Energy and entropy of adsorption", in *The Chemical Physics of Solid Surfaces and Heterogeneous Catalysis, Volume 2*, eds D. A. King and D. P. Woodruff (Elsevier, Amsterdam, 1983) p. 1.

[2] M. C. Desjonquères and D. Spanjaard, *Concepts in Surface Physics*, 2nd edn (Springer, Berlin, 1996).

[3] T. L. Hill, *An Introduction to Statistical Thermodynamics* (Dover Publications, New York, 1986).

[4] A. W. Adamson and A. P. Gast, *Physical Chemistry of Surfaces*, 6th edn (John Wiley, New York, 1997).

[5] Q. Ge, R. Kose and D. A. King, "Adsorption energetics and bonding from femtomole calorimetry and from first principles theory", in *Advances in Catalysis, Volume 45*, eds B. C. Gates and H. Knözinger (Academic Press, Boston, MA, 2000) p. 207.

[6] E. Swan and A. R. Urquhart, *J. Phys. Chem.* **31** (1927) 251.

[7] S. K. Sinha, *Ordering in Two Dimensions* (Elsevier, New York, 1980).

[8] T. L. Einstein, *Crit. Rev. Solid State Mater. Sci.* **7** (1978) 261.

[9] B. Hammer and J. K. Nørskov, "Theoretical surface science and catalysis – Calculations and concepts", in *Advances in Catalysis, Volume 45*, eds B. C. Gates and H. Knözinger (Academic Press, Boston, MA, 2000) p. 71.

[10] W. A. Brown and D. A. King, *J. Phys. Chem. B* **104** (2000) 2578.

[11] G. Ertl, *Langmuir* **3** (1987) 4.

[12] J. S. Wang, *Proc. R. Soc. London, Ser. A* **161** (1937) 127.

[13] D. L. Adams, *Surf. Sci.* **42** (1974) 12.

[14] D. A. King, *Surf. Sci.* **47** (1975) 384.

[15] H. Pfnür, P. Feulner, H. A. Engelhardt and D. Menzel, *Chem. Phys. Lett.* **59** (1978) 481.

[16] M. Boudart and G. Djéga-Mariadassou, *Kinetics of Heterogeneous Catalytic Reactions* (Princeton University Press, Princeton, NJ, 1984).

[17] K. J. Laidler, *Chemical Kinetics* (HarperCollins, New York, 1987).

[18] J. A. Serri, J. C. Tully and M. J. Cardillo, *J. Chem. Phys.* **79** (1983) 1530.

[19] C. G. Goymour and D. A. King, *J. Chem. Soc. Faraday Trans.* **69** (1973) 749.

[20] A. Cassuto and D. A. King, *Surf. Sci.* **102** (1981) 388.

[21] E. G. Seebauer, A. C. F. Kong and L. D. Schmidt, *Surf. Sci.* **193** (1988) 417.

[22] B. Meng and W. H. Weinberg, *J. Chem. Phys.* **100** (1994) 5280.

[23] J. W. Niemantsverdriet and K. Wandelt, *J. Vac. Sci. Technol. A* **6** (1988) 757.

[24] P. J. Estrup, E. F. Greene, M. J. Cardillo and J. C. Tully, *J. Phys. Chem.* **90** (1986) 4099.

[25] V. P. Zhdanov, *Surf. Sci. Rep.* **12** (1991) 183.

[26] E. K. Grimmelmann, J. C. Tully and E. Helfand, *J. Chem. Phys.* **74** (1981) 5300.

[27] K. J. Laidler, S. Glasstone and H. Eyring, *J. Chem. Phys.* **8** (1940) 659.

[28] P. Stoltze and J. K. Nørskov, *Phys. Rev. Lett.* **55** (1985) 2502.

[29] D. Menzel, "Thermal Desorption", in *Chemistry and Physics of Solid Surfaces IV, Volume 20*, eds R. Vanselow and R. Howe (Springer, New York, 1982) p. 389.

[30] H. Pfnür and D. Menzel, *J. Chem. Phys.* **79** (1983) 2400.

[31] H. Ibach, W. Erley and H. Wagner, *Surf. Sci.* **92** (1980) 29.

[32] K. W. Kolasinski, *Int. J. Mod. Phys. B* **9** (1995) 2753.

[33] J. J. Boland, *Adv. Phys.* **42** (1993) 129.

[34] M. P. D'Evelyn, Y. L. Yang and L. F. Sutcu, *J. Chem. Phys.* **96** (1992) 852.

[35] M. P. D'Evelyn, S. M. Cohen, E. Rouchouze and Y. L. Yang, *J. Chem. Phys.* **98** (1993) 3560.

[36] Y. L. Yang and M. P. D'Evelyn, *J. Vac. Sci. Technol. A* **11** (1993) 2200.

[37] M. P. D'Evelyn, Y. L. Yang and S. M. Cohen, *J. Chem. Phys.* **101** (1994) 2463.

[38] U. Höfer, L. Li and T. F. Heinz, *Phys. Rev. B* **45** (1992) 9485.

[39] G. A. Reider, U. Höfer and T. F. Heinz, *J. Chem. Phys.* **94** (1991) 4080.

[40] K. Cho, E. Kaxiras and J. D. Joannopoulos, *Phys. Rev. Lett.* **79** (1997) 5078.

[41] F. Healey, R. N. Carter and A. Hodgson, *Surf. Sci.* **328** (1995) 67.

[42] S. J. Lombardo and A. T. Bell, *Surf. Sci. Rep.* **13** (1991) 1.

[43] I. Langmuir, *J. Am. Chem. Soc.* **38** (1916) 2221.

[44] I. Langmuir, *J. Am. Chem. Soc.* **40** (1918) 1361.

[45] J. C. Tully, *Surf. Sci.* **111** (1981) 461.

[46] C. W. Muhlhausen, L. R. Williams and J. C. Tully, *J. Chem. Phys.* **83** (1985) 2594.

[47] D. A. King and M. G. Wells, *Proc. R. Soc. London, Ser. A* **339** (1974) 245.

[48] J. B. Taylor and I. Langmuir, *Phys. Rev.* **44** (1933) 423.

[49] P. Kisliuk, *J. Phys. Chem. Solids* **3** (1957) 95.

[50] C. R. Arumainayagam, M. C. McMaster and R. J. Madix, *J. Phys. Chem.* **95** (1991) 2461.

[51] H. C. Kang, C. B. Mullins and W. H. Weinberg, *J. Chem. Phys.* **92** (1990) 1397.

[52] L. A. Pétermann, *Prog. Surf. Sci.* **3** (1974) 1.

[53] T. E. Madey and J. T. Yates Jr, *Surf. Sci.* **63** (1977) 203.

[54] M. W. Roberts and C. S. McKee, *Chemistry of the Metal–Gas Interface* (Clarendon Press, Oxford, 1978).

[55] J. T. Yates Jr, "Thermal desorption of adsorbed species", in *Solid State Physics: Surfaces. Methods of Experimental Physics*, Volume 22, eds R. L. Park and M. G. Lagally (Academic Press, New York, 1985) p. 425.

[56] C. T. Campbell and S. M. Valone, *J. Vac. Sci. Technol. A* **3** (1985) 408.

[57] A. Winkler and J. T. Yates Jr, *J. Vac. Sci. Technol. A* **6** (1988) 2929.

[58] J. T. Yates Jr, *Experimental Innovations in Surface Science: A Guide to Practical Laboratory Methods and Instruments* [AIP Press (Springer), New York, 1998].

[59] G. Ehrlich, *Adv. Catal.* **14** (1963) 255.

[60] P. A. Redhead, *Vacuum* **12** (1962) 203.

[61] M. D. Alvey, K. W. Kolasinski, J. T. Yates Jr and M. Head-Gordon, *J. Chem. Phys.* **85** (1986) 6093.

[62] M. Head-Gordon, J. C. Tully, H. Schlichting and D. Menzel, *J. Chem. Phys.* **95** (1991) 9266.

[63] A. M. de Jong and J. W. Niemantsverdriet, *Surf. Sci.* **233** (1990) 355.

[64] J. B. Miller, H. R. Siddiqui, S. M. Gates, J. N. Russell Jr, J. T. Yates Jr, J. C. Tully and M. J. Cardillo, *J. Chem. Phys.* **87** (1987) 6725.

[65] J. L. Falconer and R. J. Madix, *Surf. Sci.* **48** (1975) 393.

5

Complex Surface Reactions: Catalysis and Etching

In this chapter we treat a number of complex reactive systems which have been chosen both for their technological relevance and for their scientific relevance. We investigate not only catalysis but also etching of surfaces.

From the outset of ultrahigh vacuum (UHV) surface science studies there have been legitimate concerns about whether a pressure gap and/or a materials gap exists that would make UHV studies on model systems irrelevant for the high-pressure world of industrial catalysis. The pressure gap signifies the uncertainties in extrapolating kinetic data over as many as 10 orders of magnitude. One method to surmount this barrier has been to combine UHV methods with high-pressure reaction vessels [1]. The materials gap is posed by the uncertainties derived from using single crystals to model the highly inhomogeneous materials, often composed of small metal clusters on oxide substrates, that compose industrial catalysts. These gaps have now been breached. The oxidation of CO over platinum-group metals [2, 3] and the ammonia synthesis reaction [4, 5] are arguably the two best-understood heterogeneous catalytic reactions. Using kinetics parameters derived from UHV studies on model catalysts it is possible to model the reaction rates observed on high-surface-area catalysts at high pressures [5–7]. Even complex reactions involving the formation and conversion of hydrocarbons over metal catalysts can be understood across these gaps [6, 8, 9]. Hence, pressure and materials gaps exist only insofar as gaps exist in our knowledge of how to apply properly the lessons learned from UHV surface science studies. These gaps are not intrinsic barriers to the understanding of heterogeneous catalysis in terms of elementary reactions and fundamental principles of dynamics.

5.1 Measurement of Surface Kinetics and Reaction Mechanisms

The measurement of surface kinetics follows many of the same strategies as the measurement of kinetics in other phases [10], the basis of which always relates to the measurement of concentrations as a function of time. The difference is that in surface kinetics the measurements of the rate of consumption of gas-phase reactants and the appearance of gas-phase products are insufficient to determine the reaction mechanism and kinetics. This is because a complex set of surface reactions leads to quite complex reaction orders for the gas-phase species. These reaction orders often depend on temperature and gas composition and therefore cannot unambiguously be interpreted in terms of elementary steps. A final reaction mechanism cannot be determined without measurement of surface coverages and an identification of which species are bound to the surface. For instance, in something as simple as a CO desorption spectrum, a two-peak spectrum can be interpreted in at least three ways: (1) CO binds in two distinct binding sites, (2) strong lateral interactions account for the splitting of the desorption peak from a single binding site, or (3) CO partially dissociates, the low-temperature peak is attributable to simple molecular desorption, and the high-temperature peak is attributable to recombinative desorption. This again highlights that a temperature-programmed desorption (TPD) peak area is not uniquely proportional to the *initial* coverage of any given binding site. A combination of TPD with, for example, vibrational spectroscopy on the layer before desorption and as a function of heating would deliver not only the absolute rate of the process but also its proper interpretation in terms of a reaction mechanism.

Temperature-programmed desorption applied to surface reactions is sometime called temperature-programmed reaction spectrometery (TPRS). An example of TPRS peaks is shown in Figure 5.1 [11]. For the interpretation of a bimolecular reaction, this method

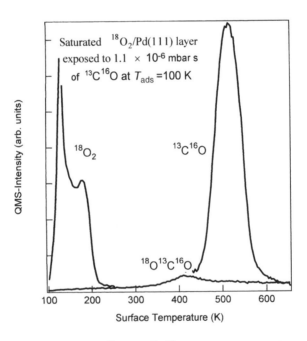

Figure 5.1   A co-adsorbed layer of $^{18}CO_2 + {}^{13}C^{16}OPd(111)$ is prepared at surface temperature $T_s = 100$ K. When heated, three products are observed in temperature-programmed reaction spectrometry: $^{18}O_2$, $^{13}C^{16}O$ and $^{13}C^{16}O^{18}O$. The CO_2 arises from the Langmuir–Hinshelwood reaction of $CO_{(a)} + O_{(a)}$. Reproduced from K. W. Kolasinski, F. Cemič, A. de Meijere and E. Hasselbrink, "Interactions in co-adsorbed CO + O_2/Pd(111) layers", *Surf. Sci.* **334** (1995) p. 19, with permission from Elsevier Science

requires four types of thermal desorption measurements. For the $CO + O_2$ reaction this would be a set of thermal desorption spectra conducted as a function of coverage for (1) pure CO, (2) pure O_2, (3) mixed $CO + O_2$ adlayers, and (4) pure CO_2 (the product). The interpretation of Figure 5.1 is then straightforward. The low-temperature O_2 desorption peaks are similar irrespective of the presence of CO. CO desorption is identical for $CO + O_2$ and CO overlayers. The differences in the mixed phase are the formation of CO_2 and the suppression of a high-temperature O_2 desorption peak.

The desorption of CO_2 arises from reaction-limited desorption caused by the Langmuir–Hinshelwood (L–H) reaction of $CO_{(a)} + O_{(a)}$ (use of subscript '(a)' indicates the adsorbed state). This conclusion is arrived at as follows. Electron energy loss spectroscopy (EELS) measurements [12] show that $O_{2(a)}$ dissociates on Pd(1 1 1) for $T_s \geq 180$ K, and above 250 K only atomic $O_{(a)}$ is left on the surface. CO_2 is only formed above 300 K, therefore the reaction must be between $CO_{(a)}$ and $O_{(a)}$ and not $CO_{(a)}$ and $O_{2(a)}$. O_2 dissociation is a prerequisite for reaction, and the reaction proceeds between two chemisorbed species. The desorption of CO_2 is limited by its rate of formation, not its rate of desorption. Pure CO_2 would desorb from Pd(1 1 1) at less than 100 K, thus at $c.$ 400 K, the surface lifetime of adsorbed CO_2 is extremely short. As soon as the CO_2 is created by reaction, it desorbs from the surface. There is no high-temperature recombinative desorption of O_2 because $O_{(a)}$ reacts completely with $CO_{(a)}$, which is present in excess for the conditions in Figure 5.1. The CO desorption peak matches that from a surface dosed only with CO because by the time the surface has reached 500 K, all $O_{2(a)}$ and $O_{(a)}$ have either desorbed or reacted, leaving the surface clean apart from $CO_{(a)}$. Thus, CO oxidation follows a Langmuir–Hinshelwood mechanism described by the following reactions:

$$CO + * \rightleftharpoons CO{-}* \qquad [5.1]$$

$$O_2 + 2* \rightleftharpoons 2O{-}* \qquad [5.2]$$

$$O{-}* + CO{-}* \rightleftharpoons CO_2 + 2* \qquad [5.3]$$

where $*$ represents an empty site.

Note that the experiment that produced Figure 5.1 utilized isotopic labelling. Isotopic labelling plays three roles in surface kinetics measurements. First, it is used to obtain better signal-to-noise ratios in the data. This is particularly important for CO and CO_2, which are often present in the background gases. Second, it is used to distinguish the products. N_2 and CO both appear at mass 28. Introduction of ^{18}O, ^{15}N or ^{13}C into the reactants shifts the N_2 and CO mass spectral peaks accordingly when the isotopically labelled species is incorporated into the desorbed product. Isotopic labelling facilitates the identification of reaction pathways by explicating which bonds are breaking. When $^{18}O_2 + {}^{13}C^{16}O$ are dosed onto Pd(1 1 1) the products observed are $^{18}O{-}^{18}O$, $^{18}O{-}^{13}C{-}^{16}O$ and $^{13}C^{16}O$ (Figure 5.1). The lack of $^{18}O{-}^{16}O$ and $^{16}O{-}^{13}C{-}^{16}O$ demonstrates that CO adsorption is nondissociative and that all of the atomic oxygen arises from the dissociation of O_2. Isotopic substitution can be coupled with vibrational spectroscopy [13] since many vibrational peaks shift sufficiently between isotopomers to facilitate spectral assignments.

Although we know that the $CO + O_2$ reaction occurs via an L–H mechanism, we can ask whether the reaction has a greater propensity to occur at certain sites on the surface. Under the conditions used for the above $CO + O_2/Pd(1\,1\,1)$ reaction, we do not expect a great sensitivity to a particular type of site. However, this is not always the case. The concept of active sites was first proposed by Taylor [14] and has been the subject of numerous studies since. For instance, Somorjai [15] has shown that step and kink sites can be particularly active in breaking C–C and C–H bonds in hydrocarbons. Consequently, vicinal surfaces of platinum are much more reactive for these reactions than are flat surfaces. Ertl and co-workers [16] also have identified steps to be the active site in the dissociation of NO on $Ru(0\,0\,1)$. This surface exhibits two different types of steps, which differ in geometric structure and reactivity. On one type of step, the oxygen atom tends to chemisorb and remain at the step. This deactivates the step, that is, this step is self-poisoned by its dissociation of NO. Oxygen atoms diffuse away from the other type of step leaving them clean and active for further dissociation. Hence, the first type of step is not an active site unless the temperature is sufficiently high to allow for the diffusion of $O_{(a)}$ away from it. The kinetics of reactions that depend on a certain active site, therefore, depends on the concentration of these sites and whether or not they become blocked during the course of the reaction. There need not be just one active site for a reaction. In a structure-sensitive reaction pronounced differences exist between the reactivities of different active sites. In a structure-insensitive reaction the difference in the reactivity of different types of active sites is negligible.

This demonstrates the fine balancing act that is the magic of catalysis. The surface must be reactive enough to break the appropriate bonds and hold adsorbates on the surface but not so reactive that it inactivates the products. This balancing act is broadly applicable to heterogeneous catalysis and has been demonstrated directly, for instance, during the oxidation of CO on $RuO_2(1\,1\,0)$ surfaces [17]. When $Ru(0\,0\,1)$ is exposed to stoichiometric mixtures of $CO + O_2$, the conversion probability to form CO_2 is extremely low. However, when CO oxidation is performed in a large excess of O_2, the reactivity is superior to that of palladium, which as we have seen above is a very efficient CO oxidation catalyst. In excess O_2, a crystalline film of $RuO_2(1\,1\,0)$ grows. This surface reveals three types of surface atoms: a ruthenium atom, a twofold coordinated bridging oxygen atom (O_{br}) and a threefold coordinated oxygen atom (O_{3f}). The binding site of CO as well as the transition state for this reaction have been determined [18]. CO binds, as expected on the Ru atom. It reacts with the O_{br} atom to form CO_2. To form the transition state on an oxygen-covered $Ru(0\,0\,1)$ surface, the Ru–O and Ru–CO bonds are significantly weakened. However the O_{br}–CO transition state is formed with little cost to the Ru–O_{br} bond strength. This significantly reduces the activation energy and leads to the site specificity. In this case the site specificity is not associated with a defect site. The O_{br} site is part of the ideal $RuO_2(1\,1\,0)$ surface. However, reaction with O_{br} is strongly favoured over O_{3f} or with oxygen chemisorbed on $Ru(0\,0\,1)$.

Molecular beam techniques have also been applied to the measurement of surface kinetics and have been reviewed extensively by Madix and co-workers [19–21]. These studies are generally performed at isothermal conditions and can involve one, two or even three molecular beams [22, 23]. The use of pulsed molecular beams and so-called molecular beam relaxation spectrometry (MBRS) allows for reaction products to be measured with moderate temporal resolution. Thus surface residence times can be established. This is particularly useful, for instance, to distinguish between L–H and Eley–Rideal (E–R) kinetics. Procedures for the analysis of data waveforms can be found elsewhere [19, 24–27].

5.2   Haber–Bosch Process

Ammonia is an exceedingly useful and important chemical on an industrial basis. Annual global production exceeds 100 million tons [28]. From it, nitrogen fertilizers are produced as well as a host of other nitrogen-containing chemicals. The production of ammonia is responsible for more than 1% of global energy consumption [29]; therefore, the discovery of a more efficient and lower-temperature ammonia synthesis would have profound implications not only for fossil-fuel consumption but also the worldwide economy. Ammonia is synthesized according to the reaction

$$N_2 + 3H_2 \rightleftharpoons 2NH_3, \quad \Delta H^0_{298} = -46.1 \text{ kJ mol}^{-1} \qquad [5.4]$$

where ΔH^0_{298} is the standard enthalpy change at 298 K. The reason why ammonia is so important industrially is essentially the same reason why Reaction [5.4] has to be performed catalytically: the triple bond in N_2 is so strong that N_2 is essentially inert. Consequently, N_2 is a poor source of nitrogen whereas ammonia represents a usefully reactive source of nitrogen for the industrial chemistry.

From the above, we anticipate that N_2 dissociation is the rate-determining step in ammonia synthesis. In other words, the dissociative sticking coefficient of N_2 is low and an effective ammonia synthesis catalyst is one that efficiently dissociates N_2.

The ammonia synthesis reaction is conceptually simple. NH_3 is not only thermo-dynamically stable, it is by far the most stable nitrogen hydride. Therefore, selectivity is not an issue for the catalyst. The role of the catalyst is to produce efficiently and rapidly an equilibrium distribution of products and reactants. The key to the Haber–Bosch process is that the catalyst efficiently breaks the N≡N bond. This releases nitrogen atoms onto the surface where they collide with adsorbed hydrogen atoms and eventually form NH_3. The elementary steps of the reaction can be written as [30]

$$N_{2(g)} + * \rightleftharpoons N_2 - * \qquad [5.5]$$

$$N_2 - * + * \rightleftharpoons 2N - * \qquad [5.6]$$

$$N - * + H - * \rightleftharpoons NH - * + * \qquad [5.7]$$

$$NH - * + H - * \rightleftharpoons NH_2 - * + * \qquad [5.8]$$

$$NH_2 - * + H - * \rightleftharpoons NH_3 - * + * \qquad [5.9]$$

$$NH_3 - * \rightleftharpoons NH_{3(g)} + * \qquad [5.10]$$

$$H_{2(g)} + 2* \rightleftharpoons 2H - * \qquad [5.11]$$

In principle, any one of the reactions can act as the rate-determining step if it has a rate that is significantly slower than all of the other elementary steps. Reaction [5.6] is the slow step under normal conditions with all subsequent reactions ensuing rapidly.

Iron is effective as a catalyst because it lowers the barrier to N_2 dissociation. It does this by gently coaxing the $N{\equiv}N$ bond to break while simultaneously forming $Fe-N$ chemisorption bonds. The activation barrier for N_2 dissociation is much lower than the $N{\equiv}N$ bond strength [941 kJ mol^{-1} (10.9 eV)] precisely because these two processes occur simultaneously. Now we arrive at the second important property of iron surfaces. They are able to break $N{\equiv}N$ bonds but in doing so they make an $Fe-N$ bond that has just the right strength. If the metal-nitrogen ($M-N$) bond is extremely strong, as it is for early transition metals, then the $N_{(a)}$ is rendered inert because it forms a surface nitride and does not react further. If, however, the $M-N$ bond is too weak, then the surface residence time of $N_{(a)}$ becomes so short that N_2 may be formed instead of NH_3. This would lower the efficiency of the catalyst.

It would appear, then, that the activation energy for N_2 dissociation (E_{ads}) and the heat of dissociative N_2 adsorption (q_{ads}) are linked and vary systematically across the periodic chart. This is an expression of the Brønsted–Evans–Polanyi relation, which states that the activation energy and reaction energy are linearly related for an elementary reaction [31, 32]. The linearity of this relationship has been confirmed for N_2 dissociation on a variety of transition metals [33]. Furthermore, the relation holds for different classes of sites. The less-reactive close-packed sites on Mo(1 1 0), Fe(1 1 1), Ru(0 0 1), Pd(1 1 1) and Cu(1 1 1) exhibit a linear relationship between E_{ads} and q_{ads}, whereas more reactive step sites on these surfaces exhibit a parallel linear relationship with correspondingly lower values of E_{ads}. Consequently, the trend of NH_3 synthesis reactivity, which peaks for iron and ruthenium on single-crystal surfaces, carries through to practical catalysts because the linear relationship is valid for ideal as well as for defect sites.

The Haber–Bosch process operates at 200–300 bar and 670–770 K over an $Fe/K/CaO/Al_2O_3$ catalyst. To make the catalyst, Fe_3O_4 is fused with a few percent of $K_2O + CaO + Al_2O_3$. The mixture is then reduced (activated) by annealing in a H_2/N_2 mixture at c. 670 K so that metallic iron particles form on the high-surface-area oxide substrate provided by Fe_3O_4/Al_2O_3. Al_2O_3 is the preferred substrate additive because it acts as a structural promoter, ensuring high dispersion of the iron clusters and hindering their tendency to sinter into larger particles. The CaO may assist in this process. The potassium acts as an electronic promoter. We shall discuss below how promoters act. For now it is enough to know that promoters do exactly what the name implies: they promote the formation of the desired product.

Inspection of Reaction [5.4] reveals an exothermic reaction with a quite favourable entropy factor. Low temperatures and high pressure should therefore favour the forward reaction. High pressure is used in the industrial process, but the reason for the high temperature is not immediately obvious. As we shall soon see, the activation barrier for N_2 dissociation is small or even negative [34, 35], thus the high temperature is not required to activate nitrogen. High temperatures ensure rapid diffusion and reaction of adsorbed intermediates and the rapid desorption of product NH_3 so that sufficient free surface sites are available to accept adsorbing N_2.

The composition of the industrial catalyst has changed little since its introduction. Haber first demonstrated the viability of the catalytic production of NH_3, but it was Mittasch who, in a demonstration of brute force combinatorial chemistry, performed over 6500 activity determinations on roughly 2500 different catalysts while developing the Haber–Bosch process [36]. Some 100 tons of such a catalyst are required for a 1000-ton-per-day plant. By definition, a catalyst is not consumed in the reaction. Nonetheless, the lifetime of the catalyst is finite. The Haber–Bosch catalyst is particularly robust with roughly a 10-year lifetime. N_2 is provided by purified air from which the O_2 has been removed by reaction with H_2. The combustion of H_2, which is generated by the water gas shift reaction (Section 5.5), provides heat for ammonia synthesis.

The ammonia synthesis catalyst is a complex mixture of metals and oxides. What is the structure of the working catalyst and how does this affect its activity? Ertl, Schlögl and co-workers have developed the following picture of the working catalyst [36]. As Figure 5.2 demonstrates, the Haber–Bosch catalyst has a complex structure. The porous oxide substrate is covered with an inhomogeneous layer of metallic iron clusters. Scanning Auger microscopy reveals that potassium preferentially segregates to the surface of the iron clusters. The potassium is bound in a type of oxide that is not representative of a known bulk phase. This potassium oxide compound is chemisorbed on iron and covers about a third of the surface. Oxygen enhances the thermal stability of potassium. Transmission electron microscopy (TEM) unambiguously reveals that the iron clusters are crystalline and that they preferentially expose the (1 1 1) face. We now need to determine whether the geometric structure of the particles and the presence of potassium are critical for the performance of the catalyst. These questions are best answered by direct surface science experiments carried out on well-characterized model catalysts.

First, let us tackle the question of whether the presence of (1 1 1) crystallites is important. Heterogeneous catalytic reactions can be usefully classified as either structure-

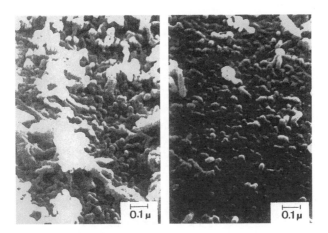

Figure 5.2 The ammonia synthesis catalyst as revealed by high-resolution scanning electron microscopy. Reproduced with permission from G. Ertl, D. Prigge, R. Schloegl and M. Weiss, *J. Catal.* **79** (1983) 359. ©1983 Academic Press

sensitive or structure-insensitive reactions [37]. Boudart and Djéga-Mariadassou [38] discuss a number of examples of both classes of reaction. Classically, the method of determining structural sensitivity is to measure the turnover number ($\equiv$ rate per catalyst atom) of a reaction as a function of catalyst particle size. Such studies convolute numerous factors and a definitive approach is to measure the rate of reaction as a function of the exposed crystal face on single-crystal model catalysts.

Somorjai and co-workers have taken precisely this approach [39, 40]. They prepared single-crystal iron samples of various orientations under standard UHV conditions and then followed the reaction of stoichiometric mixtures of N_2 and H_2. The use of a special reaction vessel allowed them to study the reaction at high pressure (20 bar) and a temperature of 773 K. The reaction rate depends strongly on the crystallographic orientation, varying by over two orders of magnitude between the three low-index planes. The reaction rate increases in the order Fe(1 1 0) < Fe(1 0 0) < Fe(1 1 1). This demonstrates unequivocally the structure sensitivity of the reaction. The clean Fe(1 1 1) surface is the most active; nonetheless, it still makes a poor catalyst. The conversion efficiency of turning N_2 into NH_3 is only of order 10^{-6}.

If dissociation of N_2 is the rate-determining step in ammonia synthesis, we expect to observe a low sticking coefficient and that the sticking coefficient should increase in the order Fe(1 1 0) < Fe(1 0 0) < Fe(1 1 1). Bozso, Ertl and co-workers confirmed this trend and measured a dissociative sticking coefficient that is of order 10^{-6} [34, 41]. Further studies [42] demonstrated that on Fe(1 1 1) the adsorption process is best described by

$$N_{2(g)} \underset{k_1}{\overset{k_{-1}}{\longleftrightarrow}} N_{2(a)} \underset{k_2}{\overset{k_{-2}}{\longleftrightarrow}} 2N_{(a)} \qquad [5.12]$$

and is actually characterized by a small negative activation barrier (-0.034 eV) when the surface temperature is varied and the pre-exponential factor is assumed constant. Nonetheless, high vibrational and translational energies are effective at promoting adsorption into the molecular precursor and, therefore, dissociation of N_2 on Fe(1 1 1) [43–45]. This behaviour is characteristic of a direct adsorption process that proceeds over an activation barrier. However, when the surface temperature is varied, the rate of N_2 dissociation decreases. The negative apparent activation barrier indicates that dissociative N_2 adsorption does not occur via a simple elementary reaction. Consistent with Reaction [5.12], dissociation of N_2 incident upon the surface with thermal energies proceeds via (at least) a two-step sequence that involves a molecularly chemisorbed intermediate (precursor). The competition between the desorption of N_2 (k_{-1}) and the dissociation of adsorbed N_2 (k_2) results in a negative apparent activation energy. The direct dissociative pathway is accessible only for N_2 molecules that have extremely high energies. Recent calculations show that even Reaction [5.12] is an oversimplification; see Mortensen *et al.* [46]. Four molecularly adsorbed states are found and dissociation proceeds successively through these states. In addition, Mortensen *et al.* have shown that the sticking data can equally well be fitted by a small positive activation barrier ($+0.03$ eV) if the temperature dependence of the pre-exponential is treated explicitly. Usually it is a good approximation to assume that A is constant; however, for such small barriers and large temperature ranges the temperature dependence of A should be taken into account.

The behaviour of N_2 on Fe(1 1 1) shows certain parallels to what we have seen for O_2/Pd(1 1 1). In both cases, a direct transition from the gas-phase molecule to the dissociative adsorbed phase does not occur readily for molecules with thermal incident energies. Adsorption into a molecular chemisorbed state opens a dissociation pathway that has a significantly lower or even vanishing barrier. These two examples illustrate the multidimensional nature of the gas–molecule potential energy hyper surface. Furthermore, they demonstrate how a surface can open new low-activation reaction pathways by first accommodating a molecule into the adsorbed phase. Once chemisorbed onto the surface, the O_2 or N_2 molecules are able to traverse regions of configuration space that are not directly accessible from the gas phase and in doing so they can dissociate via a lower barrier path.

That the ammonia synthesis reaction is structure sensitive has practical implications for a practical catalyst. Our first impression of an effective catalyst might be that we would want a catalyst with a high dispersion. The dispersion is the ratio of surface atoms to total atoms in the cluster. A dispersion of one indicates that all atoms reside on the cluster surface. In this way, no atoms would be 'wasted' in the bulk below the surface. However, as the particle size changes, the geometric structure of the cluster changes. In particular, the number of sites of different coordinations is a function of the cluster size. If a particular type of site were required for catalysis then the major concern would be to make a catalyst with a maximum of these sites rather than a maximum of surface atoms of all types. Consistent with the strong structural sensitivity, work in Somorjai's laboratory has demonstrated the importance of sevenfold-coordinated (C_7) iron atoms for efficient ammonia synthesis [39]. The C_7 sites on iron are the active sites for ammonia synthesis.

If potassium acts to enhance the reactivity of the catalysts, it should accomplish this by increasing the dissociative sticking coefficient of N_2. This is found for Fe(1 1 1) and Fe(1 0 0) surfaces [47]. Additionally, in the presence of a sufficient amount of pre-adsorbed potassium, the two iron surfaces no longer exhibit the structural sensitivity of the clean surface. Potassium-covered Fe(1 1 1) and Fe(1 0 0) are roughly equally efficient at dissociating N_2. Thus, potassium acts not only to promote the sticking coefficient but also to assist in making all exposed surfaces of the iron particles equally reactive.

The pursuit of new ammonia synthesis catalysts remains an active area of scientific endeavour. A new ruthenium-based catalyst has been introduced. Enzymatic fixation of nitrogen occurs in nature at ambient temperatures. The active part of the enzyme is believed to be a $MoFe_7S_9$ cluster embedded among organic moieties. Fundamental theoretical studies have been carried in Nørskov's group [35] to investigate whether an understanding of the enzymatic process may lead to an economically viable low-temperature synthetic route. Jacobsen [29] has shown that ternary nitrides (Fe_3Mo_3N, Co_3Mo_3N and Ni_2Mo_3N) can exhibit high activity for ammonia synthesis. A caesium-promoted Co_3Mo_3N catalyst has an activity that exceeds that of a commercial catalyst by $c.$ 30%. From these results we can predict a continuing interest in the surface chemistry of ammonia synthesis and the search for new catalyst formulations derived from a fundamental understanding of catalysis.

5.3   From Microscopic Kinetics to Catalysis

One of the great aims of chemical studies of surfaces is to describe at the microscopic level the elementary steps in complex catalytic reactions and to determine the kinetic parameters that describe the rate of the overall process. This has been accomplished for several reactions. Two of the most important are the oxidation of CO by either O_2 or NO on platinum, and the ammonia synthesis reaction. Here we describe in detail the kinetic model of ammonia synthesis developed by Stoltze and Nørskov [5, 7]. The model combines the experimental results from UHV single-crystal studies and quantum mechanical calculations. With no adjustable parameters, the model predicts rates in good agreement with high-pressure measurements made over an industrial catalyst. The results show unequivocally that the pressure gap can be overcome. First, the model begins with a set of elementary steps as proposed by Ertl [30] and described in Reactions [5.5]–[5.11].

An essential difference between surface kinetics and kinetics in other phases is that the reactions occur with a limited number of surface sites. An empty surface site is denoted * and an occupied site −*. The number of surface sites is included explicitly within the model and the total of occupied plus unoccupied sites is constant. As indicated by the equilibrium sign, each step is assumed to be reversible and all reactions, apart from the rate-determining step, are treated as equilibria. Further assumptions in the model are that the gas phase is ideal and that all sites are equivalent.

The rate-determining step is the dissociation of adsorbed N_2, depicted in Reaction [5.6] [step (2)]. The rate of ammonia synthesis is, therefore, the net rate of step (2):

$$r_2 = k_2 \theta(N_2*)\theta(*) - k_{-2}[\theta(N*)]^2 \tag{5.1}$$

where $\theta(X)$ is the coverage of species X. The net rate rather than just the forward rate $k_2\theta(N_2*)\theta(*)$ must be used because the step is assumed to be reversible. The rate constants can be written in Arrhenius form, for instance,

$$k_2 = A_2 \exp\left(\frac{-\Delta^{\ddagger}H}{RT}\right) \tag{5.2}$$

Next, we need expressions that relate all of the surface species concentrations so that we can substitute into Equation (5.1) to obtain a final rate expression. The final rate expression relates the experimental parameters ($p_{(N_2)}$, $p_{(H_2)}$, $p_{(NH_3)}$ and T) and the thermodynamic parameters to the ammonia synthesis rate. As each reaction step is assumed to be equilibrated, we can write the complete set of equilibrium constant expressions to obtain relationships for the coverages for the various species.

$$K_1 = \theta(N_2*)\left[\left(\frac{p(N_2)}{p_0}\right)\theta(*)\right]^{-1} \tag{5.3}$$

$$K_3 = \frac{\theta(NH*)\theta(*)}{\theta(N*)\theta(H*)} \tag{5.4}$$

$$K_4 = \frac{\theta(NH_2*)\theta(*)}{\theta(NH*)\theta(H*)} \tag{5.5}$$

$$K_5 = \frac{\theta(NH_3*)\theta(*)}{\theta(NH_2*)\theta(H*)} \tag{5.6}$$

$$K_6 = \frac{p(NH_3)}{p_0}\theta(*)\frac{1}{\theta(NH_3*)} \tag{5.7}$$

$$K_7 = [\theta(N*)]^2\left\{\left[\frac{p(H_2)}{p_0}\right][\theta(*)]^2\right\}^{-1} \tag{5.8}$$

where p_0 is the standard state pressure of 1 bar. The equilibrium constant for the overall reaction (Reaction [5.4]) is

$$K_g = K_1 K_2^2 K_3^2 K_4^2 K_5^2 K_6^2 K_7^3 \tag{5.9}$$

From statistical mechanics, each equilibrium constant can be calculated from the molecular partition functions of the reactants and products via Equation (4.31) (page 170). The molecular partition function is calculated from the product of the translational, vibrational, rotational and electronic partition functions as in Equation (4.33). Finally, we need an expression for the conservation of sites. In terms of fractional coverages of the adspecies and the fractional coverage of free sites, $\theta(*)$, this yields

$$\theta(N_2*) + \theta(N*) + \theta(NH*) + \theta(NH_2*) + \theta(NH_3*) + \theta(H*) + \theta(*) = 1 \tag{5.10}$$

From Equations (5.3)–(5.8) and Equation (5.10), expressions for the fractional coverages of each surface intermediate can be calculated (see Exercise 5.1). Substitution of the expressions for $\theta(N2*)$, $\theta(*)$ and $\theta(N*)$ into Equation (5.1) yields

$$r_2 = 2k_2K_1\left\{\frac{p(N_2)}{p_0} - \frac{[p(NH_3)]^2 p_0}{K_g[p(H_2)]^3}\right\}[\theta(*)]^2 \tag{5.11}$$

The rate at which NH_3 is formed is $2r_2$ molecules per second per surface site. The coverage of empty sites is given by

$$\theta(*) = \left\{ 1 + \frac{K_1 p(N_2)}{p_0} + \frac{p(NH_3) p_0^{1/2}}{K_3 K_4 K_5 K_6 K_7^{3/2} [p(H_2)]^{3/2}} + \frac{p(NH_3)}{K_4 K_5 K_6 K_7 p(H_2)} \right.$$

$$\left. + \frac{p(NH_3)}{K_5 K_6 K_7^{1/2} [p(H_2)]^{1/2} p_0^{1/2}} + \frac{p(NH_3)}{K_6} + K_7^{1/2} \frac{[p(H_2)]^{1/2}}{p_0^{1/2}} \right\}^{-1} \qquad (5.12)$$

Equations (5.11) and (5.12) contain thermodynamic and extensive variables that can be determined by experiment and theory. Thermodynamic data for the gas-phase species is readily available in the JANAF tables [48]. The equilibrium constants are calculated from the partition functions from the vibrational properties of the adsorbates. Measurements of the initial sticking coefficient of N_2 into $2N-*$ and its activation energy are used to determine A_2 and $\Delta H_2^{\ddagger}$ [47]. To compare with a working catalyst, also the surface area of the catalyst is required. The calculations of Stoltze and Nørskov [5] show that apart from the pressures, temperature, surface area and K_g, the only critical parameters in the model are A_2, $\Delta^{\ddagger}H + \varepsilon_{elec,N_2^*}$ and $\varepsilon^*_{elec,N*}$, where the terms $\varepsilon_{elec,x}$ are the electronic ground-state energies, as in Equation (4.34). All of these can be determined directly from experiments.

Figure 5.3 compares the calculated and measured outputs from a working catalytic reactor. As it is clear to see the agreement is remarkably good. The calculations are based

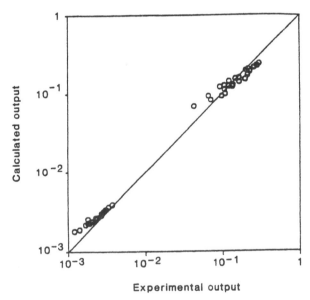

Figure 5.3 The measured NH_3 mole fraction at the reactor outlet (data points) is compared with the calculations of Stoltze and Nørskov (solid line). The Topsøe KM1R catalyst was operated at 1–300 atm and 275°C–500°C. Reproduced with permission from P. Stoltze and J. K. Nørskov, "An interpretation of the high-pressure kinetics of ammonia synthesis based on a microscopic model", *J. Catal.* **110** (1988) 1. ©1988 Academic Press

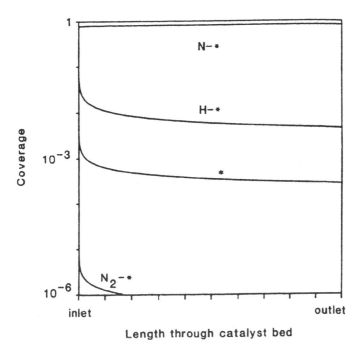

Figure 5.4 Coverages calculated by Stoltze and Nørskov for a potassium-promoted iron catalyst operating at $400°C$ with an initial mixture (at inlet) of 25% N_2, 75% H_2 and 0% NH_3. Percentage of NH_3 at outlet: 20.6%. The curves are for adsorbed nitrogen (N–∗), hydrogen (H–∗), free sites (∗) and molecular nitrogen (N_2–∗). Reproduced with permission from P. Stoltze and J. K. Nørskov, "An interpretation of the high-pressure kinetics of ammonia synthesis based on a microscopic model", *J. Catal.* **110** (1988) 1. ©1988 Academic Press

on values derived from UHV single-crystal surface studies. The experiments were performed at 1–300 atm on a practical catalyst. The extremely good agreement demonstrates that the mechanism of ammonia synthesis is understood and quantitatively describable on a molecular scale.

Figure 5.4 depicts the state of the catalyst under operating conditions. The reactor has an inlet at one end, a catalyst bed and an outlet at the other end. The gas-phase composition changes from a 1 : 3 (stoichiometric) mixture of $N_2 : H_2$ to an equilibrium mixture at the end. Nonetheless, the surface coverage trend is uniform over the length of the catalyst. The surface of the catalyst is essentially covered with adsorbed nitrogen. N–∗ is the most abundant reactive intermediate (MARI) [38]. The number of free sites is extremely small, of the order of 10^{-3}.

5.4 Fischer–Tropsch Synthesis and Related Chemistry

Several catalytic processes are built around the use of synthesis gas, or 'syn gas', a mixture of CO + H_2. Fischer–Tropsch (F–T) synthesis is the production of hydrocarbons and oxygenated hydrocarbons (oxygenates) from synthesis gas. Closely related are the methanation reaction (production of CH_4 from syn gas), methanol synthesis and the Mobil process, which converts methanol (often produced from syn gas) into transportation fuels.

Syn gas is produced from oil, natural gas, coal or other carbonaceous mineralogical deposits by steam reforming. Carbon can also be supplied in a renewable form by the use of biomass. Steam reforming is the reaction of hydrocarbons with water to form CO and H_2. Specifically, for natural gas (which is primarily methane), this is written

$$CH_4 + H_2O \leftrightharpoons CO + 3H_2, \quad \Delta H^0_{298} = -207 \text{ kJ mol}^{-1} \qquad [5.13]$$

This is performed over a K_2O-promoted nickel catalyst at 700°C–830°C, 15–40 bar pressure, on an alumina or calcium aluminate substrate [49]. The reverse reaction is strongly favoured in the low-temperature range of 250°C–350°C because the entropy term of the forward reaction is favourable. The catalyst is easily poisoned by sulfur, arsenic, halogens, phosphorus, lead and copper. These must be removed from the feedstock prior to exposure to the catalyst. This is a much less significant problem for natural gas than it is, for instance, for coal or oil shale.

The water gas shift reaction is the reaction of water gas ($CO + H_2O$) to form CO_2 and H_2:

$$CO + H_2O \leftrightharpoons CO_2 + H_2, \quad \Delta H^0_{298} = -42 \text{ kJ mol}^{-1} \qquad [5.14]$$

This reaction is used to increase the H_2 content of syn gas and is important in automotive catalysis. An Fe_3O_4 catalyst supported on Cr_2O_3 is used as a high-temperature shift catalyst (400°C–500°C). This catalyst is rather robust with respect to sulfur poisoning; indeed, the sulfide, Fe_3S_4, also acts as a catalyst for the reaction, albeit with a lower activity. If the feedstock has a low sulfur content then a low-temperature shift catalyst can be used. These catalysts consist of oxides of $Cu + Zn$ or $Cu + Zn + Al$ and operate at 190°C–260°C. From the stoichiometry of the reaction, the change in the moles of gas molecules is $\Delta n_{gas} = 0$. Therefore, the entropy term is not significant, a factor favourable for a low-temperature catalyst. However, the selectivity of the low-temperature catalyst then becomes of paramount importance because the production of methane and higher hydrocarbons is thermodynamically favoured. Because of the presence of copper, low-temperature shift catalysts are highly susceptible to sulfur poisoning.

Methanol is made industrially by the ICI low-pressure methanol process. Low pressure is in the eye of the beholder as the process is normally run at 50–100 bar. As an exothermic reaction with a highly unfavourable entropy factor,

$$CO + 2H_2 \leftrightharpoons CH_3OH, \quad \Delta H^0_{298} = -92 \text{ kJ mol}^{-1} \qquad (5.15)$$

high pressure and low temperatures favour the process. The temperature must not be too low as the catalyst is deactivated if, for example, the methanol, which is a liquid at room temperature (boiling point 64.7°C at 1 bar), does not rapidly desorb from the catalyst. The use of $CuO + ZnO + Al_2O_3$ or $CuO + ZnO$ catalyst allows the process to be run at the relatively mild temperature of 230°C–270°C. Again, the copper-based catalyst is highly susceptible to poisoning from sulfur as well as chlorine. The selectivity of the catalyst is remarkably high. Although higher alcohols, ethers and alkanes are thermodynamically preferred, the selectivity for methanol is often greater than 99%. In these catalysts the copper acts as the active site. The ZnO provides a matrix into which it can dissolve, and the Al_2O_3 ensures a high surface area (dispersion) for the catalyst. The reaction is strongly structure sensitive. This fact is essential for accurate modelling of the kinetics observed under industrial conditions as the structure of the catalyst depends on the reactant gas composition [50].

Fischer–Tropsch chemistry proceeds via a complex set of reactions that consume CO and H_2 and produce alkanes (C_nH_{2n+2}), alkenes (C_nH_{2n}), alcohols ($C_nH_{2n+1}OH$) and other oxygenated compounds, aromatics as well as CO_2 and H_2O. Some examples of these reactions are

$$nCO + (2n + 1)H_2 \rightarrow C_nH_{2n+2} + nH_2O \qquad [5.16]$$

$$nCO + 2nH_2 \rightarrow C_nH_{2n} + H_2O \qquad [5.17]$$

$$nCO + 2nH_2 \rightarrow C_nH_{2n+1}OH + (n-1)H_2O \qquad [5.18]$$

$$2nCO + (n+1)H_2 \rightarrow C_nH_{2n+2} + nCO_2 \qquad [5.19]$$

$$2nCO + nH_2 \rightarrow C_nH_{2n} + nCO_2 \qquad [5.20]$$

$$(2n-1)CO + (n-1)H_2 \rightarrow C_nH_{2n+1}OH + (n-1)CO_2 \qquad [5.21]$$

All of these reactions are accompanied by negative free-energy changes and are exothermic [28]. It is important, however, that the catalyst does not produce an equilibrium mixture of products. If this were to happen, the reaction products would be an unwieldy molasses, which would be of little economic value. Therefore, selectivity is perhaps the most important requirement of an F–T catalyst. Most group 8, 9 and 10 metals are good for these reactions; however, the product distributions obtained depend on the metal. A cobalt or iron catalyst is best for producing the low molecular weight hydrocarbons preferred as liquid fuels (both petrol and diesel) whereas a ruthenium catalyst is selective for producing high molecular weight waxy hydrocarbons. Rhodium-based catalysts give increased yields of oxygenates, particularly methanol and ethanol. A nickel catalyst is the best for formation of methane. The SASOL process uses potassium as a promoter. In addition, a small amount of copper added to the catalyst aids in the activation of the catalyst and likely helps to maintain a high surface area.

Two environmentally attractive aspects of F–T chemistry are that a renewable source of carbon (biomass) can be used as the source of syn gas and that syn gas contains no sulfur, phorphorus or nitrogen, and is low in aromatics. The liquid fuels obtained by this method are, therefore, very clean burning. The challenge in F–T chemistry remains an increase in the selectivity. A breakthrough in catalytic chemistry that would allow for the selective formation of specific alkenes or oxygenates would be of tremendous impact. To achieve this, a greater understanding of the mechanisms involved in F–T synthesis is required.

The mechanisms involved in Reactions [5.16]–[5.21] are obviously complex. Further obscuring the matter, numerous reactions are occurring along parallel paths. Nonetheless, progress has been made in understanding some mechanistic aspects of F–T chemistry. First, one needs to realize that the product distribution and therefore the reaction mechanisms depend sensitively on the surface temperature and the composition of the catalyst. Not one unique mechanism operates at all temperatures and for all catalysts. This is to be expected as a result of the interwoven nature of the numerous parallel reactions that occur.

Maitlis and co-workers studied F–T synthesis over $Rh/CeO_2/SiO_2$ and Ru/SiO_2 catalysts at 433–473 K and a $CO:H_2$ ratio of $1:2$ [51, 52]. The primary products are 1-alkenes, with propene ($CH_2=CH-CH_3$) the most abundant. They have established two important pathways that are components of F–T chemistry by tracking the incorporation of ^{13}C-containing reactants. The first is the dissociation of CO with subsequent hydrogenation to form an adsorbed methylene species (CH_2). The second series of reactions corresponds to the polymerization of CH_2 moieties with adsorbed alkenyl fragments. These two reaction cycles are depicted in Figures 5.5 and 5.6, respectively.

Figure 5.6 depicts a series of non-stoichiometric schematic reactions [51, 52]. The initiating steps of F–T synthesis are dissociative H_2 adsorption and the chemisorption of CO. Adsorbed CO then dissociates and the resulting carbide is sequentially hydrogenated to CH, CH_2 and CH_3, the presence of which have been confirmed in surface science experiments. There is general agreement that what ensues is a stepwise polymerization of methylene (CH_2 groups). Shown in Figure 5.6 is that the beginning of the hydrocarbon formation cycle is the reaction of $CH + CH_2$ to form an adsorbed vinyl species. A CH_2 species then adds to this unit followed by an isomerization to reproduce a structure similar to that of the adsorbed vinyl group. Subsequent additions of CH_2 groups then compete with the addition of H. When H is added, the cycle terminates and a 1-alkene desorbs. That $H_2^{13}C=^{13}CHBr$ is an efficient initiator of the reaction confirms this mechanism. $H_2^{13}C=^{13}CHBr$ adsorbs as $H_2^{13}C=^{13}CH$ as required by the mechanism. However, $H_3^{13}C-^{13}CH_2Br$, which does not readily form this species, is not an efficient initiator of reaction.

Several comments must be made about these mechanistic steps. Note that the exact structure of the adsorbed intermediates is unknown. Nonetheless, we know that the

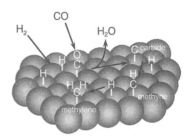

Figure 5.5   The dissociation of CO followed by hydrogenation to form an adsorbed methylene (CH_2) species. The formation of methylene is an essential step in Fischer–Tropsch synthesis. CH_2 is the product of sequential hydrogen addition steps that follow the dissociative adsorption of H_2 and CO. Oxygen is removed from the surface via H_2O formation

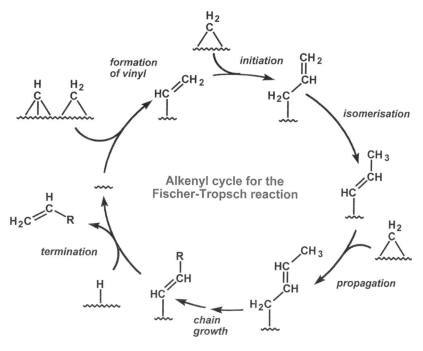

Figure 5.6   The alkenyl carrier cycle. The cycle begins in the top left-hand side of the figure with the formation of adsorbed vinyl (HC=CH$_2$) from CH and CH$_2$. Chain growth ensues, initiated by the addition of CH$_2$. Isomerization forms an adsorbed allyl (H$_2$C–CH=CH$_2$). Subsequently, further addition of CH$_2$ (propagation and chain growth) competes with addition of H (termination). Reproduced with permission from P. M. Maitlis, H. C. Long, R. Quyoum, M. L. Turner and Z.-Q. Wang, *Chem. Commun.* (1996) 1, by permission of the Royal Society of Chemistry

^{13}C=^{13}C bond is not severed completely and that the two isotopically labelled carbon atoms remain neighbours throughout, regardless of how many CH$_2$ additions occur to the molecule that contains them. This supports the isomerization/addition mechanism of Figure 5.6. In addition, desorption of the 1-alkene after the addition of a hydrogen atom to the adsorbed alkenyl species (hydrogenolysis) competes at each step with the addition of a further CH$_2$ group. This type of kinetic competition makes the product distribution particularly sensitive to temperature and catalyst activity. Importantly, the catalyst treats adsorbed double bonds gently. At 433 K the C=C found in step one is not completely broken and the two ^{13}C atoms remain bound to each other.

When the conditions of the reaction are changed, the reaction product distribution can change. The product distribution for the Ru/SiO$_2$ catalyst is much different at 463 K compared with 433 K. At 463 K significant ^{13}C=^{13}C bond scission occurs and the two ^{13}C do not often remain neighbours. Surprisingly, a Rh/CeO$_x$/SiO$_2$ catalyst at 463 K exhibits similar behaviour to the Ru/SiO$_2$ catalyst at 433 K. In summary, F–T synthesis remains a complicated problem, but better control of the product distributions should be possible with our increasingly sophisticated understanding of this catalytic system.

5.5   The Three-way Automotive Catalyst

The goal of achieving clean-burning internal combustion engines represents both a challenge to and a triumph of catalytic chemistry. Beginning in the 1970s in the United States, governments have imposed progressively stricter regulations on the release of hydrocarbons, nitrogen oxides (NO_x) and CO in automotive emissions.

The strategy of designing an automotive catalyst is much different from that of the industrial processes that we have considered above. The reaction conditions of temperature (673–773 K) and pressure (c. 1 bar) are determined, within narrow limits, by the operation of the engine and the desire for high fuel economy. The automotive catalyst aims to take the products of a near stoichiometric mixture of air and fuel after it has been combusted in the engine and ensure that the emissions consist of an equilibrium mixture containing only CO_2, H_2O and N_2. The reactant mixture that reaches the catalyst consists of H_2, H_2O, O_2, N_2, NO_x, CO, CO_2 and hydrocarbons (alkanes, alkenes and aromatics containing 1–8 carbon atoms). A certain level of impurities is derived from the fuel, engine oil and fuel or lubrication additives. The most important impurities are sulfur (as SO_2), phosphorus and lead (as tetraethyl lead). The use of alcohol in fuels leads to the introduction of aldehydes into the exhaust stream.

As summarized by Taylor [53], the pertinent chemical reactions to be considered are

$$CO + \tfrac{1}{2}O_2 \rightarrow CO_2 \qquad \qquad [5.22]$$

$$hydrocarbons + O_2 \rightarrow H_2O + CO_2 \qquad \qquad [5.23]$$

$$H_2 + \tfrac{1}{2}O_2 \rightarrow H_2O \qquad \qquad [5.24]$$

$$NO + CO \rightarrow \tfrac{1}{2}N_2 + CO_2 \qquad \qquad [5.25]$$

$$NO + H_2 \rightarrow \tfrac{1}{2}N_2 + H_2O \qquad \qquad [5.26]$$

$$hydrocarbons + NO \rightarrow N_2 + H_2O + CO_2 \qquad \qquad [5.27]$$

$$NO + \tfrac{5}{2}H_2 \rightarrow NH_3 + H_2O \qquad \qquad [5.28]$$

$$CO + H_2O \rightarrow CO_2 + H_2 \qquad \qquad [5.29]$$

$$hydrocarbons + H_2O \rightarrow CO + CO_2 + H_2 \qquad \qquad [5.30]$$

$$3NO + 2NH_3 \rightarrow \tfrac{5}{2}N_2 + 3H_2O \qquad \qquad [5.31]$$

$$2NO + H_2 \rightarrow N_2O + H_2O \qquad \qquad [5.32]$$

$$2N_2O \rightarrow 2N_2 + O_2 \qquad \qquad [5.33]$$

$$2NH_3 \rightarrow N_2 + 3H_2 \qquad \qquad [5.34]$$

These are grouped more or less in order of importance. Both *activity* and *selectivity* are important for the operation of the catalyst. The importance of activity is obvious. The catalyst must efficiently oxidize CO and hydrocarbons and reduce NO regardless of whether the exhaust mixture is rich in oxidizing agents or reducing agents. The selectivity of the catalyst is exemplified by the desire to produce N_2 not NH_3 from NO. Furthermore, O_2 should be used for complete oxidation of hydrocarbons and of CO to CO_2 rather than for the production of H_2O from H_2. Given these requirements and constraints we can now set out on the search for a perfect catalyst.

The threeway automotive catalyst, so named because it removes the three unwanted products CO, HC and NO_x, is the result of wide-ranging scientific activity that has ranged from high-pressure engineering studies on high-surface-area supported metal catalysts to UHV surface science on single-crystal model catalysts. The general composition consists of rhodium, platinum, and lead dispersed on Al_2O_3 with CeO_2 added as a type of promoter.

Rhodium is almost the perfect material for an automotive catalyst. It is the best catalyst for the reduction of NO to N_2. Rhodium is particularly active in promoting the reactions of NO + CO and NO + H_2, even at low temperatures. Rhodium also acts as an efficient steam reforming catalyst, though it does have a tendency to oxidize hydrocarbons only partially to CO. Although increasing the need for CO oxidation, this is not a severe constrain as rhodium functions as an efficient CO oxidation catalyst as well, again even at relatively low temperatures. The one great drawback of rhodium is its price. It is estimated to comprise only 0.001 ppm of the Earth's crust (compared with 0.005 ppm for platinum, and 0.01 ppm for palladium [54]. Major ore-bearing deposits are located only in South Africa and Russia as well as a lower-concentration reserve in Canada. Hence rhodium is quite expensive and its supply has been subject to political volatility. Research continues to find replacements for it.

Platinum is not normally considered an inexpensive metal unless one compares it with rhodium. Platinum is added to the threeway catalyst for its ability to oxidize CO and hydrocarbons. The warm-up time of an engine represents by far the dirtiest phase of combustion and presents the most severe problems for the catalytic conversions of CO and hydrocarbons. Platinum has relatively high activity during this phase. Platinum makes little contribution to the reduction of NO_x. Though it can reduce NO in isolation, in the presence of SO_2 and high concentrations of CO the activity of platinum for NO reduction is quite low.

Before 2000, palladium was the least expensive of the three platinum-group metals used in the threeway catalyst. However, high demand for automotive catalysts and volatility in the Russian supply have for the first time made it more expensive than platinum. Its primary role is as an oxidation catalyst for CO and hydrocarbons. It is frequently used in a separate part of the catalyst, away from the rhodium, so that oxidation and NO reduction occur in separate parts of the overall automotive catalyst. The short-term trend is for decreased palladium loadings and increased platinum loadings in automotive catalyst because of the prices of the metals. This illustrates that catalyst composition cannot be based solely on performance but is also sensitive to cost. The prices, production and uses of platinum, palladium and rhodium are detailed in Table 5.1.

Alumina is the preferred support material. It combines high surface area, favourable pore structure and strong mechanical stability. Furthermore, Al_2O_3 is an inexpensive and readily available material.

Ceria (CeO_2) is added at the level of $c.$ 1% as a promoter. Ceria acts as a structural promoter. It appears to stabilize the substrate with respect to surface area loss as well as enhancing the dispersion of the catalyst. Ceria also helps promote the water gas shift reaction (Reaction [5.29]). It is unclear whether this is due to reaction on the CeO_2 itself

Table 5.1 Production and demand for platinum-group metals. All figures are reported in thousands of ounces for the year 2000 unless otherwise noted. All figures for rhodium are for 1999. Catalysis is the exclusive use for the category 'automotive' and is the primary use for categories 'chemical' and 'petroleum'. Figures are taken from The Johnson Matthey website, at www.platinum.matthey. com/applications/autocatalysis.html. Prices are as of February 2001

Production		Demand	
source	weight	category	weight
Platinum (US$616 oz^{-1}):			
South Africa	3920	Jewellery	2940
Russia	1100	Automotive	1800
Recovered	460	Industrial	1460
North America	285	Chemical	315[a]
Other	105	Petroleum	115[a]
Palladium (US$1094 oz^{-1}):			
Russia	5200	Automotive	5160
South Africa	1960	Electronics	2070
North America	665	Dental	870
Recovered	230	Chemical	240[a]
Other	95		
Rhodium (US$2125 oz^{-1}):			
South Africa	400	Automotive	502
Russia	80	Chemical	37
Recovered	66	Glass	30
North America	20	Electronics	6
Other	10	Other	11

[a]1999.

or whether this results from some type of electronic promotion. In addition, ceria acts as a reservoir for oxygen. The lattice of ceria is unusually forgiving with respect to the amount of oxygen contained in it, much more so than alumina. O_2 can dissociate on the metal particles and then diffuse into the surrounding ceria. This effect is known as spillover. The oxygen is not permanently trapped in the ceria; rather, it remains available for future reaction. Oxygen storage capacity is an important characteristic because the air/fuel ratio is not constant in a working engine. The presence of ceria allows the catalyst/substrate system to store oxygen when it is too plentiful and release it for reaction when the oxygen concentration is too low.

Poisoning is an ever-present problem for catalysts, and the threeway catalyst is no exception. Fortunately for the environment and its inhabitants, the automotive catalyst is poisoned by lead. Tetraethyl lead has been added to automotive fuels as an octane enhancer as well as for its beneficial tribological properties. Researchers quickly realized, however, that lead has a detrimental effect on the performance of the threeway catalyst. Removing the lead from fuel easily solves the problem, though this has necessitated some changes in engine design. This solution was rapidly adopted in the United States but has taken significantly longer to implement in Europe. The reduced lead burden on the environment is an unintended beneficial aspect of the catalytic converter.

5.6 Promoters

The aim of catalysis is to enhance the rate of formation of a desired product. A catalyst does so by lowering the appropriate activation barriers along the reactive path. In the case of ammonia synthesis, the problematic activation barrier is that of breaking the $N\equiv N$ bond. An iron surface enhances the reactivity of N_2 by accommodating it into a molecularly chemisorbed state that has a relatively low barrier toward dissociation into the atomic fragments. The barrier to dissociation out of the molecularly chemisorbed states is lower than the barrier to dissociation of the gas-phase molecule because chemisorption has weakened the $N-N$ bond, making it more susceptible to scission.

Nonetheless, a clean iron surface is still not as reactive as we would like for an effective catalyst. The industrial catalyst also contains potassium on the surface of the iron particles. Potassium significantly increases the activity of the catalyst and is therefore called a promoter. The effect of an electronic promoter is clear: it lowers the barrier to reaction. The mechanism by which a promoter achieves this is less unambiguous.

In Chapter 3 we investigated the dissociative adsorption of H_2 (Section 3.5) and the molecular chemisorption of CO (Section 3.4.2). I both of these cases, the filling of an antibonding electronic state led to an increased binding of the molecular species to the surface and a concomitant reduction of the intramolecular bond strength. This suggests one mechanism by which a promoter can act. If the promoter can facilitate charge transfer from the substrate into antibonding electronic states of the adsorbate, the adsorbate is more likely to dissociate. In other words, increased substrate-to-adsorbate charge transfer can reduce activation barriers. Molecular states near the Fermi energy, E_F, the most sensitive to changes in the electronic structure of the substrate because slight shifts in the positions of these states relative to E_F can change their occupation and the extent of mixing with substrate levels. The donation of electrons from an electropositive adsorbate, such as an alkali metal, to the substrate significantly perturbs the surface density of states near E_F. The greater availability of electrons near E_F leads to an enhanced ability of the substrate to donate electrons into the adsorbate, which lowers the barrier to dissociation. This mechanism of promotion has been developed theoretically by Feibelman and Hamann [55].

Nørskov, Holloway and Lang [56] have pursued an alternative mechanism for promotion. They have concentrated on the transition state to dissociation. If the promoter stabilizes the transition state relative to the initial state of the reactants, it lowers the activation energy. Because of the extended bonds and distorted electronic structure of transitions states compared with stable molecules, transition states tend to exhibit larger dipole moments than do ground-state molecules. This makes them susceptible to the influence of electrostatic fields. Once an alkali metal has donated an electron to the substrate it is effectively an ion adsorbed on the surface. This ion is strongly screened on a metal; nonetheless, a local electrostatic field is associated with the alkali metal. The interaction of this field with the transition state can reduce the energy of the transition state and thereby reduce the activation barrier to reaction through this transition state. Whether this effect is attractive or repulsive depends on the relative sign of the electrostatic fields on the two adsorbates. For most molecular adsorbates this field is related to electron transfer into the antibonding molecular orbitals.

It is still not clear which one of these promotion mechanisms predominates. In all likelihood, both are possible and the mechanism of promotion is a system-specific property rather than being represented by only one global mechanism. On the one hand, calculations suggest that the electrostatic stabilization of the transition state is responsible for the alkali-metal promotion of N_2 dissociation on ruthenium surfaces [57]. On the other hand, stabilization of chemisorbed $N_{2(a)}$ appears to account for a large fraction of the promotion effect on potassium-promoted iron [58]. It therefore appears that the mechanism of promotion must be considered on a case-by-case basis and that there may even be circumstances in which both models contribute to the overall effect.

Promotion is most commonly thought of in terms of the electronic promotion that is described above, but it is important to remember that an effective catalyst is one that stably and selectively produces the desired product. Thus, promoters are also added to a catalyst to affect the stability and selectivity of the catalyst and not just the activity. In the ammonia synthesis catalyst, Al_2O_3 and CaO are added as structural promoters. They enhance the surface area of the dispersed iron particles and work against the sintering of these into larger particles. In F–T synthesis, potassium is added to the catalyst when an increased selectivity for higher molecular weight products is desired.

5.7 Poisons

Poisons are adsorbed species that lower the activity of a catalyst. The above case studies show that poisoning of catalysts is a ubiquitous phenomenon. Poisons can sometimes be avoided simply by changing the reactive feed, but this is not always possible. Poisons can sometimes be removed by side-reactions. This represents a type of self-cleaning catalyst. Oxygen can often be removed from a catalyst if the reactive mixture is strongly reducing, as in the ammonia synthesis. Carbon can be removed under strongly oxidizing conditions. It is the formation of tenacious deposits, that cannot be removed chemically that must be avoided.

Poisons can reduce the rate of a surface reaction by two distinctly different mechanisms. The first is an electronic mechanism, which is an indirect, long-range effect. This is essentially the inverse of the process described above for promotion in which the poison acts to change activation energies in an unfavourable manner. The presence of a poison can destabilize either a chemisorbed reactive intermediate or the transition state of an elementary step. Again, electron donation or electrostatic effects may be involved in this process. Reaction rates can be suppressed either because an activation barrier is increased or because the concentration of one of the reactants is reduced. If a poison acts to reduce the chemisorption bond strength, the affected adsorbate is more likely to desorb than to proceed in the reaction.

The second mechanism is called site blocking. This is a direct, geometric effect. If a poison occupies all of the adsorption sites, there are none left over to take part in the catalytic reaction. For instance, when a three-way catalyst is covered with lead it is inactivated because is not an effective catalyst for a catalytic converter.

Site blocking is a localized effect that can be utilized in the nanoscale modification of surfaces. Hydrogen adsorption on silicon surfaces greatly reduces the reactivity of these surfaces toward oxidation by O_2 and H_2O. If the $H_{(a)}$ is removed from the surface, the uncovered sites become reactive and can readily be oxidized in the presence of O_2 or H_2O. Lyding, Avouris, Quate and co-workers have taken advantage of this poisoning of silicon oxidation to pattern silicon surfaces [59–61]. Using a scanning probe tip to desorb hydrogen, it is possible to modify the silicon surface with near atomic resolution (see Figure 4(a), page xviii).

Poisons are not always undesirable. They are sometimes intentionally added to a catalyst to stabilize their performance. Thus, a calculated reduction of the initial reactivity of a catalyst can pay off if the long-term activity is higher on a specifically poisoned catalyst than on an unintentionally poisoned one. Another reason for intentionally poisoning a catalyst is that the poison may be specific for a certain reaction. Step sites can be particularly effective at breaking bonds. Depending on the reaction, we may want to suppress this function. By decorating the steps with a poison, the reactivity of the steps can be turned off so that no reaction occurs there. Hence, poisons are sometimes added to a catalyst to enhance its selectivity.

In F–T chemistry, coking, the formation of graphite on the surface, must be avoided. One solution is the intentional partial poisoning of the catalyst with sulfur by adding H_2S to the feed. An ingeniously novel suggestion to avoid graphite formation comes from Besenbacher *et al.* [62]. The addition of a small amount of gold to the nickel catalyst leads to the formation of a surface alloy. This alloyed surface has a slightly lower activity than the pure nickel surface. The formation of graphite, however, is completely suppressed. Therefore, although the initial reactivity is somewhat lower for the Au/Ni catalyst no long-term degradation of the catalytic activity occurs. This strategic implementation of poisoning represents a great victory for the fundamental surface science approach to heterogeneous catalysis. The poisoning effect of gold was predicted based on an understanding of the reactivity of metal surfaces gleaned from theoretical studies of dissociative hydrogen adsorption (Section 3.6). The successful implementation of this knowledge in a practical catalyst demonstrates that the principles governing the interactions of small molecules under UHV conditions can, indeed, be used to develop an understanding of real-world catalysts.

A catalyst can be deactivated in a number of other ways apart from poisoning. If the metal clusters in the active catalyst agglomerate into larger particles a net reduction of surface area occurs. If the reaction is structure sensitive, the loss of reactivity may be larger than the loss of surface area would indicate. Some catalysts have materials added to them that hinder the sintering of the active phase. Al_2O_3 and CaO perform this function in the Haber–Bosch process and are hence known as structural promoters. The catalyst may actually be lost if it is volatile, reacts to form volatile products, diffuses into the support or reacts with the support to form nonactive phases. The oxidation of potassium on the ammonia synthesis reduces the volatility of potassium. The choice of substrate and control of the reaction temperature are critical to avoid volatilization and reactions between the catalyst and support. Most heterogeneous catalysts exist in porous matrices. If the pores become blocked by reaction products or by-products, the reactivity suffers.

5.8   Rate Oscillations and Spatiotemporal Pattern Formation

'Nonlinear dynamics' is a term used to describe a broad range of phenomena that includes not only oscillating chemical reactions and pattern formation but also chaos and turbulence. Nonlinear dynamics has been applied to the study of phenomena as diverse as traffic jams and heart arrhythmia. Oscillating chemical reactions have been known for a long time, with the most celebrated example being the Belousov–Zhabotinskii reaction. The first observation of oscillating kinetics in heterogeneous catalysis was made in the group of Wicke [63]. Imbihl and Ertl [64] have reviewed the field with particular emphasis on studies carried out on single-crystal surfaces. They point out that, compared with pattern formation in homogeneous systems, surface reactions provide three unique aspects. First, anisotropic diffusion is possible on surfaces. Second, mass transport in the gas phase provides a means of communication across the surface that can lead to global synchronization. Last, since the reactions occur on a surface we can use a library of techniques that have high spatial resolution to characterize the spatial and temporal concentration gradients that are formed.

Imbihl and Ertl [64] list 16 surface reactions that exhibit oscillatory kinetics. Single-crystal studies have concentrated on CO oxidation on platinum and palladium, and on NO reduction by CO, H_2 and NH_3 on Pt(1 0 0) and Rh(1 1 0). In general, oscillating reactions are accompanied by spatiotemporal pattern formation, and surface reactions are no exception. We concentrate on CO oxidation on platinum as it exhibits particularly rich chemical behaviour.

CO oxidation follows the Langmuir–Hinshelwood mechanism described in Section 5.1. Particularly important for the observation of oscillations is that CO oxidation exhibits asymmetric inhibition. Inhibition occurs when a reactant poisons the reaction. In this case, CO forms a densely packed layer upon which O_2 cannot dissociatively adsorb. Conversely, oxygen atoms form an open adlayer into which CO readily adsorbs. Therefore, the reaction is poisoned only by high coverages of CO, and the reaction rate exhibits two branches. On the high-rate branch at low p_{CO} (partial pressure of CO), the CO_2 production rate increases linearly with p_{CO}. On the low-rate branch at high p_{CO}, the reaction rate decreases with p_{CO}. In other words, the effective reaction order relative to p_{CO} changes from $n = 1$ to $n < 0$ even though there is no change in the reaction mechanism. This highlights the need to understand what is occurring on the surface in order to understand effective reaction orders whereas the converse cannot be done unambiguously.

The reaction dynamics described above are sufficient for explaining the bistability of the reaction rate but are not yet sufficient to explain oscillations. A further mechanistic element is required to introduce a nonlinearity into the reaction kinetics. The first is the adsorbate-induced reconstruction of the Pt(1 0 0) and Pt(1 1 0) surfaces. The second is that the sticking coefficient of O_2 changes dramatically on the different reconstructions.

Rate oscillation are observed under conditions where O_2 adsorption is rate limiting. To illustrate, consider a CO-covered (but not saturated) Pt(1 1 0)-(1×1) surface. O_2 adsorbs readily on this surface and the reaction rate increases as more O_2 adsorbs. Eventually, the rate becomes too high for adsorption of CO to replenish the surface. Consequently, the CO coverage starts to drop and the rate passes through a maximum. When the CO coverage, θ_{CO}, drops below c. 0.2 ML, the surface reconstructs into the (1×2) phase. O_2 does not stick well on this surface but CO adsorption is much less structure sensitive. The reaction rate drops to a minimum and the CO coverage begins to build up on the (1×2) surface until the critical coverage is exceeded. The surface reconstructs and the system returns to the initial conditions so that the next oscillation can begin.

Reconstructions are not a required element of oscillating surface reactions. A more general statement is that periodic site blocking leads to oscillations. Thermokinetic oscillations can be related to a periodic site-blocking mechanism. In this mechanism, the heat of reaction can lead to either heating or cooling of the surface depending on whether the reaction is exothermic or endothermic. Because reaction rates depend exponentially on temperature, a change in the surface temperature dramatically changes the balance between adsorption and desorption. A decreasing temperature caused by an endothermrc reaction leads to the build up of site-blocking adsorbates, killing off reaction and forming a deactivated phase. After deactivation, the temperature rises, the site blocker desorbs, the activated state is restored and oscillations ensue. Alternatively, an exothermic reaction can lead to a heating of the surface that accelerates either desorption or reaction involving the blocking species. If the reaction rate exceeds the rate of replenishment of reactants, the catalyst is eventually cleaned off, at which point the temperature drops and the site blocker again begins to build up.

Autocatalysis can also provide the mechanism for oscillations. In the reaction of $NO + CO$, NO dissociation requires two empty sites. The formation of N_2 and CO_2 leads to the formation of four empty sites. Under conditions in which NO dissociation is rate limiting, the formation of products increases the rate of NO dissociation. The rate can accelerate so rapidly that a surface explosion occurs in which products desorb in extremely narrow peaks only 2–5 K wide.

Rate oscillations are closely coupled to some mechanism of synchronization. If there were none, different reaction rates across the surface would simply average out to a uniform rate. Synchronization can occur via heat transfer, partial pressure variations or surface diffusion. Only the latter two are relevant in low-pressure ($p < 10^{-3}$ mbar) single-crystal studies.

These couplings can result in pattern formation as well. Imbihl, Ertl and co-workers [65, 66] were the first to identify propagating waves on single-crystal surfaces. Several classes of patterns can form solely on the basis of diffusional coupling both in fluid phases and on surfaces. These include target patterns, spiral waves and pulses. In addition to these travelling waves, stationary waves known as Turing structures can form. Anisotropic diffusion and lateral interactions lead to deformations in patterns, altering what would have been circular patterns into elliptical patterns. Gas-phase coupling leads to either stabilization or destabilization of a uniformly oscillating surface depending on conditions. An example of travelling spiral waves observed for CO oxidation on Pt(1 1 0) is shown in Figure 5.7.

Advanced topic: cluster assembled catalysts

Techniques now exist for the production of beams of atomic and molecular clusters [67–69]. Furthermore, the cluster can be size selected, in favourable circumstances, for clusters ranging from two to thousands of atoms. These clusters can be deposited onto a suitable substrate for further study. Alternatively, organometallic chemistry can be used to create multicentre cluster compounds of metals [70]. The use of size-selected, deposited metallic clusters represents an alternative route to the formation of model catalysts [70, 71].

Gas-phase clusters are known to exhibit size-dependent reactivity in specific reactions, for example for H_2 dissociation on metals [68] or for O_2 on silicon [72]. Thus, we might also expect supported clusters to exhibit size-dependent reactivity. Size-dependent

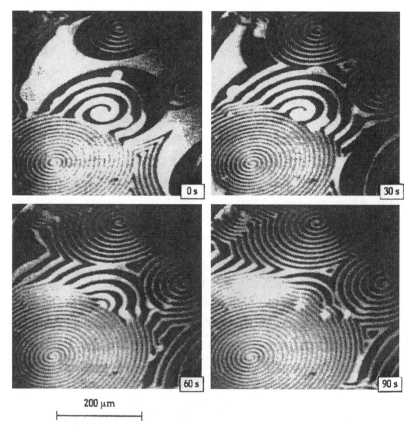

200 μm

Figure 5.7 Spatiotemporal pattern formation in CO oxidation over platinum. Reproduced with permission from R. Imbihl and G. Ertl, *Chem. Rev.* **95** (1995) 697. ©1995 American Chemical Society

reactivity, of course, ties into the concepts of structure-sensitive and structure-sensitive reactions that were mentioned in Section 5.1. First, we note that gas-phase reactivity has little predictive relevance for reactivity of supported clusters. The interaction of the cluster with the support not only changes the electronic and geometric structure of a cluster, but also opens up new energy relaxation pathways that are not available to gas-phase clusters, such as loss of thermal energy to substrate phonons. Therefore, the reactivity of clusters is greatly modified by interactions with the substrate. For instance, the dissociation of O_2 on silicon clusters exhibits strong size dependence in the gas phase but virtually none for supported clusters [72]. Nonetheless, size-dependent reactivity has been observed for supported clusters as in the case of CO oxidation on small gold clusters [71].

The study of size-selected supported clusters promises to bring new insight to catalytic studies. When coupled with structural probes such as scanning tunnelling microscopy (STM) it addresses important questions regarding the interactions between catalyst particles and substrates. Supported metal clusters with diameters less than $10\,\text{Å}$ are smaller than those that have generally been used in practical catalysts [70]. The study of this neglected size range may lead to developments in catalytic chemistry.

5.9   Etching

Etching represents a completely different type of surface reaction because the surface itself is not only involved in the reaction, but also is being volatilized or otherwise removed from the substrate. Etching is important in a number of technological processes. Below we survey several of these.

5.9.1   Coal gasification and graphite etching

We have already encountered one type of surface etching reaction. Coal gasification is the reaction of an inhomogeneous carbonaceous solid with H_2O to form H_2, CO, CO_2 and CH_4. The water gas shift and methanation reactions discussed in Sections 5.4 and 5.5 are of obvious importance to this class of reactions. Conditions can be optimized so that either methane or syn gas is the primary product. Impurities within the coal itself may act as catalysts for gasification. The demands put upon a catalyst are extremely severe because of poisoning and because it is inherently more difficult to catalyse a reaction involving the etching of a solid. One way to avoid mass transport problems is to use a catalyst that is liquid under the reaction conditions. K_2CO_3 has been used to catalyse coal gasification in a process developed by Exxon [49]. The reaction is run at $1000\,\text{K}$, at which K_2CO_3 is liquid. The reactions are by no means understood. Various surface oxides are formed and these appear to exert control over the reaction [73]. In addition, numerous hydrocarbon decomposition reactions occur simultaneously with the reactions of solid carbon.

The reaction of graphite with H_2O [74], O_2 [75, 76] and H_2 [77] have been studied under well-defined conditions. Low-temperature adsorption of H_2O is nondissociative and does not lead to etching of graphite. The addition of potassium to the surface catalyses the formation of CO_2 at a temperature of just $750\,\text{K}$, consistent with its use in the Exxon process. The reaction of modulated molecular beams of O_2 with pyrolytic graphite depends strongly on the surface structure. CO is the primary product, with much less CO_2 being formed. The reaction has an appreciable rate only above 1000 K and then increases slowly with temperature. As etching proceeds the surface is roughened significantly. Nonetheless, the reaction probability remains low ($\leq 10^{-2}$). This implies that only certain defect sites act as active sites for the reactions. The reaction probability is insensitive to changes in the O_2 energy, which indicates that the reaction is not activated in the coordinates of the molecule. Activation occurs only by increasing the surface temperature. Graphite is not etched by molecular hydrogen. Even a beam of atomic hydrogen is only able to etch pyrolytic graphite with a reaction probability of less than 10^{-2}. Methane is produced below 800 K, whereas acetylene (C_2H_2) is the major etch product above 1000 K. Between these temperatures, the hydrogen atoms recombine to form H_2 without etching the surface. The basal plane is much less reactive and remains relatively smooth after etching. The prism plane is more reactive and forms ridges parallel to the basal plane much as is found after etching in O_2.

The etching of highly oriented pyrolitic graphite (HOPG) in O_2 or air can lead to the formation of circular etch pits [78], which have been used as molecular corrals for the study of self-assembled monolayers (Section 6.8.2). When etched for short periods at *c.* 920 K, flat-bottomed, one-monolayer-deep etch pits with diameters of 50 Å to 5000 Å are formed. Again, defects are of great importance in the etching process. The defects have much higher reactivity than have the perfect terraces; therefore, defects act as nucleation sites from which the circular pits grow. Etching in air can be accompanied by the introduction of particulates onto the surface. Particles that land on the surface can act as further nucleation centres that lead to enhanced etching via a catalytic process. The particles often appear to be mobile. This leads to the formation of channels formed by etching in the wake of the diffusing particle.

5.9.2   Silicon etching in aqueous fluoride solutions

The etching of silicon surfaces in aqueous fluoride solutions is of great scientific and technological interest [79–82]. Depending on the reaction conditions, etching can produce either perfectly flat, ideally terminated surfaces or else an exceedingly complex high-surface-area material known as porous silicon (por-Si). Solutions of HF are one of the few that can etch glass. Glass is largely composed of amorphous SiO_2 (a-SiO_2). A native oxide layer covers a silicon wafer because of air oxidation of the silicon single crystal after cutting and polishing. The native oxide is a SiO_2 layer of about 4 nm thickness. HF is used in semiconductor processing to remove the native oxide [83]. There is great interest in obtaining the flattest possible silicon surface after cleaning as surface roughness affects device performance [84, 85]. Por-Si is of great interest because it is an inexpensive means for production of nanostructured silicon that emits visible light because of quantum confinement effects [86]. Por-Si has numerous other interesting characteristics that make it interesting for a variety of other applications in optics and chemical sensing [82].

The Si$-$F bond is significantly stronger than the Si$-$H bond. Therefore, it was assumed for years that the surface obtained by etching in $HF_{(aq)}$ (or other solutions obtained from NH_4F etc.) was terminated by chemisorbed fluorine atoms. Chabal and co-workers [87] demonstrated with infrared (IR) spectroscopy that the surface is actually terminated with hydrogen atoms. Furthermore, a nearly perfect $Si(1\,1\,1)$-(1×1) surface can be formed under appropriate condition. The $Si(1\,0\,0)$ surfaces produced by fluoride etching can be quite flat but do not exhibit the same degree of perfection [80, 88–91].

The mechanism of silicon etching in fluoride solutions has received a great deal of attention in the literature and is not yet completely understood [79, 81, 92–95]. Nevertheless, we can explain a number of pertinent details about the reaction dynamics and, in so doing, we illustrate a number of important concepts and show how UHV surface science can be used to help understand complex reactions that occur at the liquid/solid interface. We use the model of Gerischer, Allongue and Costa-Kieling [93] as the basis of our discussion. The mechanism is not complete in every detail nor for all reaction conditions. Its simplicity and intuitive nature, however, make it a good pedagogical tool and foundation for understanding this system. This mechanism was proposed to explain the electrochemical etching of the surface, in which a hole is captured by the Si—H bond to initiate the reaction. Nevertheless, we can use it to illustrate the chemical reactions that occur after initiation.

The first question to be addressed is why the surface is hydrogen-terminated. The answer is related to the electronegativities of fluorine and hydrogen. Although the Si—F bond is much stronger and therefore more stable on thermodynamic grounds it is, in addition, highly electron withdrawing. The adsorption of fluorine leads to an induction effect whereby the Si—Si bonds, in particular the backbonds of surface atoms to the underlying silicon atoms, are polarized. The induction effect weakens the Si—Si bonds and makes them susceptible to further chemical attack. Thus, the monofluoride-covered surface is unstable on kinetic grounds as the addition of a fluorine atom makes the surface more reactive. The addition of a second fluorine atom increases the inductive effect, making the addition of a third fluorine atom that much easier (Figure 5.8). The resulting $HSiF_3$ is liberated as the final etch product. In solution, this is hydrolized to H_2SiF_6, as silicon (hydrogen) fluorides are unstable in water.

From a purely thermodynamic viewpoint the system Si + HF is unstable. Although the reaction to form $SiH_4 + 2F_2$ is endothermic by $c.$ $70\,kJ\,mol^{-1}$, the reaction yielding $SiF_4 + 2H_2$ is exothermic by $c.$ $10\,kJ\,mol^{-1}$ and is accompanied by a highly favourable entropy factor. We might then wonder why the etching of silicon in HF stops instead of proceeding completely to SiF_4 until either all of the silicon or all of the fluorine is consumed. The answer is that a kinetic constraint stops the reaction. The kinetic constraint arises from the hydrogen termination of the surface. The origin of the hydrogen termination can be observed by inspection of Figure 5.8. In the first step, a surface hydrogen is replaced by a fluorine atom. This occurs preferentially at a defect site occupied by a dihydride species. In the two subsequent steps, HF adds to the silicon surface. The fluorine atom attaches to the silicon atom that already has fluorine bound to it. The hydrogen atom adds to the silicon atom that has no fluorine atoms bound to it. The fluorine atom preferentially attaches to the silicon with other fluorine atoms attached because this silicon atom is deficient in electrons. Since fluorine is a strong nucleophile it prefers to attach to this silicon atom rather than to the one bound in the lattice. The result is that the surface is left hydrogen-terminated after the etched silicon atom departs.

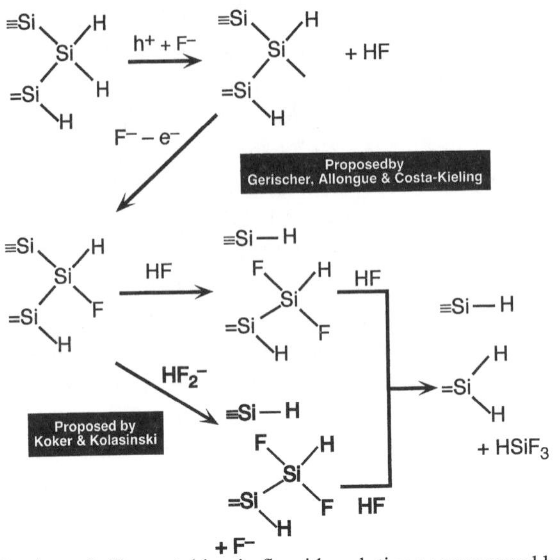

Figure 5.8 The mechanism of silicon etching in fluoride solutions as proposed by Gerischer, Allongue and Costa-Kieling [93] and as modified by Koker and Kolasinski [95]. Electrochemical or photo-assisted etching is initiated by hole capture at the surface. This is followed by rapid replacement of adsorbed hydrogen, $H_{(a)}$, by adsorbed fluorine, $F_{(a)}$, and slow addition of HF or HF_2^-. A final HF addition completes the etch cycle and leaves the surface hydrogen terminated

The Si–H bond is close to being nonpolar. The electrons are shared nearly equally and there is no polarization of backbonds induced by $H_{(a)}$. The nonpolar bond results in a hydrophobic surface. The surface relaxes in the presence of chemisorbed hydrogen such that the silicon atoms approach the bulk-terminated structure. Because of this lack of strain in the silicon lattice, the lack of dangling bonds and the nonpolar nature of the Si–H bond, the hydrogen-terminated silicon surface is chemically passivated. Under vacuum the surface remains clean indefinitely. Even when exposed to the atmosphere, the surface resists oxidation. While hydrocarbons physisorb readily on the hydrogen-terminated surface, water and O_2 take weeks to oxidize completely the hydrogen-terminated surface even at atmospheric pressure. It is this passivation effect that protects the silicon crystal, and the etch rate of silicon in $HF_{(aq)}$ in the dark is less than 1 Å min^{-1} even though the system is thermodynamically unstable.

The extreme passivation of perfect Si(1 1 1)-(1×1):H surface also explains why $HF_{(aq)}$ solutions of the proper pH can lead to the formation of nearly perfect hydrogen-terminated Si(1 1 1)-(1×1):H surfaces. The step sites have a much higher reactivity than the terraces. Therefore, etching occurs preferentially at the steps. The steps are stripped away whereas the terraces are virtually immune to pitting. Expansive ideal terraces are formed, the size of which are ultimately limited by the miscut of the crystal from the true (1 1 1) plane. Two essential characteristics of etching lead to ideal surfaces: (1) the initiation step is not random but starts preferentially at the step; (2) the etching is highly anisotropic, starting at the step and proceeding laterally instead of down into the surface.

The Si(1 1 1)-(1×1):H surface is a strain-free surface in which the surface atoms occupy positions close to the bulk-termination values. A Si(1 1 1) surface with the same degree of perfection cannot be produced. This can be understood based on adsorbate geometry and strain. There is no stable Si(1 0 0)-(1×1) dihydride phase that corresponds to the bulk structure with the two dangling bonds terminated by hydrogen atoms. In such a structure, the hydrogen atoms are so close to their neighbours that they interact repulsively. Because of this, the Si(1 0 0)-(1×1) dihydride surface is strained and reactive. UHV STM studies by Boland [96] have shown that although the low-energy electron diffraction (LEED) pattern from such a surface appears (1 × 1), the surface is actually rough with monohydride and dihydride units present. The (1 × 1) pattern corresponds to the diffraction from the ordered subsurface layers rather than to diffraction from an ordered dihydride phase, the disordered surface causing only diffuse scattering and an increase in the background of the LEED pattern.

If a silicon crystal is left for an extended period in $HF_{(aq)}$, the surface spontaneously roughens. A Si(1 0 0) surface develops pyramidal facets that have Si(1 1 1) faces. This is because the Si(1 1 1) surface is more stable than the Si(1 0 0) surface in relation to etching; effectively, the Si(1 1 1) surface is a survivor whereas the Si(1 0 0) surface is etched at a higher rate and disappears. Nonetheless, por-Si does not spontaneously form. Por-Si can be produced under appropriate electrochemical conditions [79, 97] or under laser irradiation [98, 99]. Essentially, the same etch chemistry is involved but now an extremely complex and nonplanar structure is formed. The difference between planar etching and por-Si formation can be traced back to the initiation step. In electrochemical and laser-assisted etching, a hole is driven to the surface and captured at the surface. Unlike the case of chemical etching, initiation occurs randomly across the surface and pitting of the terraces can occur. Moreover, once a pit has formed, the electronic structure of the silicon responds in a way that preferentially directs holes to the bottom of the pits rather than to the pit walls [100–102]. Subsequent initiation is at the bottom of the pore rather than at the sidewalls. Thus, once a pore is nucleated, it continues to propagate into the surface.

5.9.3   Selective area growth and etching

In the formation of integrated circuits (ICs) and microelectromechanical systems (MEMS), the substrate must be etched in a precise manner to create two-dimensional and three-dimensional structures. To achieve this, the surface must be patterned in some way so that its resistance to etching can be manipulated in a controlled manner. Lithography [103, 104] is a technique of pattern transfer into a substrate. The general strategy of lithography is that the substrate to be patterned is coated with a resist. The resist is a material that is sensitive to irradiation by photons, electrons or ions. In the lithographic step, the resist is irradiated to transfer the pattern. Irradiation is often performed through a stencil mask. However, it is also possible to perform maskless lithography in which the irradiating beam is moved across the surface to expose the resist in a predetermined pattern. In Section 5.8 we briefly mentioned how hydrogen can be used as a resist for maskless lithography using electron beam irradiation from an STM tip.

In conventional lithography, the resist is a thin polymer film that is applied from a solution onto the substrate by spin coating. Irradiation of the resist leads to chemical changes in the polymer, which lead to a change in the solubility between the irradiated and nonirradiated regions. In a negative resist such as poly(methylmethacrylate) (PMMA) or cyclized poly(*cis*-isoprene) with bisazide cross-linkers, irradiation leads to cross-linking of the polymer chains. The irradiated area is rendered insoluble in the next step, which is development in a solvent. A positive resist such as diazonaphthoquinone (DNQ)/novolak resins are insoluble. Irradiation transforms the hydrophobic DNQ into indenecarboxylic acid. This renders the exposed regions soluble in the development step. Various compounds have been used as resists. These and the issues involved in resist performance are treated by Gutmann [103]. Advanced lithographic techniques involving wavelengths below 250/nm require chemically amplified (CA) resists. Irradiation does not change the solubility of CA resists directly but it does release a chemical component into the thin film of resist. In a subsequent baking step, this component reacts with the thin film to form a region of altered solubility [104].

In the development step, the latent image from lithography is transformed into a real physical image. Development usually consists of introducing a solvent to the exposed resist. The chemical identity of the solvent depends on the resist. It may be an alkaline solution [NaOH, KOH or $N(CH_3)_4OH$] or a polar or nonpolar organic solvent (butyl-acetate, xylene or ethylpyruvate, etc.). The most important feature is that the solvent must selectively dissolve either only the exposed region (positive tone resist) or only the unexposed region (negative tone resist). The process of lithography and development is illustrated in Figure 5.9.

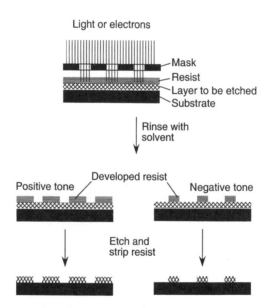

Figure 5.9 Lithographic pattern transfer. Irradiation through a mask transfers a pattern into the resist. Development removes the irradiated area of a positive tone resist, whereas the nonirradiated region is removed from a negative tone resist. Etching transfers the pattern into the layer beneath the resist

The substrate is now ready for selective area modification. One type of modification is selective area growth. For instance, if the patterned and unpatterned regions alternate between hydrophobic and hydrophilic tendencies, then Langmuir–Blodgett films and self-assembled monolayers (Chapter 6) can be used to transfer patterns of assorted organic molecules onto the surface. The grafting of controlled patches of biomolecules (proteins, peptides, DNA, etc.) onto silicon surfaces forms the basis of biochips used in immunoassays and the study of genomics and proteomics.

Chemical vapour deposition can be used to transfer inorganic films into the patterned regions. To be effective the patterned region must have a different reactivity from that of the unpatterned region. This is the basis of a technique for the selective area growth of carbon nanotubes that is explored in Chapter 6. In a resistless approach, iron is evaporated through a mask to transfer a pattern of catalytically active regions onto a silicon substrate. Alternatively, one can pattern a resist, develop it and then fill the developed regions with a solution containing an iron compound [105]. The iron compound is then activated by annealing to form patterned regions that are catalytically active for the production of carbon nanotubes.

After development, the resist may be called upon to function as a mask. There are two distinct ways in which the resist may act as a mask: (1) as a chemical mask, and (2) as an etch mask. In the first mode, the developed substrate is exposed to a reactive chemical environment. This could be oxygen, ammonia, an evaporated metal or a laser-ablated insulator such as SiO_2. The mask does not react with the chemicals but the bare substrate does. After reaction, the resist is removed with an appropriate solvent, leaving behind a substrate that is patterned with, for example, an oxide, nitride or metal. In the second mode, the substrate is exposed to a reactive chemical environment that can etch the surface, such as KOH (for silicon), ion radiation or a halogen-containing plasma. The resist is chosen such that its etch behaviour is opposite to that of the substrate. After removal of the resist, the substrate is left with a pattern of troughs. By successive alternation of various etching and growth steps, the intricate structures of ICs and MEMs are constructed. Selective area growth and etching allow for the incorporation of materials of various compositions into the final structure.

Advanced topic: silicon pillar formation

To illustrate these principles, we refer to Figure 5.10. First, a silicon wafer surface is coated with an unconventional 'resist' [106]. This resist consists of polystyrene spheres with a diameter of 500 nm dispersed in water. Upon drying of the solvent, the polystyrene spheres form a self-assembled closed-packed hexagonal array, as shown in Figure 5.10(a). This layer is then irradiated with evaporated metal atoms (e.g. silver or gold)). The metal atoms fill the interstices. Rinsing with chloroform or toluene then develops the film. This dissolves the polystyrene spheres, leaving an ordered array of metal clusters. More conventionally, the polystyrene spheres would be called a shadow mask, but the analogy to classical lithography should not be lost. The rinsing step develops a negative tone image.

The array of metal clusters can now be used as an etch mask. Reactive ion etching with use of a mixture of SF_6 and CF_4 preferentially etches the bare silicon surface, the clusters inhibiting the etching of the silicon beneath them. The result is the formation of an

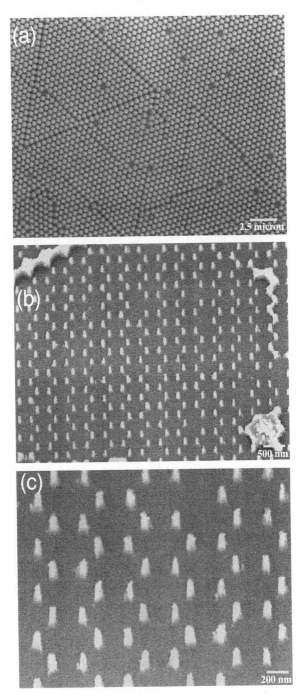

Figure 5.10 The creation of nanoscale silicon pillars via reactive ion etching. (a) Polystyrene spheres deposited from solution self-assemble into a close-packed lattice. Various vacancy and domain boundary defects are visible in the lattice. (b) After irradiation with metal vapour and reactive ion etching, the silicon pillars are formed. (c) A close-up demonstrates that the pillars are not only ordered but also exhibit a small size distribution. A. Wellner is thanked for providing these electron micrographs

ordered array of silicon nanopillars. The height and width of the pillars depends on the details of the lithographic and etching procedures. Heights of more than 100 nm and widths of less than 50 nm are readily achieved with this method, as seen in Figures 5.10(b) and 5.10(c).

5.10   Summary of Important Concepts

- The pressure gap and materials gap are not intrinsic barriers to understanding catalysis on the basis of UHV surface studies.

- Some catalytic reactions exhibit kinetics that depends strongly on the presence of certain types of sites or crystallographic planes. These are structure-sensitive reactions. Particularly reactive sites are known as active sites.

- Structure-insensitive reactions have kinetic parameters that are similar across a range of sites.

- The kinetics of a series of reactions is often controlled by a single reaction – the rate-determining step.

- The rate of an elementary step (or an irreversible one-step reaction) is given by

$$r = k[\theta_A^n \theta_B^m \ldots] \tag{5.13}$$

  whereas the rate of a reversible elementary reaction is given by the net rate of reaction

$$r = k_1[\theta_A^m \theta_B^n \ldots] - k_{-1}[\theta_A^{m'} \theta_B^{n'} \ldots] \tag{5.14}$$

  where k_1 is the rate constant of the forward reaction and k_{-1} that of the reverse reaction.

- An accurate count of empty sites must be taken into account in the kinetics of surface reactions.

- Both activity and selectivity are important characteristics of industrial catalysts.

- A practical catalyst is usually composed of an active phase and a support that helps to maintain the catalytic material in the active phase.

- A promoter enhances the reactivity and/or selectivity of the active phase either by electronic effects or by direct interactions with the reactants.

- Poisons reduce activity either through site blocking or through electronic effects.

- Poisons are sometimes strategically added to practical catalysts.

- Nonlinearities in surface kinetics can lead to rate oscillations and spatiotemporal pattern formation.

- Etching involves reactions that consume the surface.

- The region of the surface that is etched can be controlled by the use of a mask.

- In a positive tone resist the exposed portion of the resist is removed during development whereas in a negative resist, the unexposed region is removed.

Exercises

5.1   Show that the following expressions [5] hold for the equilibrium surface coverages during ammonia synthesis:

$$\theta(N_2*) = K_1 \frac{p_{N_2}}{p_0} \theta_* \tag{5.15}$$

$$\theta(N*) = \frac{p_{NH_3} p_0^{0.5}}{K_3 K_4 K_5 K_6 K_7^{1.5} p_{H_2}^{1.5}} \theta(*) \tag{5.16}$$

$$\theta(NH*) = \frac{p_{NH_3}}{K_4 K_5 K_6 p_{H_2}} \theta(*) \tag{5.17}$$

$$\theta(NH_2) = \frac{p_{NH_3}}{K_5 K_6 K_7^{0.5} p_{H_2}^{0.5} p_0^{0.5}} \theta(*) \tag{5.18}$$

$$\theta(NH_3) = \frac{p_{NH_3}}{K_6 p_0} \theta_* \tag{5.19}$$

$$\theta(H*) = K_7^{0.5} \frac{p_{H_2}^{0.5}}{p_0^{0.5}} \theta(*) \tag{5.20}$$

5.2   Consider the dissociative adsorption of N_2 through a molecularly bound state:

$$N_{2(g)} + * \rightleftharpoons N_2 - * \tag{5.35}$$
$$N_2 - * + * \rightleftharpoons 2N - * \tag{5.36}$$

Show that the rate of formation of adsorbed nitrogen atoms is

$$\frac{d\theta(N*)}{dt} = 2k_2 K_1 \frac{p_{N_2}}{p_0} [1 - \theta(N*)]^2 \left[1 + K_1 \frac{p_{N_2}}{p_0}\right]^{-2} - \frac{2k_2}{K_2} [\theta(N*)]^2 \tag{5.27}$$

Assume that the system is at equilibrium.

5.3   Calculate the equilibrium $H_{(a)}$ coverage on an iron surface for $p_{H_2} = 225$ bar and $T = 723$ K. Calculate the same for $N_{(a)}$ for $p_{N_2} = 225$ bar and $T = 723$ K. How do these compare with coverage on a practical ammonia catalyst? Discuss the effect of the reaction on the equilibrium coverages. Take $q_{ads}(H_2) = 100$ kJ mol^{-1} and $q_{ads}(N_2) = 240$ kJ mol^{-1}.

5.4 A number of heterogeneous reactions, including NH_3 formation, exhibit a maximum in the rate in the middle of the row of transition metals [33]. Plots of reactivity against periodic table group number with such a maximum are called volcano plots because of their shape [38]. Discuss the origin of this trend for ammonia synthesis.

5.5 Calculate algebraically the effective activation energy for N_2 dissociation on Fe(1 1 1).

5.6 The following results are observed in co-adsorption studies of CO, C_6H_6 or PF_3 with potassium on Pt(1 1 1) [107]: (1) $CO + K/Pt(1 1 1)$: the CO—Pt bond is significantly strengthened; (2) C_6H_6, which adsorb molecularly at low surface temperature, T_s, is less likely to dissociated upon heating in the presence of co-adsorbed potassium; (3) the amount of PF_3 that can adsorb drops roughly linearly with $\theta(K)$.

(a) Explain these observations.

(b) If the rate-determining step of a catalytic reaction is

(A) CO dissociation

(B) C_6H_6 desorption

how does the presence of potassium affect the reaction rate?

5.7 The threeway catalytic converter used in automobiles catalyses the oxidation of unburned hydrocarbons and CO while reducing simultaneously NO to N_2. Consider only the reactions of H_2 and NO. (Reactions [5.26] and [5.28]). The desired products are N_2 and H_2O. NH_3 formation must be suppressed. Write out a complete set of elementary reactions for which H_2O, N_2 and NH_3 are the final products. Pinpoint particular elementary steps that are decisive for determining the selectivity of the catalyst for N_2 over NH_3.

5.8 A reaction is carried out on a single-crystal Pt(1 1 1) sample. The reaction requires the scission of a C—H bond. The reaction is completely poisoned by just a few percent of a monolayer of preadsorbed oxygen at low temperature, but, at high temperature, oxygen coverage up to about a tenth of a monolayer has little effect on the reaction rate. Discuss the poisoning behaviour of preabsorbed oxygen.

5.9 The reaction

$$NO + CO \rightarrow \tfrac{1}{2}N_2 + CO_2 \qquad\qquad [5.37]$$

is a useful reaction in an automotive catalyst. It follows Langmuir–Hinshelwood kinetics. Consider the following to be a complete set of reactions for this system:

$$NO_{(g)} \leftrightharpoons NO_{(a)} \qquad\qquad [5.38]$$

$$CO_{(g)} \leftrightharpoons CO_{(a)} \qquad\qquad [5.39]$$

$$NO_{(a)} \leftrightharpoons N_{(a)} + O_{(a)} \qquad\qquad [5.40]$$

$$2N_{(a)} \leftrightharpoons \tfrac{1}{2}N_{2(g)} \qquad\qquad [5.41]$$

$$CO_{(a)} + O_{(a)} \rightarrow CO_{2(g)} \qquad\qquad [5.42]$$

(a) A trace of sulfur weakens the NO—surface bond. If sulfur co-adsorption substantially affects only $NO_{(a)}$, explain what the effect of the presence of a trace of $S_{(a)}$ has on the rate of production of N_2 and CO_2.

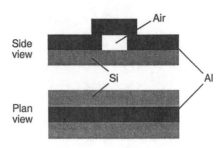

Figure 5.11   Structure required: see Exercise 5.11

(b) Assume that potassium co-adsorption promotes N_2 dissociation but affects no other reaction. Does potassium co-adsorption change the rate of CO_2 production? Explain.

(c) NO dissociates only on the steps of Pt(1 1 1). Assume that the resulting oxygen atoms are immobile at 770 K – the reaction temperature in a typical automatic catalytic converter. Given that CO desorption is first order and characterized by an activation energy, E_{des} of 135 kJ mol^{-1}, and an Arrhenius pre-exponential factor A, 3.00×10^{13} s^{-1} and that CO diffusion is characterized by an activation E_{dif}, of 19.8 kJ mol^{-1}, and a diffusion pre-exponential factor, D_0, of 5.00×10^{-6} cm^2 s^{-1}, would it be wise to use Pt(1 1 1) single crystals with terrace widths of 3500 Å to eliminate CO from the exhaust gas? Justify your answer with calculations.

5.10   For methanation of CO over nickel, CO dissociation is not the rate-determining step. Potassium co-adsorption increases the CO dissociation rate. Explain how in this case potassium acts as a poison.
Hint: consider the net effect on H_2 adsorption [108].

5.11   Design a scheme involving lithography, deposition and etching that will create the structure shown in Figure 5.11.
Hint: the etch rate of SiO_2 is so high compared with that of crystalline silicon (c-Si) that a sacrificial layer of SiO_2 can be removed by $HF_{(aq)}$ without attacking the c-Si.

Further Reading

M. Boudart and G. Djéga-Mariadassou, *Kinetics of Heterogeneous Catalytic Reactions* (Princeton University Press, Princeton, NJ, 1984).

B. C. Gates and H. Knözinger (eds), *Impact of Surface Science on Catalysis* (Academic Press, Boston, 2001).

D. A. King and D. P. Woodruft (eds), *The Chemical Physics of Solid Surfaces and Heterogeneous Catalysis: Fundamental Studies of Heterogeneous Catalysis, Volume 4*, (Elsevier, Amsterdam, 1982).

J. H. Larsen and I. Chorkendorff, "From fundamental studies of reactivity on single crystals to design catalyst", *Surf. Sci. Rep.* **35** (1999) 163.

S.J. Lombardo and A. T. Bell, "A review of theoretical models of adsorption, diffusion, desorption, and reaction of gases on metal surfaces", *Surf. Sci. Rep.* **13** (1991) 1.

I. Mochida and K. Sakanishi, "Catalysis in coal liquefaction", in *Advances in Catalysis, Volume 40*, eds D. D. Eley, H. Pines and W. O. Haag (Academic Press, Boston, MA, 1994), p. 39.

B. E. Nieuwenhuys "Toward understanding automotive exhaust conversion catalysis at the atomic level", in *Advances in Catalysis, Volume 44*, eds. W. O. Haag, B. C. Gages and H. Knözinger (Academic Press, Boston, MA, 1999), p. 260.

J. K. Nørskov, "Electronic Factors in Catalysis", *Prog. Surf. Sci.* **38** (1991) 103.

R. Pearce and M. V. Twigg, "Coal- and natural gas-based chemistry", in *Catalysis and Chemical Processes*, eds R. Pearce and W. R. Patterson, (Blackie & Sons, Glasgow, 1981), p.114.

G. A. Somorjai, *Introduction to Surface Chemistry and Catalysis* (John Wiley, New York, 1994).

P. Stoltze, "Microkinetic simulation of catalytic reactions", *Prog. Surf. Sci.* **65** (2000) 65.

J. M. Thomas and W. J. Thomas, *Principles and Practice of Heterogeneous Catalysis* (VCH, Weinheim, 1996).

G. M. Wallraff and W. D. Hinsberg, "Lithographic imaging techniques for the formation of nanoscopic features", *Chem. Rev.* **99**, 1801 (1999).

H. F. Winters and J. W. Coburn, "Surface science aspects of etching reactions", *Surf. Sci. Rep.* **14** (1992) 161.

References

[1] G. A. Somorjai, *Chem. Rev.* **96** (1996) 1223.

[2] T. Engel and G. Ertl, "Oxidation of carbon monoxide", in *The Chemical Physics of Solid Surfaces and Heterogeneous Catalysis: Fundamental Studies of Heterogeneous Catalysis*, Volume 4, eds D. A. King and D. P. Woodruff (Elsevier, Amsterdam, 1982) p. 73.

[3] T. Engel and G. Ertl, *Adv. Catal.* **28** (1979) 1.

[4] M. Grunze, "Synthesis and decomposition of amonia", *The Chemical Physics of Solid Surfaces and Heterogeneous Catalysis: Fundamental Studies of Heterogeneous Catalysis* Volume 4, eds D. A. King and D. P. Woodruff (Elsevier, Amsterdam, 1982) p. 143.

[5] P. Stoltze and J. K. Nørskov, *J. Catal.* **110** (1988) 1.

[6] D. W. Goodman, *Chem. Rev.* **95** (1995) 523.

[7] P. Stoltze and J. K. Nørskov, *Phys. Rev. Lett.* **55** (1985) 2502.

[8] D. W. Goodman and J. E. Houston, *Science* **236** (1987) 403.

[9] S. M. Davis and G. A. Somorjai, "Hydrocarbon conversion over metal catalysts", in *The Chemical Physics of Solid Surfaces and Heterogeneous Catalysis: Fundamental Studies of Heterogeneous Catalysis*, Volume 4, eds D. A. King and D. P. Woodruff (Elsevier, Amsterdam, 1982) p. 217.

[10] K. J. Laidler, *Chemical Kinetics* (HarperCollins, New York, 1987).

[11] K. W. Kolasinski, F. Cemič, A. de Meijere and E. Hasselbrink, *Surf. Sci.* **334** (1995) 19.

[12] K. W. Kolasinski, F. Cemič and E. Hasselbrink, *Chem. Phys. Lett.* **219** (1994) 113.

[13] J. T. Yates Jr and K. Kolasinski, *J. Chem. Phys.* **79** (1983) 1026.

[14] H. S. Taylor, *Proceedings of the Royal Society of London A* **108** (1925) 105.

[15] G. A. Somorjai, *Adv. Catal.* **26** (1977) 1.

[16] T. Zambelli, J. Wintterlin, J. Trost and G. Ertl, *Science* **273** (1996) 1688.

[17] H. Over, Y. D. Kim, A. P. Seitsonen, S. Wendt, E. Lundgren, M. Schmid, P. Varga, A. Morgante and G. Ertl, *Science* **287** (2000) 1474.

[18] Z.-P. Liu, P. Hu and A. Alavi, *J. Chem. Phys.* **114** (2001) 5956.

[19] J. A. Schwarz and R. J. Madix, *Surf. Sci.* **46** (1974) 317.

[20] M. P. D'Evelyn and R. J. Madix, *Surf. Sci. Rep.* **3** (1983) 413.

[21] C. R. Arumainayagam and R. J. Madix, *Prog. Surf. Sci.* **38** (1991) 1.

[22] D. F. Padowitz and S. J. Sibener, *J. Vac. Sci. Technol. A* **9** (1991) 2289.

[23] D. F. Padowitz and S. J. Sibener, *Surf. Sci.* **254** (1991) 125.

[24] R. H. Jones, D. R. Olander, W. J. Siekhaus and J. A. Schwarz, *J. Vac. Sci. Technol.* **9** (1972) 1429.

[25] D. R. Olander and A. Ullman, *Int. J. Chem. Kinet.* **8** (1976) 625.

[26] D. R. Olander, *J. Colloid Interface Sci.* **58** (1977) 169.

[27] H. H. Sawin and R. P. Merrill, *J. Vac. Sci. Technol.* **19** (1981) 40.

[28] J. M. Thomas and W. J. Thomas, *Principles and Practice of Heterogeneous Catalysis* (VCH, Weinheim, 1996).

[29] C. J. H. Jacobsen, *Chem. Commun.* (2000) 1057.

[30] G. Ertl, "Reaction mechanisms in catalysis by metals", in *Critical Reviews in Solid State and Materials Science*, Volume 10 (CRC Press, Boca Raton, FL, 1982) p. 349.

[31] N. Brønsted, *Chem. Rev.* **5** (1928) 231.

[32] M. G. Evans and N. P. Polanyi, *Trans. Faraday Society* **34** (1938) 11.

[33] A. Logadottir, T. H. Rod, J. K. Nørskov, B. Hammer, S. Dall and C. J. H. Jacobsen, *J. Catal.* **197** (2001) 229.

[34] F. Bozso, G. Ertl and M. Weiss, *J. Catal.* **50** (1977) 519.

[35] T. H. Rod, A. Logadottir and J. K. Nørskov, *J. Chem. Phys.* **112** (2000) 5343.

[36] G. Ertl, *Angew. Chem. Int. Ed. Engl.* **29** (1990) 1219.

[37] M. Boudart, *Adv. Catal.* **20** (1969) 153.

[38] M. Boudart and G. Djéga-Mariadassou, *Kinetics of Heterogeneous Catalytic Reactions* (Princeton University Press, Princeton, NJ, 1984).

[39] D. D. Strongin, J. Carrazza, S. R. Bare and G. A. Somorjai, *J. Catal.* **103** (1987) 213.

[40] N. D. Spencer, R. C. Schoonmaker and G. A. Somorjai, *J. Catal.* **74** (1982) 129.

[41] F. Bozso, G. Ertl, M. Grunze and M. Weiss, *J. Catal.* **49** (1977) 18.

[42] G. Ertl, S. B. Lee and M. Weiss, *Surf. Sci.* **114** (1982) 515.

[43] C. T. Rettner and H. Stein, *J. Chem. Phys.* **87** (1987) 770.

[44] C. T. Rettner, H. E. Pfnür, H. Stein and D. J. Auerbach, *J. Vac. Sci. Technol. A* **6** (1988) 899.

[45] C. T. Rettner and H. Stein, *Phys. Rev. Lett.* **59** (1987) 2768.

[46] J. J. Mortensen, L. B. Hansen, B. Hammer and J. K. Nørskov, *J. Catal.* **182** (1999) 479.

[47] G. Ertl, S. B Lee and M. Weiss, *Surf. Sci.* **114** (1982) 527.

[48] M. W. Chase, *JANAF thermochemical tables* (National Institute of Standards and Technology, Gaithersburg, MD, 1998).

[49] R. Pearce and M. V. Twigg, "Coal- and natural gas-based chemistry", in *Catalysis and Chemical Processes*, eds R. Pearce and W. R. Patterson (Blackie & Sons, Glasgow, 1981) p. 114.

[50] H. Topsøe, C. V. Qvesen, G. S. Clausen, N.-Y. Topsøe, P. E. Højlung Nielsen, E. Tørnqvist and J. K. Nørskov, "Importance of dynamics in real catalyst systems", in *Dynamics of Surfaces and Reaction Kinetics in Heterogeneous Catalysis*. eds G. F. Froment and K. C. Waugh (Elsevier, Amsterdam, 1997) p. 121.

[51] P. M. Maitlis, H. C. Long, R. Quyoum, M. L. Turner and Z.-Q. Wang, *Chem. Commun.* (1996) 1.

[52] B. E. Mann, M. L. Turner, R. Quyoum, N. Marsih and P. M. Maitlis, *J. Am. Chem. Soc.* **121** (1999) 6497.

[53] K. C. Taylor, "Automobile catalytic converters", in *Catalysis: Science and Technology*, Volume 5, eds J. R. Anderson and M. Boudart (Springer, Berlin, 1984) p. 119.

[54] C. A. Hampel and G. G. Hawley, *The Encyclopedia of Chemistry* (Van Nostrand Reinhold, New York, 1973).

[55] P. J. Feibelman and D. R. Hamann, *Surf. Sci.* **149** (1985) 48.

[56] J. K. Nørskov, S. Holloway and N. D. Lang, *Surf. Sci.* **137** (1984) 65.

[57] J. J. Mortensen, B. Hammer and J. K. Nørskov, *Phys. Rev. Lett.* **80** (1998) 4333.

[58] J. K. Nørskov and P. Stoltze, *Surf. Sci.* **189/190** (1987) 91.

[59] S. C. Minne, S. R. Manalis, A. Atalar and C. F. Quate, *J. Vac. Sci. Technol. B* **14** (1996) 2546.

[60] T.-C. Shen, C. Wang, G. C. Abeln, J. R. Tucker, J. W. Lyding, P. Avouris and R. E. Walkup, *Science* **268** (1995) 1590.

[61] J. W. Lyding, T.-C. Shen, J. S. Hubacek, J. R. Tucker and G. C. Abeln, *Appl. Phys. Lett.* **64** (1994) 2010.

[62] F. Besenbacher, I. Chorkendorff, B. S. Clausen, B. Hammer, A. M. Molenbroek, J. K. Nørskov and I. Stensgaard, *Science* **279** (1998) 1913.

[63] P. Hugo, *Ber. Bunsenges. Phys. Chem.* **74** (1970) 121.

[64] R. Imbihl and G. Ertl, *Chem. Rev.* **95** (1995) 697.

[65] R. Imbihl, M. P. Cox, G. Ertl, G. Müller and W. Brenig, *J. Chem. Phys.* **83** (1985) 1578.

[66] M. P. Cox, G. Ertl and R. Imbihl, *Phys. Rev. Lett.* **54** (1985) 1725.

[67] H. Haberland, "Clusters of Atoms and Molecules", in *Springer Series in Chemical Physics, Volume 52* (Springer, Berlin, 1993).

[68] H. Haberland, "Clusters of Atoms and Molecules II", in *Springer Series in Chemical Physics, Volume 56* (Springer, Berlin, 1994).

[69] G. Scoles, *Atomic and Molecular Beam Methods*, Volume 1 (Oxford University Press, New York, 1988).

[70] B. C. Gates, *Chem. Rev.* **95** (1995) 511.

[71] U. Heiz and W.-D. Schneider, *J. Phys. D* **33** (2000) R85.

[72] M. F. Jarrold, *Science* **252** (1991) 1085.

[73] M. L. Gorbaty, F. J. Wright, R. K. Lyon, R. B. Long, R. H. Schlosberg, Z. Baset, R. Liotta, B. G. Silbemagel and D. R. Neskora, *Science* **206** (1979) 1029.

[74] D. V. Charkarov, L. Österlund and B. Kasemo, *Langmuir* **11** (1995) 1201.

[75] D. R. Olander, W. Siekhaus, R. Jones and J. A. Schwarz, *J. Chem. Phys.* **57** (1972) 408.

[76] D. R. Olander, R. H. Jones, J. A. Schwarz and W. J. Siekhaus, *J. Chem. Phys.* **57** (1972) 421.

[77] M. Balooch and D. R. Olander, *J. Chem. Phys.* **63** (1975) 4772.

[78] H. Chang and A. J. Bard, *J. Am. Chem. Soc.* **113** (1991) 5588.

[79] R. L. Smith and S. D. Collins, *J. Appl. Phys.* **71** (1992) R1.

[80] Y. J. Chabal, A. L. Harris, K. Raghavachari and J. C. Tully, *Int. J. Mod. Phys. B* **7** (1993) 1031.

[81] L. A. Jones, G. M. Taylor, F.-X. Wei and D. F. Thomas, *Prog. Surf. Sci.* **50** (1995) 283.

[82] A. G. Cullis, L. I. Canham and P. D. J. Calcott, *J. Appl. Phys.* **82** (1997) 909.

[83] G. J. Pietsch, *Appl. Phys. A: Mater. Sci. Process.* **60** (1995) 347.

[84] R. I. Hegde, M. A. Chonko and P. J. Tobin, *J. Vac. Sci. Technol. B* **14** (1996) 3299.

[85] T. Ohmi, M. Miyashita, M. Itano, T. Imaoka and I. Kawanabe, *IEEE Trans. Electron Devices* **39** (1992) 537.

[86] L. T. Canham, *Appl. Phys. Lett.* **57** (1990) 1046.

[87] G. S. Higashi, Y. J. Chabal, G. W. Trucks and K. Raghavachari, *Appl. Phys. Lett.* **56** (1990) 656.

[88] P. Jakob and Y. J. Chabal, *J. Chem. Phys.* **95** (1991) 2897.

[89] P. Jakob, P. Dumas and Y. J. Chabal, *Appl. Phys. Lett.* **59** (1991) 2968.

[90] P. Dumas, Y. J. Chabal and P. Jakob, *Surf. Sci.* **269/270** (1992) 867.

[91] P. Jakob, Y. J. Chabal, K. Raghavachari, R. S. Becker and A. J. Becker, *Surf. Sci.* **275** (1992) 407.

[92] G. W. Trucks, K. Raghavachari, G. S. Higashi and Y. J. Chabal, *Phys. Rev. Lett.* **65** (1990) 504.

[93] H. Gerischer, P. Allongue and V. Costa-Kieling, *Ber. Bunsenges, Phys. Chem.* **97** (1993) 753.

[94] E. S. Kooij and D. Vanmaekelbergh, *J. Electrochem. Soc.* **144** (1997) 1296.

[95] L. Koker and K. W. Kolasinski, *J. Phys. Chem. B* **105** (2001) 3864.

[96] J. J. Boland, *Adv. Phys.* **42** (1993) 129.

[97] A. Uhlir, *Bell Syst. Tech. J.* **35** (1956) 333.

[98] N. Noguchi and I. Suemune, *Appl. Phys. Lett.* **62** (1993) 1429.

[99] L. Koker and K. W. Kolasinski, *Phys. Chem. Chem. Phys.* **2** (2000) 277.

[100] M. I. J. Beale, J. D. Benjamin, M. J. Uren, N. G. Chew and A. G. Cullis, *J. Crystal Growth* **73** (1985) 622.

[101] R. T. Collins, M. A. Tischler and J. H. Stathis, *Appl. Phys. Lett.* **61** (1992) 1649.

[102] S. Frohnhoff, M. Marso, M. G. Berger, M. Thönissen, H. Lüth and H. Münder, *J. Electrochem. Soc.* **142** (1995) 615.

[103] A. Gutmann, "Photolithography", in *Electronic Materials Chemistry*, ed. H. B. Pogge (Marcel Dekker, New York, 1996) p. 199.

[104] G. M. Wallraff and W. D. Hinsberg, *Chem. Rev.* **99** (1999) 1801.

[105] H. T. Soh, C. F. Quate, A. F. Morpurgo, C. M. Marcus, J. Kong and H. Dai, *Appl. Phys. Lett.* **75** (1999) 627.

[106] K. Seeger and R. E. Palmer, *J. Phys. D* **32** (1999) L129.

[107] E. L. Garfunkel, J. J. Maj, J. C. Frost, M. H. Farias and G. A. Somorjai, *J. Phys. Chem.* **87** (1983) 3629.

[108] C. T. Campbell and D. W. Goodman, *Surf. Sci.* **123** (1982) 413.

6

Growth and Epitaxy

Up to this point we have concentrated on adsorption at the level of one monolayer or less. We need to go beyond thinking about only a single monolayer. What happens as the coverage approaches and then exceeds one monolayer? We have seen that strong chemisorption often leads to ordered overlayers. Particularly at high coverage, chemisorbates often take on structures that are related to the structure of the substrate. The factors that influence the order of a monolayer are the strength of the adsorption interaction, the strength of lateral interactions, the relative strength of adsorption versus lateral interactions, corrugation in the adsorbate/substrate interaction potential and mobility in the layer. In short, we need to know how the adsorbate fits into the template of the substrate and how it gets there.

When we consider the adsorption of a second layer on top of the first, we again need to ask how the subsequent layers fit onto the layers below them. In order to quantify this fit and to understand the different modes of layer growth, we need to understand strain and surface tension.

6.1 Stress and Strain

Consider two separate single crystal solids, A and B, taken to be two Si(1 1 1) surfaces. If two unreconstructed Si(1 1 1) surfaces are brought into contact, the dangling bonds match up perfectly and, once the new bonds between the two surfaces are formed, the two crystals form an interface that is indistinguishable from the rest of the crystal. The interface is perfectly matched geometrically. If we were to bring two unreconstructed (1 1 1) surfaces of silicon and germanium into contact they would not match. The lattice constant of germanium is larger than that of silicon – 5.658 Å compared with 5.43 Å. The lattice mismatch, ε, expressed in terms of the lattice constants a_A and a_B of the two materials, is

$$\varepsilon = \frac{a_A - a_B}{a_B} \tag{6.1}$$

For germanium (A) on silicon (B) this amounts to 0.042. If we force the two surfaces to form bonds, these bonds are highly perturbed and this perturbation must be relaxed in some way if the crystal is to lower its energy. Far from the interface, the two pure materials have the structures of the pure materials, but what does the new interface look like? Will the perturbation be relaxed gradually? Will the interface assume the geometry of silicon, germanium or something completely new?

A stress is a force per unit area. Imagine an arbitrary interface drawn through a body. Three types of forces can exist at the interface. If the material on either side of the interface would prefer to expand into the neighbouring region, the stress is called compression. A compressive force pushes on the interface. If the material on either side of the interface has the tendency to contract away from the interface, the stress is called tension. A tensile forces pulls away from the interface. Both compression and tension are directed perpendicular to the interface. A force acting parallel to the interface is called a shearing stress.

When a stress acts upon a nonrigid lattice, the lattice is deformed. The deformation of a stressed material is called strain. Strain depends on stress in both character and magnitude. Each type of stress leads to a certain type of strain. For instance, compression or tension will lead to changes in the length of a wire, whereas a change in pressure on a three-dimensional solid leads to a volume change. Stress is defined by

$$\text{stress} = \frac{F}{A} \tag{6.2}$$

where F is force, and A is area. If the stress were applied to a wire, changing the length of the wire from l to $l + \Delta l$, the strain – compressive or tensile, depending on the sign of Δl – is given by

$$\text{strain} = \frac{\Delta l}{l} \tag{6.3}$$

Hooke's law, which is valid in the limit of not too large a stress, states that stress is proportional to strain for solid materials. This allows us to define the Young's modulus for the case of tension as

$$Y = \frac{\text{stress}}{\text{strain}} = \frac{Fl}{A\Delta l} \tag{6.4}$$

The amount of strain at the interface of two materials, A and B, depends on several factors. The relative sizes of the two materials, that is the lattice constants, are important factors. If the two atoms have the same size then they fit together. If one atom is much larger than the other then even if the two materials have the same lattice symmetry they do not fit into a lattice of the same size. The lattice symmetry of the two materials must be taken into account as well. Something obviously has to give if one tries to fit a hexagonal close-packed (hcp) lattice onto a diamond lattice. The relative strengths of the A−A, A−B and B−B interactions as well as the temperature can be decisive. If two materials have different coefficients of thermal expansion, the stress experienced at the interface is temperature dependent.

If germanium is deposited on a Si(1 1 1)-(1×1) surface and assumes a (1×1)-overlayer structure, the overlayer is subjected to a large compressive stress. The overlayer is strained with respect to bulk germanium because the Ge−Ge distance in the overlayer is much smaller than the corresponding bulk value. Below we will investigate how stress and strain relaxation affect the properties of interfaces.

6.2 Types of Interfaces

Before considering the dynamics of growth, let us consider how strain relaxation affects the types of interfaces that are formed. These are depicted in Figure 6.1. If there is no mixing of the two materials, a sharp interface is formed [Figure 6.1(a)]. When mixing occurs a nonabrupt interface is formed [Figure 6.1(b)]. The mixing may take two different forms. If one of the materials is soluble in the other, then it can diffuse into the other material and create a region of variable composition. The dissolved material may act as a substitutional impurity or reside in interstitial sites. If A and B can form a stable compound, a reactive interface is formed [Figure 6.1(c)]. In this case, a new compound is formed between the two pure phases; hence, two interfaces rather than one interface are created by the growth process. When growth occurs on a crystalline substrate – a type of growth that is of particular interest for semiconductor processing – the overlayer may grow either in a crystalline state or in an amorphous state. A special case of growth is epitaxial growth, in which the overlayer takes on the same structure as the substrate and grows in registry with the substrate. Such films are called epitaxial layers or epilayers.

The situation shown in Figure 6.1 is often oversimplified in that the transition from one material may be abrupt from the standpoint of composition; nonetheless, from the standpoint of structure, interfaces are quite often found to be more slowly varying beasts. By this I mean that when making the transition from A to B the composition may change from pure A to pure B within one or two atomic layers. The structure, on the other hand, changes from the structure of pure A to the structure of pure B over a much longer length scale. This is frequently the case for materials of similar structure but with different lattice constants. The long-range change of structure allows for a gradual easing of the strain associated with the formation of an interface between two materials with lattices that do not match exactly.

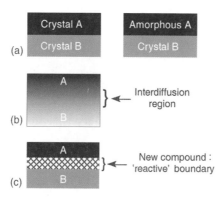

Figure 6.1   The variety of interfaces that can be formed at the interface of two materials: (a) sharp interface; (b) nonabrupt interface; (c) reactive interface

The type of interface obtained depends not only on the materials involved but also on the *manner in which they are grown*; in other words, the morphology and composition of layered structures depends on the balance between kinetics and thermodynamics. Many important structures are intentionally grown to be nonequilibrium or quasi-equilibrium structures. The reaction conditions – in particular the temperature, the rate of deposition and the composition of the gas phase – are regulated to grow interfaces of the desired structure. Because of the importance of the manner in which growth is performed we will investigate below the dynamics of growth and the techniques used to create designer films of controlled composition and thickness.

6.3 Surface-Energy and Surface Tension

6.3.1 *Liquid surfaces*

To help us understand the interactions that occur at growing interfaces, we need to define and utilize surface energy and surface tension. For a liquid, these two terms are synonymous and we use them interchangeably. For largely historical reasons, the term surface tension is commonly used to refer to solid surfaces. As we shall see below, this terminology is somewhat ambiguous for solids, therefore we confine the use of the term surface tension to liquids.

We begin by discussing liquid surfaces. Surface tension, γ, is defined by the relationship between the amount of work performed in enlarging a surface, δW^s, and the surface area created, dA. At constant temperature and pressure the amount of work done to expand reversibly a volume by an amount dV is

$$\delta W_{T,P} = PdV \tag{6.5}$$

By direct analogy, the amount of work required to expand reversibly a surface is given by

$$\delta W^s_{T,P} = \gamma dA \tag{6.6}$$

Hence, the surface tension is the surface analogue of pressure. Whereas three-dimensional pressure is perpendicular to the surface (the direction of expansion), the surface tension is parallel to the surface. Surface tension, unlike the pressure of an ideal gas, is a material-dependent property. Surface tension scales with the force required to expand the surface area of a material. The surface energy of a material in the solid state is larger than the surface energy of that material in the liquid phase. Surface energy tends to scale with the heat of sublimation. Typical values of surface energy range from 2.90×10^4 N m^{-1} for $W_{(s)}$ to 721.3 N m^{-1} for $H_2O_{(l)}$, to 3.65 N m^{-1} for $He_{(l)}$. Equivalent units are J m^{-2}. As confirmed in Table 6.1 [1], there is a general tendency for surface energy to decrease with increasing temperature.

The meaning of surface tension does not end with an analogy to two-dimensional pressure. The change in Gibbs energy upon a change in surface area for a pure liquid is given by

$$dG = -SdT + VdP + \gamma dA \tag{6.7}$$

Table 6.1   Selected values of surface tension or surface energy. Source: Adamson and Gast [1]

	$T(^\circ C)$	$\gamma(\text{mN m}^{-1})$
Liquid–vapour interface		
Perfluoropentane	20	9.89
Heptane	20	20.14
Ethanol	20	22.39
Methanol	20	22.50
Benzene	30	27.56
	20	28.88
Water	25	72.13
	20	72.94
Mercury	25	485.5
	20	486.5
Silver	1100	878.5
Copper	1083[a] (T_f)	1300
Platinum	1772[a]	1880

Note: T, temperature; γ, surface tension or surface energy
[a]At temperature of fusion.

At constant temperature and pressure, this reduces to

$$dG_{T,P} = \gamma dA \tag{6.8}$$

In a liquid the surface area is increased by adding atoms to the surface from the bulk, thus

$$dG_{T,P} = G^S dA = \gamma dA \tag{6.9}$$

where surface Gibbs energy, G^s, is equal to the surface tension:

$$G^s = \gamma \tag{6.10}$$

From Equation (6.10) we see that the surface Gibbs energy is always positive. In other words, the formation of a surface is an energetically unfavourable venture. Dynamically, this can be understood by considering that an increase of surface area by the addition of bulk atoms to the surface is accomplished by moving fully coordinated bulk atoms to surface sites of lower coordination. This process is energetically unfavourable both because the atom must be moved to the surface and because of the less favourable bonding environment at the surface compared with the bulk.

A system at equilibrium has attained a minimum value of the Gibbs energy. For a surface at a constant temperature, pressure and composition, the condition for minimum Gibbs energy is found by minimizing the surface Gibbs energy according to

$$\int \gamma(\mathbf{m}) dA = \text{minimum} \tag{6.11}$$

In Equation (6.11) we write the surface tension as a function of $\mathbf{m}$, a unit vector along the surface normal. This allows for any potential orientation dependence in the surface energy, as is essential for a solid. For a material in which γ is constant, that is, in which the surface energy does not depend on the surface orientation, the minimum of the integral in Equation (6.11) is obtained when the surface area of the material is minimized. Consequently, liquids and other isotropic materials contract into a sphere (in the absence of other forces such as gravity that can distort the particle shape) because a sphere has the minimum surface area.

To handle growth, we need to generalize Equation (6.7) to include the presence of more than one material. The Gibbs energy is then

$$dG = -SdT + VdP + \sum_i \gamma_i dA_i + \sum_j \mu_j dn_j \qquad (6.12)$$

for a system with i interfaces and n_j moles of j components of chemical potential μ_j. At constant temperatures and pressure

$$dG_{T,P} = \sum_i \gamma_i dA_i + \sum_j \mu_j dn_j \qquad (6.13)$$

and the equilibrium configuration of the system is obviously much more complex than a simple sphere. The system now attempts to maximize the areal fraction of low-surface-tension components while simultaneously maximizing the concentration of low-chemical-potential components. The system must balance the tendency toward minimum surface area with the drive to form substances with the lowest chemical potential.

6.3.2 Solid surfaces

For a liquid, the creation and the stretching of the surface are equivalent processes. The energetics are therefore governed by one parameter, γ, and surface energy (associated with surface creation) is identical to surface tension (associated with surface deformation). For a solid, these two processes are inequivalent because increased solid surface area can be achieved not only by adding atoms from the bulk but also by increasing the distance between surface atoms. Furthermore, deformations of a solid surface change the intrinsic properties of the surface, and this leads to a further differentiation between solid and liquid surfaces. These differences were first noted by Gibbs [2].

The energy required to create a unit of surface area is given by the surface-orientation-dependent scalar quantity $\gamma(\mathbf{m})$. This is the surface energy of a solid. The bonding of surface atoms depends on the crystallographic orientation of the surface, from which it follows that $\gamma(\mathbf{m})$ is, in fact, dependent upon the surface orientation. The crystalline structure of a surface means that the material can no longer be treated as isotropic, as we did for a liquid. The energy to move an atom depends on the direction in which the atom is moved. Therefore, the energy of deformation is no longer a scalar quantity. The intrinsic surface stress tensor, $\tau_{\alpha\beta}(\mathbf{m})$, describes the energy change upon deformation according to

$$\Delta E = \int \tau_{\alpha\beta}(\mathbf{m})\varepsilon_{\alpha\beta}\,dA \tag{6.14}$$

Equation (6.14) is valid for small deformations $\varepsilon_{\alpha\beta}$ in which the change in surface energy is linear in strain. Indices α and β are two-dimensional indices in the surface plane, that describe the direction of the deformation. Positive values of $\tau_{\alpha\beta}$ correspond to tensile stress, which is associated with a surface that tends to contract. Negative values of $\tau_{\alpha,\beta}$ correspond to compressive stress for which the surface favours expansion. As noted in Chapter 1, most surfaces exhibit tensile stress. An exception is the Si(1 0 0)-(2×1) surface that exhibits compressive stress perpendicular to the dimers [3]. For more on the tensor properties of the intrinsic surface stress tensor, see the review of Koch [4].

The presence of lattice strain leads to a dependence of the surface energy on the direction of the deformation. Including only a linear term in the intrinsic stress tensor we can write

$$\gamma(\mathbf{m}, \varepsilon_{\alpha\beta}) = \gamma_0(\mathbf{m}) + \tau_{\alpha\beta}(\mathbf{m})\varepsilon_{\alpha\beta} \tag{6.15}$$

A lattice mismatch between the substrate and an epilayer results in lattice strain. Thus, Equations (6.14) and (6.15) depict thermodynamic control of the growth mode and of the equilibrium structure of epitaxial layers.

Crystalline materials spontaneously form particles of particular shapes that reflect the symmetry of the lattice and that expose the surfaces that exhibit the lowest surface energies. Hence, cubic materials tend to form cubic crystallites whereas hexagonal materials have a tendency to pack themselves into crystallites with angled faces. Wulff first described the formation of single crystals with an equilibrium crystal shape (ECS) [5]. The exact thermodynamic formulation can be found elsewhere [6, 7].

Small crystallites supported on a substrate are particularly important in catalysis. In many cases, only small energy differences exist for crystallites that expose different crystallographic planes. Thus, a distribution of shapes and exposed crystal faces exists at equilibrium. The distribution of shapes affects the reactivity of a catalyst for structure-sensitive reactions such as methanol formation on Cu/ZnO_2. The Wulff construction is a method for determining the distribution of crystallite shapes [8].

6.4 Growth Modes

6.4.1 Liquid-on-solid growth

Consider the case in which one material, B, is deposited on a second, A, and there is no mixing between the two components. A lack of mixing may occur because the two components only form compounds that have a higher chemical potential than either of the two pure components. Alternatively, a lack of mixing can occur if there is a large activation barrier to the formation of a compound or to penetration of the deposited component into the bulk of the substrate. As long as deposition is carried out at a sufficiently low temperature, B cannot overcome the activation barrier and it stays on the surface.

Consider a liquid placed on top of a nondeformable solid. In other words, we neglect small changes in γ that may arise in Equation (6.13) because of liquid–solid interactions. As long as the interaction between the liquid and the substrate is not too strong, this is a good approximation. In Figure 6.2 we sketch a liquid on a substrate and use this to define the contact angle, ψ. At equilibrium the forces on the three components (substrate, liquid, gas phase) must be balanced. That is, the tangential components of the force exerted by the substrate–gas interface is equal and opposite to the forces exerted by the substrate–liquid interface and the liquid–gas interfaces. Balancing of the forces lead to the Young equation

$$\gamma_{\text{lg}} \cos \psi + \gamma_{\text{sl}} = \gamma_{\text{sg}} \tag{6.16}$$

where subscripts s, l and g relate to the substrate, liquid and gas, respectively.

We can define two interesting limits to Equation (6.16). When $\psi = 0°$, the deposit spreads across the surface and coats it uniformly – it wets the surface. This is the familiar situation of water on (clean) glass. For $\psi > 0°$, the deposit forms a droplet and does not wet the surface. This describes the interaction of mercury with glass, or water on grease-covered glass. In terms of the surface tensions, these two limits correspond to

$$\psi = 0° \Rightarrow \gamma_{\text{sg}} \geq \gamma_{\text{sl}} + \gamma_{\text{lg}} \tag{6.17}$$

$$\psi > 0° \Rightarrow \gamma_{\text{sg}} < \gamma_{\text{sl}} + \gamma_{\text{lg}} \tag{6.18}$$

Surface tension is a bulk property that does not give us an idea of the microscopic processes that lead to wetting or droplet formation. The key to wetting is the balance of the strengths of the A–A, A–B and B–B interactions. If the A–B interactions are strong compared with the B–B interactions, B would prefer to bond to the substrate rather than to itself. This leads to wetting. If, on the other hand, the B–B interactions are strong compared with the A–B interactions, B would prefer to stick to itself rather than to the substrate. In this case, B attempts to make a droplet to minimize the area of the A–B interface and maximize the number of B–B interactions. Accordingly, taking $-\Phi_{\text{AA}}$ and $-\Phi_{\text{AB}}$ as the energies of A–A and A–B bonds, respectively, we may define the energy difference

$$\Delta = \Phi_{\text{AA}} - \Phi_{\text{AB}}. \tag{6.19}$$

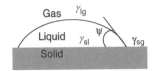

Figure 6.2 The contact angle, ψ, is defined by the tangent to the gas–liquid interface at the point of intersection with the solid. $\psi < 90°$ is indicative of attractive liquid–solid interactions, whereas $\psi > 90°$ signifies repulsive interactions. γ_{lg}, γ_{sg} and γ_{sl} are the surface energies at the liquid–gas, substrate–gas and substrate–liquid interfaces, respectively

According to Equation (6.19) we can restate the wetting condition as $\Delta < 0$. Equivalently, the non-wetting condition is $\Delta > 0$.

A glass surface is covered with $Si-OH$ moieties. These highly polar groups have a strong attractive interaction with polar molecules such as water. Hence water wets a glass surface. A greasy glass surface is coated with a hydrocarbon film. The nonpolar hydrocarbon interacts weakly with water and the interaction of hydrogen bonds between water molecules dominates over the water–hydrocarbon interaction. Hence, the water forms droplets to minimize the water–hydrocarbon interactions and maximize the water–water interactions.

As a thought experiment, consider a monolayer-thick uniform film of water covering a hydrophobic substrate. We now let the system evolve to equilibrium. The water is attracted to itself and repelled by the substrate. In order to minimize the contact area between water and the substrate, the film spontaneously breaks up into three-dimensional islands. Early on, the formation of islands is a random process. Some accrete material faster than do others, thus producing a broad distribution of island sizes.

The exact shape of the islands depends on the relative values of the substrate and liquid surface energies. In the limit of extremely repulsive liquid–substrate interactions, the equilibrium island shape would be a sphere. For this example we assume the islands are hemispherical. The nonwetting nature of the interaction means that it costs more energy to create a liquid–solid interface than it does to create the vapour–liquid interface. Furthermore, since the vapour–liquid surface area increases more rapidly with island radius than does the contact area, small islands are unstable with respect to large islands.

The relative instability of small islands with respect to large islands means that large islands grow at the expense of small islands. Given no mass transport limitations an equilibrium configuration of the system is attained in which a single large hemispherical island is formed. If, however, during island growth, the mean distance between islands exceeds the characteristic diffusion length of the liquid molecules, then the islands lose communication with one another and a quasi-equilibrium (metastable steady-state) structure of relatively few large islands is attained. The diffusion rate, the amount of material deposited and the rate of island nucleation determine the exact size distribution. The process of large-island growth at the expense of small-island growth is known as Ostwald ripening. Inherently, Ostwald ripening leads to a broad size distribution and no preferred or optimum size is achieved unless all the material is able to form just one big island.

6.4.2 Solid-on-solid growth

Life without stress would be easy. The same applies to a description of growth phenomena at equilibrium. Equations (6.17) and (6.18) would seem to indicate that there are only two modes of growth. However, when investigated on a microscopic level, the three growth modes outlined in Figure 6.3 are found. The appearance of strain, resulting from lattice mismatch, is the reason why solid-on-solid growth is more complicated than the cases considered above.

Equations (6.17) and (6.18) are valid for deposits in equilibrium with their vapour on rigid substrates. Before considering the effects of strain, we consider the influence of the

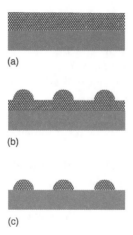

Figure 6.3 Equilibrium growth modes in solid-on-solid epitaxy: (a) layer-by-layer growth (Frank–van der Merwe); (b) layer plus island growth (Stranski–Krastanov); (c) island growth (Volmer–Weber)

vapour phase. Equations (6.17) and (6.18) are equally valid for liquid and solid deposits as long as there is sufficient mass transfer between the various phases to ensure the formation of (local) equilibrium structures. In practise, this may require temperatures high enough to form a liquid deposit; however, the fundamental requirement is that the diffusion rate during growth is high compared with the rate at which new material impinges on the interface.

As we shall see below, growth is often performed in a vacuum chamber in which the pressure of the gas phase P, does not correspond to the equilibrium vapour pressure of the depositing material, $P*$. In this case, a correction to the change in the Gibbs energy of the system must be made:

$$\Delta G = nk_{\mathrm{B}}T \ln\left(\frac{P}{P*}\right) \tag{6.20}$$

where k_{B} is the Boltzmann constant.

The ratio

$$\zeta = \frac{P}{P*} \tag{6.21}$$

is known as the degree of supersaturation. This leads to a further parameter that determines the equilibrium structure of the growing interface. Thus by controlling pressure and temperature the range of structures that can be grown as well as their growth mode can, in principle, be changed.

6.4.3 Strain in solid-on-solid growth

Even using the correction to the Young equation provided by Equation (6.20) we still cannot explain the occurrence of the Stranski–Krastanov growth mode [Figure 6.3(b)]. The observation of equilibrium structures formed by this mode is explicable only by means of the influence of strain. The three growth modes can be summarized as follows.

- Layer-by-layer (Frank–van der Merwe) growth [9]:
  - $\psi = 0 \Rightarrow \gamma_{sg} \geq \gamma_{sd} + \gamma_{dg} + Ck_B T \ln(P^*/P)$ (where subscript d represents the deposit);
  - precise lattice matching ($\varepsilon \approx 0$) and balanced interactions are required;
  - the substrate is wetted by the adsorbate.

- Layer plus island (Stranski–Krastanov) growth [10]:
  - after the growth of a wetting layer, often 1–4 atomic layers thick, the overlayer cannot continue to grow in the strained structure and instead forms less strained three-dimensional islands;
  - the adsorbate experiences a strong interaction with the surface but the overlayer experiences significant strain.

- Three-dimensional island (Volmer–Weber) growth [11]:
  - $\psi > 0 \Rightarrow \gamma_{sg} < \gamma_{sd} + \gamma_{dg} + Ck_B T \ln(P^*/P)$;
  - the adsorbate is non-wetting ($\varepsilon \neq 0$);
  - adsorbate–adsorbate interactions are far stronger than adsorbate–substrate interactions, and lattice mismatch is too great to be relieved by subtle structural relaxations.

Stranski–Krastanov growth is the most commonly observed growth mode. It is also an intriguing growth mode as it is sometimes accompanied by the observation of islands that have a small size distribution (i.e. an optimum size) and that can form ordered arrays [12]. The spontaneous formation of islands represents a type of self-assembly process for quantum dot formation. The interplay between kinetics and thermodynamics in determining the shape and size of the islands remains a topic of intense study. Island formation in semiconductor epitaxy, especially of II–VI and III–V compounds as well as the Si/Ge system, has attracted great interest because of the unique optical and electronic properties of semiconductor quantum dots [13, 14]. Quantum dots get their name from the observation of size-dependent quantum mechanical effects in nanoscale objects.

An example of Stranski–Krastanov growth is the deposition of germanium onto Si(1 0 0). After a wetting layer of *c*. 3 ML (monolayer) has coated the silicon substrate, germanium islands begin to form. Both squared-based pyramidal and dome-shaped islands are formed. Similar morphologies are found in Ge/Si alloys. The evolution of these islands has been investigated by Ross, Tromp and Reuter [15] during the chemical vapour deposition growth of Ge_xSi_{1-x} films from disilane/digermane (Si_2H_6/Ge_2H_6) mixtures at 923–973 K. Islands initially take on pyramidal shapes and these grow in size until a critical size is achieved, at which point they transform into domes. A transitional truncated pyramidal structure is also observed. The pyramids do not appear to be stable. They either grow large enough so that they can convert into domes, or else they shrink and disappear. The growth of larger islands at the expense of smaller ones is reminiscent of Ostwald ripening; however, the establishment of an optimum island size does not accompany true Ostwald ripening. The relative proportions of the three structures depend sensitively on the conditions, in particular on the temperature.

A general understanding of the *equilibrium* structure and growth modes of epitaxial systems is developed in the Advanced Topic below. From the above discussion we have learned that the geometric structure of overlayers depends critically on several parameters:

- the relative strengths of adsorbate–adsorbate and adsorbate–substrate interactions;

- lattice matching (i.e. how the lattice of the substrate fits with the lattice of the overlayer);

- strain fields (i.e. how strain relaxes in the substrate, islands and the area around the islands);

- temperature (i.e. surface energies and therefore reconstructions change with temperature).

Advanced topic: equilibrium overlayer structure and growth mode

Shchukin and Bimberg [12] laid out a theoretical framework describing the role of strain in growth processes. Daruka and Barabási [16] constructed from this a theory that describes all observed growth modes. They consider the deposition of H monolayers of atoms A on substrate B. The system is allowed to equilibrate forming n_1 monolayers of A in a wetting layer, with the remaining $H - n_1$ monolayers distributed into three-dimensional islands. The islands are assumed to take on pyramidal shapes with a constant aspect ratio.

Neglecting entropic contributions, the internal energy of the deposit can be written as

$$u(H, n_1, n_2, \varepsilon) = E_{\mathrm{ml}}(n_1) + n_2 E_{\mathrm{isl}} + (H - n_1 - n_2)E_{\mathrm{rip}}. \tag{6.22}$$

where subscripts ml, isl and rip refer to the monolayer, island and ripened islands, respectively. The first terms on the right-hand side is the contribution of the wetting layer. It represents an integral over the binding and elastic energy densities. The effects of strain must be properly accounted for in the calculation of both energy densities. The uniformly strained wetting layer has an energy density, G, given by the sum of the strain energy density and the binding energy of the layer:

$$G = C\varepsilon^2 - \Phi_{\mathrm{AA}} \tag{6.23}$$

where C is a material-dependent constant and where ε and $-\Phi_{\mathrm{AA}}$ are, as before, the lattice mismatch and the strength of the A$-$A bond, respectively. In a strained layer of finite thickness, the binding energy density of A atoms increases from $-\Phi_{\mathrm{AB}}$ in the first monolayer to $-\Phi_{\mathrm{AA}}$ far from the interface. Accounting for this, the total energy stored in the wetting layer is written

$$E_{\mathrm{ml}}(n_1) = \int_0^{n_1} dn \left\{ G + \Delta \left[\Theta(1 - n) + \Theta(1 - n)\exp\left(-\frac{n - 1}{a}\right) \right] \right\} \tag{6.24}$$

where $\Theta(1 - n) = 0$ if $(1 - n) < 0$, and $\Theta(1 - n) = 1$ if $(1 - n) > 0$. Δ is the energy difference, defined in Equation (6.19).

The second term on the right-hand side of Equation (6.22) describes the free energy per atom of the islands and the island–island interaction. Islands assemble if the strain energy density of an island is lower than that of the wetting layer. This is accounted for by introducing a form factor g that can take on values from 0 to 1. The binding energy is as before. The elastic energy of an edge of length L is proportional to $-L \ln L$. The edge energy density is therefore proportional to $-(\ln L)L^2$. In addition, account must be taken of the facet energy and of the interaction of the homoepitaxial and heteroepitaxial stress fields. Writing $x = L/L_0$ (the reduced island size) and the island–island interaction as E_{ii} yields

$$E_{isl} = gC\varepsilon^2 - \Phi_{AA} + E_0\left(-\frac{2}{x^2}\ln(e^{1/2}x) + \frac{\alpha}{x}\right) + E_{ii} \qquad (6.25)$$

where E_0 is the characteristic energy, and $\alpha = p(\gamma - \varepsilon)$; p and γ are constants that describe the coupling between the stress fields and the extra surface energy introduced by the islands, respectively. E_{ii} depends on the island spacing, shape and size as well as ε. E_{ii} is dominated by dipole–dipole interactions; thus, at low coverages of islands (the dilute limit) it can be neglected. E_{rip}, the total elastic energy per atom of the ripened islands, is obtained from Equation (6.25) by taking the limit $x \to \infty$. This yields E_{rip} equal to $gC\varepsilon^2 - \Phi_{AA}$.

Using Equation (6.25), Daruka and Barabási have calculated the phase diagram that appears in Figure 6.4. This phase diagram demonstrates that Frank–van der Merwe (FM), Stranski–Krastanov (SK) and Volmer–Weber (VW) growth modes can each represent the *equilibrium* growth mode for the appropriate combinations of deposited material and lattice mismatch. Two SK phases are found, differing as to whether the wetting layer forms before (SK$_1$) or after (SK$_2$) the islands form. Also found in Figure 6.4 are three distinct ripening phases, R$_1$–R$_3$. R$_1$ corresponds to classic Ostwald ripening in the presence of a wetting layer. R$_r$ is a modified ripening phase with a wetting layer and stable small islands. The islands are formed during the SK stage of growth. Subsequently, their growth is arrested but not all of them are lost to the ripening islands. The R$_3$ phase is similar to R$_2$ but lacks the wetting layer.

In Equation (6.25) allowance is made for the island–island interactions. Just as for atomic and molecular adsorbates, these lateral interactions become progressively more important as the coverage increases. If the strain fields that cause E_{ii} are anisotropic, they have the potential to lead to ordered arrays of islands. Shchukin and Birnberg [12] have investigated these interactions theoretically and have shown that under appropriate conditions, islands coalesce into an ordered square array.

In the construction of Equation (6.22) the entropic contribution has been neglected. Strictly speaking, the phase diagram of Figure 6.4 is valid only at $T = 0$ K. The extension of the theory outlined here to include entropic effects remains an outstanding challenge. This is required to describe the temperature dependence of growth systems. The temperature plays a vital role because surface energies are temperature dependent. The most obvious consequence of this is that the most stable reconstruction of a surface is temperature dependent. Furthermore, some surfaces are known to undergo reversible roughening transitions at a specific temperature. Additional temperature effects can arise because the two materials may have different coefficients of thermal expansion, which results in a temperature dependence for ε. This last effect, assuming all other material-dependent parameters are constant, is accounted for within the theory outlined above.

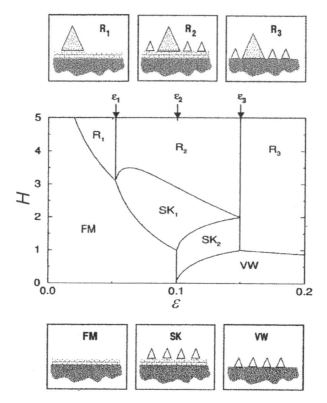

Figure 6.4   Phase diagram, calculated by Daruka and Barabási, as a function of coverage H and misfit ε. The small panels at the top and bottom illustrate the growth modes observed: small open triangles, stable islands; large shaded triangles, ripened islands; FM, Frank–van der Merwe growth; SK, Stranksi–Krastanov growth (SK_1, wetting layer forms before islands; SK_2, wetting layer forms after islands); VW, Volmer–Weber growth. Ripening phases: R_1, Ostwald ripening in the presence of a wetting layer; R_2, a modified ripening phase with a wetting layer and small stable islands; R_3, a modified ripening phase with small stable islands but without a wetting layer. Reproduced with permission from I. Daruka and A.-L. Barabási, *Phys. Rev. Lett.* **79** (1997) 3708. ©1997 by the American Physical Society

6.5   Growth Away from Equilibrium

6.5.1   Thermodynamics versus kinetics

Measurements on the InAs/GaAs system have shown that FM growth gives way to SK growth at a critical coverage of *c.* 1.7 ML [17]. This is a true phase transition; thus, the driving force of the growth mode transition is thermodynamic rather than kinetic. Other systems exhibit clear evidence of the influence of nonequilibrium processes upon layer morphology [18]. Two important parameters – temperature and pressure – are at the disposal of the experimentalist to drive a system away from equilibrium and into a regime in which dynamical and kinetic parameters determine the layer morphology. The temperature of the substrate controls the diffusion and desorption rates of adsorbates.

Sticking coefficients may also be temperature dependent. The pressure (and relative composition if more than one species is involved) of the gas phase control the impingement rate of adsorbing atoms and/or molecules. Consequently, whether growth is thermodynamically or dynamically controlled depends upon the experimental conditions.

If the temperature is low enough for islands to nucleate and if the diffusion rate is anisotropic, highly nonequilibrium structures can be formed. In this regime, especially for large amounts of deposited material, the size and shape of islands is determined by the interplay between the kinetics of adsorption and diffusion, and, often, pyramidal islands are the predominant feature of a generally rough surface profile. Atomistic understanding of the morphology in this regime is difficult but attempts have been made to understand the morphology in terms of phenomenological models and scaling relationships [19]. A general treatment of nonequilibrium growth is complex, but in the next section the special case of homoepitaxial growth of metals is treated, which because of the absence of strain is particularly revealing of the dynamical processes involved in film growth.

We also need to distinguish between the kinetics that is operative during and after growth. During growth, we must treat an open system. Matter is being added to the surface. According to the conditions, equilibrium on the surface and between the surface and the gas phase may or may not be established. After growth, the film morphology is subject to change as a result of the kinetics of diffusion as well as desorption. In addition, the chemical potential of surface species is affected by the change in pressure. The morphology of films is usually studied at temperatures lower than the growth temperature. The phase of a film is temperature dependent; therefore, it is always important to establish whether the observed film structure is the equilibrium structure corresponding to the observation temperature or whether it represents the equilibrium structure characteristic of the growth temperature. The latter may be observed if growth is interrupted and the temperature is decreased so rapidly that diffusion does not have the opportunity to equilibrate the film to its new conditions.

6.5.2 *Nonequilibrium growth modes*

When kinetics rather than thermodynamics determines the structure of growing films, the balance of various surface processes must be analysed. These processes are explicated in Figure 6.5. Under kinetic control the relative rates of terrace diffusion, accommodation of atoms at steps, nucleation, diffusion across steps and deposition determine the growth mode and film morphology. As found in Section 3.2, diffusion across a terrace is usually

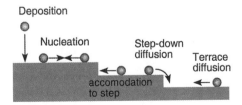

Figure 6.5   Surface processes involved in film growth

characterized by a lower activation barrier than is diffusion across a step. Furthermore, diffusion in the step-down direction has a lower barrier than diffusion in the step-up direction. Indeed, step-up diffusion generally can be neglected at normal temperatures. In Figure 6.5, desorption is neglected because it is usually not important for homoepitaxy of metals, which is the specific case that we consider below. Further details can be found in the review of Rosenfeld, Poelsema and Comsa [20].

Proof that kinetic control of growth can occur is easily demonstrated by studying homoepitaxial growth. Under thermodynamic control, homoepitaxial growth leads to layer-by-layer growth because there is necessarily no strain. Nonetheless, three distinct growth) modes, shown in Figure 6.6, are observed. Close to equilibrium, step-flow grow

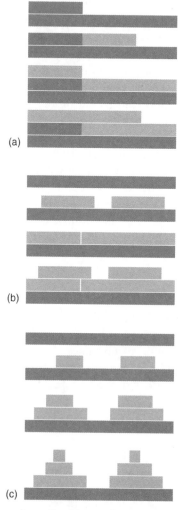

Figure 6.6 Nonequilibrium growth modes: (a) step-flow growth; (b) layer growth by island coalescence; (c) multilayer growth

is observed [Figure 6.6(a)]. In step-flow growth no interlayer transport occurs and terrace diffusion is so rapid that all atoms reach a step before nucleation of islands can occur.

Far from equilibrium, the nucleation of islands is rapid. The growth mode is decided by the extent of interlayer transport. In the limit of high interlayer transport no islands nucleate on a layer until it is complete. This is ideal layer growth (two-dimensional growth [Figure 6.6(b)]. In the limit of no interlayer transport, islands nucleate on top of incomplete layers. The higher the amount of deposition, the greater the number of islands on islands. This is multilayer growth (three-dimensional growth) [Figure 6.6(c)]. Where the activation energy of diffusion down a step tends to infinity ($E_{\mathrm{dif}}^{\mathrm{step}} \to \infty$) ideal multilayer growth occurs. However in the opposite limit ($E_{\mathrm{dif}}^{\mathrm{step}} \to 0$) ideal layer growth does not occur as long as there is a finite probability for island nucleation on an incomplete layer. Hence, ideal layer growth is never observed, though some systems do approach this behaviour. Nonetheless, after the deposition of many layers, any far-from-

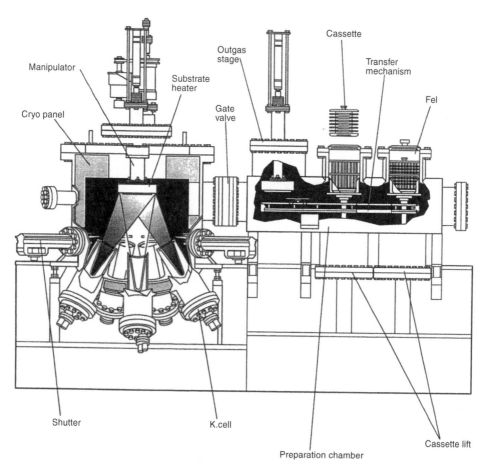

Figure 6.7 A molecular beam epitaxy machine. The main chamber contains several Knudsen cells (K-cells) for the evaporation of material onto the substrate. Reproduced from the VG Semicon manual with permission. B. Vögele of VG Semicon is thanked for providing this figure

equilibrium system will grow rough and, as noted above, will tend toward pyramidal structures.

The distinction between layer and multilayer growth can be made quantitative, as follows [20]. Let θ_c be the critical coverage at which nucleation occurs on top of the growing islands, and let θ_{coal} be the coverage at which islands coalesce to form a connected layer. Thus, the conditions defining the growth modes are

$$\theta_c > \theta_{coal} \Rightarrow \text{layer growth} \tag{6.26}$$

and

$$\theta_c < \theta_{coal} \Rightarrow \text{multilayer growth} \tag{6.27}$$

Typically, $0.5 \leq \theta_c \leq 0.8$. The critical coverage is not a constant; however, it can be increased only by increasing the temperature or decreasing the deposition rate.

Temperature and pressure are the two defining parameters that determine the growth mode. Temperature controls the rate of diffusion and pressure controls the rate of deposition, r_{dep}. As surface temperature, T_s, tends to infinity and r_{dep} tends to 0, equilibrium is approached and step-flow growth is observed. As T_s tends to 0 and r_{dep} tends to ∞, interlayer transport stops but nucleation still occurs. Multilayer growth is observed in this limit. If $E_s = 0$ (i.e. if terrace and step-down diffusion have the same barrier height; For more on E_s, see Section 3.2, page 82) then, for any r_{dep}, transitions from step-flow to layer to multilayer growth occur as the temperature decreases. If $E_s > 0$, the transitions step-flow to layer to multilayer occur only for higher values of r_{dep}, and a direct step-flow to multiplayer transition occurs for small r_{dep}.

We are now in a position to make predictions about growth modes in a real system. On a nonreconstructing face-centred cubic [fcc(1 0 0)] transition metal surface, E_s is of the order of 0.05 eV or even smaller. In contrast, E_s is of the order of a few tenths of an electron volt for fcc(1 1 1) surfaces. Consequently, growth on fcc(1 0 0) surfaces is generally much smoother than it is on fcc(1 1 1) surfaces. Higher temperatures are required for the observation of smooth fcc(1 1 1) surfaces not because step-flow growth occurs at a significantly higher temperature on these surfaces, but because these surfaces do not usually exhibit the layer growth mode. For example, the onset of step-flow growth for both Ag(1 0 0) and Ag(1 1 1) is $c.$ 500 K. However, Ag(1 0 0) exhibits layer growth down to 77 K whereas Ag(1 1 1) switches immediately to multilayer growth.

6.6   Techniques for Growing Layers

The structure of interfaces is controlled not only by thermodynamics but also by the kinetics of adsorption, desorption, diffusion across the interface and diffusion into the substrate. Therefore, the techniques used to grow interfaces play an important role in determining the types of interfaces that are formed. Below we introduce several growth techniques that are commonly used.

6.6.1 Molecular beam epitaxy

Molecular beam epitaxy (MBE) grew out of fundamental studies on the sticking coefficients of metal atoms performed at Bell Laboratories. In MBE a substrate is placed inside a vacuum chamber as in Fig. 6.7. Often this substrate is a single-crystal sample of a semiconductor such as a silicon, GaAs or CdTe. The chamber is fitted with facilities for heating the substrate and probing the surface structure and composition. The defining elements of the chamber are several Knudsen cells (K-cells). A Knudsen cell is a source from which a thermal beam emanates after the evaporation of a solid or a liquid. It is composed of a heatable crucible placed behind a small aperture from which a directional but highly divergent beam propagates. A shutter is placed in front of the aperture so that the flow of gas can be rapidly turned on and off. This allows for the precise control of the dose provided by the Knudsen cell such that the amount of deposited material can be controlled on the submonolayer level. This is important not only for the control of the layer structure but also so that precise amount of dopants can be added to the layers. The control of dopant concentrations allows one to control the electrical properties of the semiconductor films.

MBE generally involves the deposition of atoms, though dimers, trimers, and so on can constitute some fraction of the evaporated vapour based on an equilibrium distribution. Surface chemistry plays a role in MBE particularly for the growth of III–V (e.g. GaAs) and III–VI (e.g. ZnSe) materials. Whereas adsorption, diffusion and reactions in the adsorbed phase are important in MBE, desorption is generally not. The majority of atoms dosed to the surface are incorporated into the film.

Surface mobility plays a major role in MBE; thus deposition is conventionally carried out at high substrate temperatures. Although high substrate temperatures are good for ensuring sufficient surface mobility, they can also result in interdiffusion of deposited atoms into the substrate. Furthermore, dopants may have different degrees of solubility in the epilayer and the substrate or, if several epilayers of varying composition are created, the dopant may be more soluble in one layer compared with the neighbouring layer. In other words, the desired structure may not be the equilibrium structure, and high temperatures tend to push the system toward equilibrium. This is a potential problem with MBE.

6.6.2 Chemical Vapour deposition

In chemical vapour deposition (CVD) molecular precursors to film growth are dosed onto a surface from the gas phase or by using a supersonic or Knudsen molecular beam. These molecular precursors include, for example, $AsH_3 + Ga(CH_3)_3$ for GaAs, SiH_4 or $Si_2H_6 + GeH_4$ or Ge_2H_6 for SiGe alloys. The wide range of metalloorganic compounds that can be used to fabricate III–V and II–VI materials can be found elsewhere [21]. From the compositions of these molecules and the compositions of the desired epilayers it is obvious that not all of the atoms dosed onto the surface remain. Therefore, in contrast to MBE, CVD requires facile desorption to be among the reactions that occur during growth. Fundamental surface science studies have shed light on the interplay between adsorption, the stability of surface intermediates and desorption in determining the growth kinetics and outcomes [22].

For instance in the growth of homoepitaxial silicon layers or heteroepitaxial SiGe layers on silicon from silanes and germanes, the desorption of molecular hydrogen plays a vital role. The hydrogen-terminated surface is virtually inert and neither silanes nor germanes dissociate on this surface [22–25]. Therefore, at low temperature, as adsorbed hydrogen, H(a), builds up on the surface, the surface growth is choked off and eventually stops. This is an example of a self-limiting reaction. The desorption of hydrogen limits the growth kinetics and sets a lower bound to the temperature at which growth can be performed. Hydrogen desorbs from silicon at about 800 K, and at about 650 K from germanium. At higher temperatures, the dissociative adsorption of silanes and germanes limits the growth. The presence of $H_{(a)}$ also affects the diffusion of silicon and germanium atoms, which leads to distinct differences in layer morphology in CVD-grown as compared with MBE-grown layers.

High surface temperatures again bring about potential problems with the interdiffusion of dopants and other atoms. However, thermal desorption need not be the only means of removing adsorbates from the surface. Photon or electron irradiation can be combined with CVD. This opens the possibility of directly writing structures during CVD. If one of the reactants has a low sticking coefficient, energy can be put into the system to increase the sticking coefficient. This can be provided, for instance, by the high translational energy of a supersonic beam. Sticking and desorption can also be enhanced by igniting a plasma in front of the surface.

6.6.3 Ablation techniques

Some materials are not easily evaporated or they form compounds that are not well suited to CVD. Fortunately, other methods exist to volatilize refractory materials and allow them to be deposited on a substrate. Collectively, these are known as ablation techniques. Ablation occurs when high-energy particles encounter a surface and dislodge surface and possibly bulk atoms. When high-energy ions are used, the technique is called sputtering. Sputtering can be achieved with lower energy ions if a plasma, as in magnetron sputtering, assists the process. A high-power laser can also ablate atoms from a target. For the ejection of atoms from a target under laser irradiation we usually differentiate between true laser ablation – the explosive release of atoms from a target as the result of a nonequilibrium process – and laser vaporization – the evaporative release of atoms as a result of a thermal process. Details of the processes occurring during laser ablation and its use in pulsed-laser deposition can be found elsewhere [26]. Alternatively, high-energy electron irradiation can be used to induce vaporization of a target.

Advanced topic: catalytic growth of nanotubes

Growth can involve a catalytic agent. This is particularly important in the formation of nanotubes [27]. Fullerenes such as C_{60} are formed in any number of reactive systems in which carbon is volatilized and allowed to condense. Single-walled and multiwalled nanotubes (SWT and MWT, respectively) of carbon are formed when certain metals are added to graphite that is subsequently vapourized [28, 29]. Smalley and co-workers [30] have shown that extremely high yields of SWT are obtained from mixture of graphite with *c.* 1 at% of either Co or Pt. This mixture is formed into a rod that is used as a laser ablation target. The metals form clusters in the gas phase. The clusters act as catalytic agents upon which carbon atoms adsorb and subsequently react to form nanotubes. The nanotubes grow outward from the catalytic metal cluster. The size of the cluster during the initial stages of nanotube formation is likely responsible for determining whether SWT or MWT are formed.

Dai and co-workers [31] used a catalysed CVD reaction to produce ordered arrays of carbon nanotubes. They deposited iron on a porous silicon (por-Si) substrate. Evaporation of iron through a shadow mask allowed them to transfer the iron to the surface in a predetermined pattern. The patterned substrate was annealed in air so that both the por-Si and the iron are present as oxides. The sample was then heated to 973 K in flowing ethylene (C_2H_4) in a process that can be equated to chemical vapour deposition. The state of the surface-bound iron under these highly reducing conditions is unknown, though it is likely reduced (at least partially) to a metallic form. Ordered arrays of carbon nanotubes grew up away from the iron oxide islands (i.e. growth was from the bottom up). The iron acted as a catalyst for the dissociation of C_2H_4 and the subsequent growth of the nanotubes. In a reaction catalysed by gas-phase clusters, the carbon can easily reach the catalyst even though a nanotube is growing out from it on one side. At the surface, the geometry is more constrained. The role of the por-Si substrate is to ensure that the C_2H_4 can reach the iron as well as the bottom of the growing nanotubes.

Quate, Dai and co-workers have in addition demonstrated the CVD growth of individual SWT [32]. Again, an iron-based catalyst was used that was probably reduced under reaction conditions. In this case CH_4 was used as the reactive gas. The iron catalyst was supported either on a single-crystal silicon substrate or on Al_2O_3 nanoparticles. In both cases SWT with diameters of *c.* 1–5 nm and lengths of up to tens of micrometers were obtained.

Terrones, Kroto, Walton and co-workers [33] used a catalysed solid-state reaction to form SiO_x nanoflowers (see Figure 6.8), SiC and silicon nanotubes, and SiC nanotubes wrapped in a-SiO_2. A powder of SiC is heated to *c.* 1800 K in CO. The product distribution is dependent upon the catalyst that is mixed into the powder. One-dimensional SiC and silicon nanowires are obtained from an iron-catalysed reaction. Cobalt catalyses the production of three-

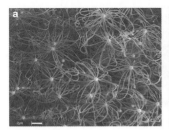

Figure 6.8 Scanning electron micrographs of: (a) a surface covered by flower-like silica nanostructures generated by heating SiC + Co under an Ar−CO atmosphere; (b) a typical three-dimensional feature radiating from a central spherical cobalt-rich particle; (c) a uniform nanoflower film formed from SiC + Co heated in pure CO. Reproduced with permission from M. Terrones, N. Grobert, W. K. Hsu, Y. Q. Zhu, W. B. Hu, H. Terrones, J. P. Hare, H. W. Kroto and D. R. M. Walton, 'Advances in the creation of filled nanotubes and novel nanowires', *MRS Bulletin* **28**(8) (1999) 43

dimensional SiO_x nanoflowers. When iron and cobalt are both combined with SiC, the result is the formation of elongated single crystals of β-SiC wrapped in a sheath of amorphous SiO_x. These nanowires have a SiC core diameter of c. 6–40 nm and an outer diameter of c. 10–60 nm. Exceptionally long nanowires of up to 100 µm are sometimes found. The fascinating chemistry involved in this system is not yet understood. As described above for the work of Dai *et al.* [31], the strongly reducing reaction conditions maintain the working catalyst in the metallic state. The metallic particles appear to be acting as catalytic sites for the dissociation of CO. They also act as tethers for the growing nanowires. The reactants (silicon, carbon and oxygen) diffuse across the surface of the metallic clusters and, again, the nanotubes grow out from the metal. Volatile SiO may play a role in the reaction as well.

6.7 Langmuir Films

When a small amount of an insoluble liquid is poured onto another liquid, it spreads out across the surface. The history of the observation of these films has been recounted by Laidler [34] and is summarized here. The water-calming effects of oil films on water were first noted by Pliny the Elder in 560 BC. The effects of these films later motivated Benjamin Franklin to investigate them. The truly modern era of the study of insoluble liquid films was ushered in by Agnes Pockels, who pursued many studies in her kitchen. She communicated her results to Lord Rayleigh, and the fascination she awakened in him led him to study further these films. Rayleigh was the first to calculate that the films had to have a width of approximately one molecule. Langmuir took up the cause and greatly improved the equipment, known as a Langmuir trough, used to study these films [35]. Because of his extensive studies of molecular films on the surface of a liquid, these films are called Langmuir films.

A particularly interesting type of Langmuir film is formed when an amphiphile is poured onto the surface of water. Amphiphiles are molecules that are polar (hydrophilic) on one end and nonpolar (hydrophobic) on the other. Examples include stearic acid. $(C_{17}H_{35}CO_2H)$, alkanethiols $(CH_3(CH_2)_nSH)$ and perfluoronated alkanethiols. Thus the hydrophobic end is generally a long alkyl chain, either hydrogenated or perfluoronated. Many different end-groups are suitable to form amphiphiles, such as $-CH_2OH$, $-COOH$, $-CN$, $-CONH_2$, $-CH=NOH$, $-C_6H_4OH$, $-CH_2COCH_3$, $-NHCONH_2$, $-NHCOCH_3$. When such a molecule is deposited on the surface of water, the polar end is attracted to the water whereas the alkyl chain attempts to avoid the water. The structure of the film and the orientation of the molecules depend on the area available to each molecule. The area can be varied by containing the molecules and applying pressure to the film with the aid of a movable barrier. The Langmuir trough provides not only a containment area with a movable barrier but also a means to measure the applied pressure.

When a three-dimensional gas is subjected to increasing pressure it eventually condenses, first into a liquid and then into a solid. Each phase exhibits progressively less compressibility and changes in the extent of ordering. Similarly, a two-dimensional Langmuir film exhibits different phases based on the applied pressure. One way to think of the effect of pressure is that the application of pressure effectively changes the area available for each molecule.

The effects of pressure are illustrated in Figure 6.9. When the area per molecule is large, a disordered gaseous phase is formed. A gas possesses neither short-range nor long-range order and there is little interaction between the molecules. As the area is decreased a phase transition occurs into a liquid-like phase. Liquids are less compressible than are gases and exhibit short-range but no long-range order. In these fluid phases, the chain is not rigid and adds to the disorder in the films. Further reduction in the area compresses the film into a solid-like phase. Langmuir determined that the molecules are standing up straight because the area per molecule is independent of chain length. This has been confirmed by X-ray and polarized infrared absorption measurements. If the area is decreased even further, the film collapses. Molecules get pushed out of the first layer and shearing leads to bilayer formation. The order in the solid state means that the molecules all point with the polar end-group in the water, and the alkyl chains stand up away from the surface.

Various condensed phases have been observed [1]. Not all phase transitions are observed for all films. The types of condensed phases that can be formed depend sensitively on the intermolecular interactions within the film and between the film and the liquid. The liquid under the Langmuir film does not present an ordered lattice to the amphiphile; therefore, lateral interactions are responsible for the formation of ordered solid-phase Langmuir films. Mobility is not a restriction, as it was in the formation of ordered layers by MBE or CVD; instead, the degree of ordering depends on the applied pressure and the strength of intermolecular interactions.

Figure 6.10 shows the interactions that occur within a Langmuir film. The head-group of the amphiphile (the most polar end) is hydrogen bonded to the liquid (usually water). Hydrogen bonding is a relatively weak interaction of order 20 kJ mol^{-1} (c. 0.2 eV) that can be equated to physisorption. In some instances, electrostatic forces rather than

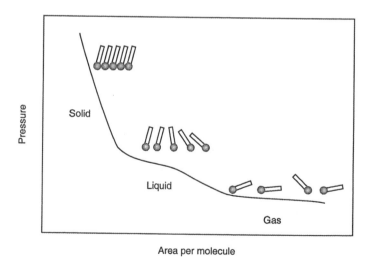

Figure 6.9 Phase diagram of surface pressure against area per molecule in a Langmuir film. The (hydrophobic) tails should only be considered rigid in the solid-like phase in which tail–tail interactions lock them into an ordered structure

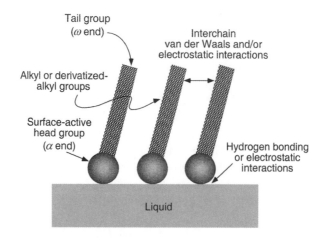

Figure 6.10 A Langmuir film is a monomolecular layer of an amphiphile on a liquid. In this example the polar head-group of the amphiphile interacts with the liquid through hydrogen bonding or electrostatic interactions. By aligning and tilting, the tail groups maximize their attractive interactions

hydrogen bonding bind the head – group to the polar liquid. The hydrophobic chains experience a repulsive interaction with the water. An upright configuration is thus favoured, but this interaction alone would not result in ordering. Ordering arises from nonpolar chain–chain interactions. These interactions, caused by van der Waals forces, are maximized when neighbouring molecules align with one another. Although van der Waals forces are weak, the chains generally contain 8–20 or more carbon atoms, and this contribution can amount to tens of kilojoules per mole. If the chain is not composed solely of methylene groups (CH_2) then hydrogen bonding may also play a role [e.g. for an ether (ROR) tail]. Further impetus toward ordering can be given by the (polar or nonpolar) interactions of end-groups.

6.8   Langmuir–Blodgett Films

6.8.1   Meniscus formation

A meniscus is formed when a solid is pushed into a liquid. The balance of surface tension forces at the liquid–solid interface determines the profile of the meniscus. Figure 6.11 illustrates how the relative surface tension of the liquid and solid lead to either concave or convex meniscuses. The surfaces of oxides tend to be covered with hydroxyl groups. These surfaces are polar and hydrophilic and therefore the attractive forces associated with the hydroxyl groups lead to upward-sloping meniscuses in water. Common hydrophilic substrates include Al_2O_3, Cr_2O_3, SnO_2, SiO_2 (glass), gold and silver. A nonpolar surface is not wet by water and forms a downward-curving meniscus. Silicon is a particularly versatile substrate in that it can be made hydrophilic when an oxide layer is present, or hydrophobic when the surface is hydrogen-terminated. A silicon surface normally has a thin oxide layer on it, known as the native oxide. This surface is readily

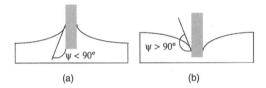

Figure 6.11   Meniscus formation: (a) a hydrophilic substrate in water; (b) a hydrophobic substrate in water. ψ, contact angle

transformed into a hydrogen-terminated surface by dipping it in acidic aqueous solutions containing fluoride.

6.8.2   Vertical deposition

Now consider the situation depicted in Figure 6.12 in which a solid substrate interacts with a Langmuir film. The formation of a meniscus leads to a gradual transition from the surface of the liquid to the surface of the substrate. The liquid becomes progressively thinner until it gives way to the solid substrate. Thus, the molecules in the Langmuir film ride up the meniscus and gradually the binding switches from a molecule–liquid interaction to a molecule–substrate interaction. A film of amphiphiles transferred onto a solid substrate is known as a Langmuir–Blodgett film.

Figure 6.12 demonstrates in addition how a film can be transferred from the surface of the liquid to the surface of the substrate. If the substrate is pulled vertically out of the liquid and a movable boundary is used to maintain the order in the Langmuir film, the film can be transferred to the substrate. The contact angle of a static system differs from

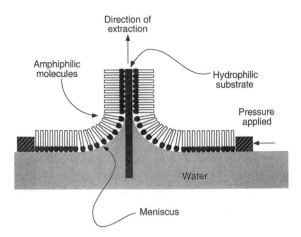

Figure 6.12   The transfer of a Langmuir film onto a solid substrate via vertical deposition. In this example the substrate is hydrophilic and interacts attractively with the head-group of the amphiphile. Movable barriers are required to maintain a constant surface pressure in the film to ensure uniform deposition

that of a dynamic system in which the substrate is moving. As long as the velocity of the substrate is not too high, the general geometry shown in Figure 6.12 is maintained and the film is transferred. When the film is retracted too rapidly, the system breaks down and transfer does not occur. In other words, deposition of a Langmuir–Blodgett film occurs only below a critical velocity.

Figure 6.12 depicts how a Langmuir-Blodgett film is deposited upon retraction of a hydrophilic surface; conversely, deposition during immersion can take place for a hydrophobic surface. The substrate need not be a 'clean' substrate; that is, the substrate can be one upon which a Langmuir–Blodgett film has already been deposited. In this way multilayer films can be built up that are composed of several layers of amphiphiles. Figure 6.13 displays several types of films that can be formed and defines the classifications of multilayer films. X-type multilayer films are built up from nonpolar–nonpolar interactions that bind the amphiphile to the surface. Subsequently, head–tail interactions bind one layer to the next. Such a film arises exclusively for amphiphiles in which the polarity does not differ greatly between the two termini. These are generally the least stable multilayer films. Y-type multilayer films contain amphiphiles in alternating head–tail/tail–head configurations. They are the most stable multilayer films and are most likely to occur for strongly polar head groups. Z-type multilayer films are an inverted version of X-type films; that is, the head-group binds to the substrate and thereafter tail–head interactions hold neighbouring layers together. These films may be more tightly bound to the substrate than X-type films but the lack of tail–tail, head–head interactions between layers results in a much less stable film than for Y-type films.

Langmuir–Blodgett films are usually weakly bound to the substrate by physisorption. Hence, these films are not stable with regard to washing in solvents. Exposure to air may gradually cause their destruction. However, particular combinations, such as RSH + Au (where R is an organic group) can lead to chemisorbed films.

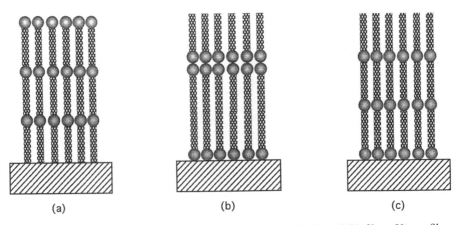

(a)   (b)   (c)

Figure 6.13   X-type, Y-type and Z-type multilayer Langmuir–Blodgett (LB) films. Y-type films are deposited successively on upstrokes and downstrokes. X-type films deposit only on downstrokes. Z-type films deposit only on upstrokes. Downstroke, insertion of the substrate into the LB trough; upstroke, retraction of the substrate from the LB trough

6.8.3   Horizontal lifting (Shaefer's method)

Rather than approaching perpendicular to the surface, the substrate can be pressed onto a Langmuir film in a parallel geometry. This method of transferring Langmuir films onto a substrate is known as horizontal lifting, or Shaefer's method. Although Langmuir and Shaefer may have been the first to perform in-depth scientific studies on the horizontal deposition method, the technique has been used in Japan for over 800 years in the decorative arts [36].* The Japanese technique of *sumi-nagashi* (meaning 'ink flow') utilizes a horizontal deposition method to transfer marbled patterns onto paper. Legend has it that this art form was bestowed upon the Hiroba family by the god Kasuga in 1151. The family continues the tradition of making *sumi-nagashi* prints today. The technique consists of placing inks, generally black, red and indigo, on top of water. The inks are suspensions of organic pigments or graphite in protein; hence they spread out over the surface of the water and do not mix with it. After completing the design with various brush strokes, and by blowing on the film the artist then transfers the pattern by placing a sheet of paper on top of the film. Here the technique deviates from that of Langmuir and Shaefer in that paper is a porous material and the ink is drawn into the fibres of the material rather than remaining on the outer surface of the substrate. Sheets of *sumi-nagashi* are traditionally used for writing poetry but recently they have also found use as endpapers, book covers (in particular, this book's cover owes its design to an example of *sumi-nagashi*) and in other decorative uses.

6.9   Self-Assembled Monolayers

Self-assembled monolayers are molecular assemblies formed spontaneously by immersion of an appropriate substrate into a solution of an active surfactant in an organic solvent. Interest in SAMs has grown exponentially in recent years but the first report by Zisman and co-workers [37] in 1946 went almost unnoticed by his contemporaries. By far the most studied system is that of an alkanethiol [$HS(CH_2)_nX$ with $X = CH_3$, CF_3, $CHCH_2$, CH_2OH, $COOH$, etc.] interacting with a gold surface, as first investigated by Nuzzo and Allara [38]. Alkanethiols are often abbreviated as RSH or HSRX, where X represents the tail group. The head group is the sulfur end of the molecule, which is denoted the α end. The tail group is denoted as the ω end. Numerous systems form SAMs as a result of adsorption either from solution or from the gas phase [39–41]. These include:

- alkanethiols on gold, silver, copper and GaAs;

- dialkyl sulfides (RSR′) and disulfides (RS−SR′) on gold;

- organosilanes [$RSiCl_3$, $RSi(OCH_3)_3$, $RSi(NH_2)_3$] on SiO_2, Al_2O_3, glass, quartz, mica, GeO_2, ZnSe and gold;

- alcohols (ROH) and amines (RNH_2) on platinum;

*I am deeply grateful to Sachiko Usui of International Research Center for Japanese Studies, Kyoto, Japan, for providing information on *sumi-nagashi*.

- carboxylic acids (fatty acids, ROOH) on Al_2O_3, CuO, AgO and silver;
- alcohols or terminal alkenes ($RC=CH_2$) on hydrogen-terminated silicon.

Multifarious organic functionalities have been tethered to silicon surfaces through the formation of Si—C bonds [42]. The formation of Si—C bonds often requires activation by irradiation, a peroxide initiator or heating. These conditions are harsher than those usually associated with the formation of SAMs, especially for the prototypical alkanethiol/gold system. This budding area of SAM research is of great interest because of the ability to pattern silicon (and por-Si) surfaces as well as the technological implications of silicon materials.

The formation of an RSH SAM generally follows Langmuirian kinetics. The rate of adsorption is dependent on the concentration of the alkanethiol in solution, which is usually of the order of 10^{-3} mol l^{-1}. Jung and Campbell [43] have determined the sticking coefficient for $C_2–C_{18}$ straight-chain alkanethiols. Langmuirian kinetics (see Chapter 4) were followed up to c. 80% of the saturation coverage. The value of the initial sticking coefficient is of the order $10^{-6}–10^{-8}$, increasing with increasing chain length. The initial sticking coefficient is of the order of unity for the adsorption of RSH from the gas phase. Clearly, the presence of a solvent hinders adsorption of the alkanethiol. It does so in two ways. First, solvent molecules (ethanol in the case studied by Jung and Campbell) adsorb on the surface and must be displaced by the incoming RSH. Second, a solvation shell surrounds a dissolved molecule. The alkanethiol must shed the solvent molecules that constitute the solvation shell before it can adsorb on the surface. The combination of these two processes leads to an adsorption barrier in the case of adsorption from the solution that is not present for adsorption from the gas phase.

A monolayer forms at room temperature in minutes to hours, though days may be required to obtain the most highly ordered, close-packed structure. A spectacular coincidence in the balance of intermolecular forces allows for formation of ordered structures at room temperature. Most RSH/Au SAMs are observed to disorder if heated above c. 380 K. Increased ordering of a completed monolayer can be achieved by cooling, particularly if a temperature below c. 250 K can be reached [44].

6.9.1 Amphiphiles and bonding interactions

For the formation of Langmuir films, it was essential to use amphiphiles to obtain ordered films. The same considerations and interactions found in Figure 6.10 are also pertinent to the formation of SAMs. There are now two major differences. First, the head-group is chemisorbed to the surface, with typical bond energies of more than 100 kJ mol^{-1} (>1 eV). Second, the substrate is a well-ordered solid. Therefore, a driving force toward ordering is provided by the directionality of chemisorption. While chemisorption may force the head-groups to seek registry with the surface, the ordering of the chains and tails is brought about predominantly by the intermolecular interactions outlined before.

The role of the substrate is also important. The substrate must be unreactive to everything but the molecule that is attempting to form a monolayer. Gold and hydrogen-terminated silicon are both rather inert surfaces. Foreign adsorbates physisorb to these surfaces. The target molecule easily displaces the physisorbed molecules. The tenacious chemisorption of impurities would poison the formation of the SAM. Another important characteristic of the substrate is the spacing of the surface atoms. The Au(1 1 1) surface provides a virtually perfect template for alkanethiols. The sulfur atoms are able to assume a close-packed structure, presumably by filling the threefold hollow sites with S–S separation of 4.99 Å. This is sufficiently far apart that the chains are not crowded while leaving them sufficient space to tilt and form an ordered structure that maximizes the chain–chain interactions.

6.9.2 Mechanism of self-assembled monolayer formation

The dynamics of SAM formation of alkanethiols on gold has been elucidated by Poirier and Pylant [45] and is depicted in Figure 6.14. They investigated the interaction of various RSH molecules with a Au(1 1 1) surface under ultrahigh vacuum (UHV) conditions. Nonetheless, the observations they made seem to be generally applicable to the growth of SAMs from solution as well as to the process that occurs during the formation of ordered Langmuir films. The description below applies to studies carried out at room temperature. The two major differences compared with adsorption out of solution are that solvent molecules need not be displaced from either the surface or the solvation shell.

The clean Au(1 1 1) is reconstructed into a herringbone structure. At low coverage, the adsorbed alkanethiol is present as a mobile two-dimensional gas. In this phase the alkyl chain is oriented roughly parallel to the surface. As the coverage is increased, ordered islands of a solid phase begin to appear. The solid and two-dimensional gas phases are in equilibrium and the appearance of the solid phase represents a first-order phase transition. The solid phase, denoted $solid_1$, is anchored to the surface through the sulfur atom. The molecular axis is still parallel to the surface. The appearance of $solid_1$ leads to local changes in the Au(1 1 1) reconstruction. That is, the substrate changes its structure only under the nucleated islands. As the coverage is increased yet further, the surface becomes saturated with $solid_1$. Continued adsorption results in a second phase transition to a second solid phase ($solid_2$). In $solid_2$ the RSH molecules stand upright and the same packing density is obtained as that achieved at saturation for a SAM deposited from solution. The driving forces of this phase transition are lateral interactions in the solid phase as well as the propensity of the alkanethiols to form as many Au–S bonds as possible.

This growth mechanism is a perfect illustration of the effects of equation (1.1) (page 9). It demonstrates how multiple (substrate and adsorbate) phases can coexist at the same temperature. It also demonstrates how a sequence of adsorbate structures can be populated as σ_A increases. The system as a whole needs to minimize its free energy and it does this by counterbalancing, for instance, the gains achieved by increased adsorption and increased chain–chain interactions versus the energetic penalty of changing the reconstruction of the substrate or the orientation of the adsorbate.

The importance of adsorbate-induced reconstruction of the substrate as a driving force behind self-assembly is confirmed by the studies of Besenbacher and co-workers [46] of hexa-*tert*-butyldecacyclene (HtBDC) on Cu(1 1 0). In this case the critical nucleation centre is two HtBDC molecules that join and simultaneously expel c. 14 copper atoms

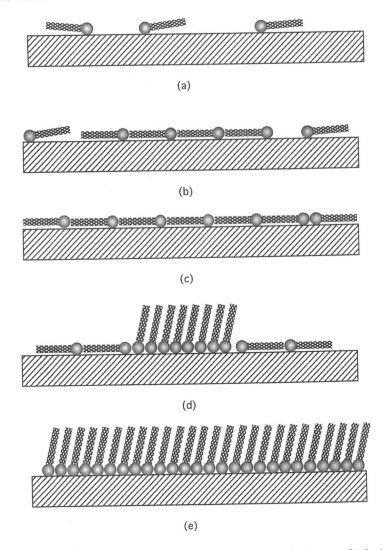

Figure 6.14 Mechanisms of self-assembly: (a) disordered, mobile lattice gas; (b) lattice gas plus ordered islands; (c) saturated surface-aligned phase; (d) nucleation of solid-phase islands; (e) saturated solid phase. Self-assembly progresses in stages, as long as a sufficient supply of adsorbates is available, until the thermodynamically most favoured final state is reached. Not depicted in the figure is a reconstruction of the surface, which often accompanies the process

from beneath them. The energetic cost of removing these atoms is balanced by the added stability of the chemisorbed molecules. Thermal energy is required to aid in this process, and adsorption below c. 250 K results neither in surface reconstruction nor self-assembly.

The thermodynamic driving force is what favours the formation of self-assembled monolayers. Given enough of the appropriate adsorbate and sufficient time, the system locks into a final structure that is well ordered. Let us look in more detail at the dynamics of the transition into the ordered phase. Growth of a new phase starts from a nucleus. The nucleus is a cluster of atoms or molecules of a critical size with the appropriate orientation. The formation of this nucleus is a random process. The nucleus is assembled by statistical fluctuations that bring together the constituent units in the proper structure. Once the critical nucleus forms, the rest of the phase assembles around it.

Beebe and co-workers [47] have studied the formation of ordered domains of 4′-octyl-4-biphenylcarbonitrile (8CB) on graphite. To facilitate their study, they used specially prepared graphite substrates with circular etch pits. The one-monolayer-deep depressions act as molecular corrals that fence off the adsorbates inside from the surrounding terraces, allowing for detailed studies of isolated surface regions. 8CB self-assembles on graphite on a timescale of hours. Figure 6.15 exhibits a scanning tunnelling microscope image of a graphite sample immersed in pure liquid 8CB at room temperature. Studies of such images reveal several interesting aspects of the dynamics of nucleation and self-assembly. Figure 6.15 illustrates three molecular corrals surrounded by regular graphite terraces. Two of the three corrals and the terraces are covered by regular arrays of 8CB. The third corral is filled with a disordered layer of 8CB. This corral appears empty because the disordered 8CB molecules are highly mobile. This image confirms the picture painted above: the adsorbate undergoes a phase transition from a disordered phase into the ordered phase. Furthermore, this phase transition is irreversible at room temperature. Once the ordered phase is formed, it does not spontaneously revert to the disordered phase.

The ordering of 8CB is much more rapid than the formation of nucleation centres. Nucleation occurs over a period of hours. Ordering, in contrast, occurs too rapidly (tens of milliseconds) to be observed by the microscope. Effectively, the disordered molecules wait around for a nucleation centre to form, but once it appears the remaining molecules snap into position. The formation of a nucleus is obviously difficult. By studying the nucleation rate in corrals with radii from roughly 50 Å to 1350 Å, Beebe and co-workers determined that nucleation is more likely to occur on terraces than on steps. Quite possibly, the disorder brought about by the steps destabilizes the nucleation centres, making them less favourable sites for nucleation. This is in marked contrast to numerous metal-on-metal and semiconductor-on-semiconductor growth systems in which islands preferential nucleate at defect sites.

Advanced topic: chemistry with self-assembled monolayers

The head-groups in a SAM are strongly anchored and therefore they are more tightly bound to the substrate than are Langmuir–Blodgett films. SAMs can be washed and rinsed with solvents. Chemical reactions with the tail or between chains can be performed without disrupting the monolayer. Once anchored to the surface with an ω-functional group exposed, the full tool chest of organic or organosilane chemistry can be brought into action to transform the termination of the SAM into a bewildering array of different molecular entities. The ω-substitution of bulky groups has the tendency to reduce the order in the SAM.

These properties of SAMs make them ideal candidates for fundamental studies on molecular recognition. A SAM makes it possible to tether an ordered array of specific functional groups onto the surface. Interactions of this group with another molecule can then

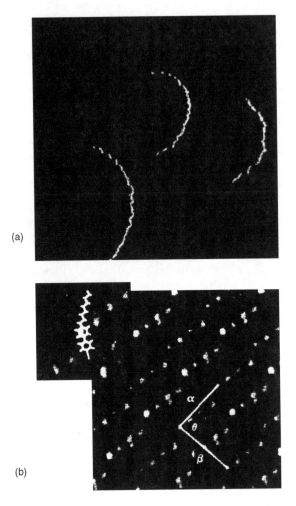

Figure 6.15 Self-assembled rows of 4'octyl-4-biphenylcarbonitrile (8CB) molecules in corrals on graphite. (a) A constant-height scanning tunnelling microscope image. Several hours after deposition, ordered monolayer films have formed on the terrace and in one corral but not the other. Note how the rows in the corral are not aligned with the rows formed on the surrounding terrace. The second corral is filled with a disordered layer of 8CB molecules. (b) A 95×95 Å^2 constant-height image with molecular resolution. The position of one molecule is marked by the model overlay. Reprinted with permission from D. L. Patrick, V. J. Cee and T. P. Beebe Jr, *Science* **256** (1994) 231. ©1994 American Association for the Advancement of Science

be studied with the advantage that the tethered molecule has a known orientation and is fixed in space. Prime and Whitesides, for example, have used SAMs to study specific protein-binding interactions [48]. As discussed in Chapter 2, SAM formation on the tip of an atomic force microscope has been used to make tips of known chemical termination [49]. These layers can then be used to study chemical interactions between the tip-bound molecules and molecules immobilized on a substrate. This has proved a fruitful method of investigating the chemistry of a variety of biomolecules. The stability and uniform chemical structure of SAMs also make them ideal surfaces to study tribology, the science of friction, and the influence of interfacial layers on electrochemistry and charge transfer at surfaces.

6.9.3 Self-assembly in inorganic thin films

Self-assembly or spontaneous ordering can occur under much different circumstances than what we have investigated in the previous section. In Section 6.4 we were introduced to Stranski–Krastanov growth and discussed how certain systems exhibit a tendency to form spontaneously islands that display a narrow size distribution. In Section 6.4, we concentrated on heteroepitaxial semiconductor systems, but these are not the only ones known to exhibit spontaneous ordering. Kern, Comsa and co-workers [50] have found periodic stripes of adsorbed oxygen atoms on Cu(1 1 0). Triangular islands of silver grown on platinum demonstrate that self-ordered growth can also occur in metal-on-metal systems [51]. Strain relief again forms the basis of the explanation of this phenomenon. The lattice mismatch between platinum and silver is large. The second monolayer of silver on platinum assumes a structure in which a trigonal network of dislocations partially relaxes the compressive strain. Careful preparation of this layer leads to long-range ordering of this strain-relief structure. Further deposition of silver leads to the formation of triangular islands. These islands preferentially nucleate away from the dislocation and, therefore, they form an ordered array with a narrow size distribution. This example illustrates not only the importance of strain relief but also how control of the nucleation process can be used to create self-organized structures.

An example of spontaneous formation of islands with a narrow size distribution is observed [52] when polycrystalline palladium films deposited on $LaAlO_3$ are annealed in O_2 at 1173 K to form PdO_2. The PdO_2 obviously must not wet the $LaAlO_3$ substrate as all of the palladium is oxidized and the $LaAlO_3$ is exposed between the islands. The oxidation of palladium to PdO_2 is accompanied by a 38% volume increase. The resulting layer expansion and lattice mismatch provide the driving force for island formation. However, an explanation based solely on thermodynamic grounds seems unfounded as island formation occurs only when one starts with polycrystalline films. An epitaxial layer of palladium oxidized under the same conditions leads only to a rough film. This result indicates that the kinetics of mass transport play a critical role in island formation and again illustrates that layer growth is a complex phenomenon that must be investigated on a case-by-case basis to determine the growth characteristics and whether kinetic or thermodynamic factors are directing the film properties.

6.10 Summary of Important Concepts

- A stress is expressed in force per unit area.

- A stress applied to a lattice causes a deformation called strain.

- Surface energy and surface tension are synonymous for liquids.

- The term surface tension only should be applied to liquids. For solids the proper term is surface energy.

- Surface energy can be thought of as being analogous to a two-dimensional pressure [compare Equations (6.5) and (6.6)] or else as the surface Gibbs energy [Equation (6.10)].

- The multilayer growth of an adsorbate on a substrate is characterized by one of three growth modes: Frank–van der Merwe (layer-by-layer), Stranski–Krastanov (layer plus island) or Volmer–Weber (island).

- At equilibrium the growth mode is determined by the effects of strain and the relative strength of the A–A, A–B and B–B interactions (A is the adsorbate atom, and B is the substrate atom).

- Under nonequilibrium conditions the dynamics of adsorption, desorption and diffusion can lead to deviations from the equilibrium growth mode.

- Langmuir films are monomolecular films on the surface of a liquid.

- When these films are transferred onto a solid substrate, they are called Langmuir–Blodgett films.

- Self-assembled monolayers are ordered monolayer films that form spontaneously on a solid substrate.

- Whereas the chemisorption interaction between the head-group and the surface accounts for the bulk of binding interaction in SAMs, it is the weak, largely noncovalent interactions between the chains and end-groups that lead to the order in the SAM.

Exercises

6.1 Derive Equation (6.16).

6.2 Consider a hemispherical liquid island of radius r with surface energies $\gamma_{ls} > \gamma_{lg}$ in equilibrium with its vapour. Calculate the island surface energy as a function of r and demonstrate that small islands are unstable with respect to large islands. Assume the substrate to be rigid and that the island energy is composed only of island–substrate and island–vapour terms.

6.3 Consider a system that for a given set of conditions exhibits step-flow growth. Discuss the effects that the adsorption of heteroatoms can have on homoepitaxial growth. Consider two low heteroatom coverage cases:

(a) the heteroatoms decorate the steps;

(b) the heteroatoms occupy isolated terrace sites.

6.4 Silicon is the most important semiconductor for electronic applications. GaAs and its III–V sister compounds are better suited than silicon as building blocks for optical devices such as light emitting diodes (LEDs) and lasers. The integration of optical components with electronics is a highly desirable manufacturing goal for improved communications, computing and display devices. Discuss the fundamental physical reasons why it is difficult to integrate GaAs circuitry silicon.

6.5   Consider the dynamics of deposition of X, Y and Z multilayer films. For each case determine whether deposition occurs on the downstroke (insertion of substrate into the Langmuir–Blodgett trough), upstroke (retraction) or in both directions. Discuss the reasons for these dependencies.

6.6   The sticking coefficient is defined as

$$s = \frac{r_{ads}}{Z_w} \qquad (6.28)$$

and represents the probability of a successful adsorption event. The collision frequency in solution is given by

$$Z_w = c_{sol}\left(\frac{k_B T}{2\pi m}\right)^{1/2} \qquad (6.29)$$

where c_{sol} is the concentration in molecules per cubic metre. The initial sticking coefficient of $CH_3(CH_2)_7SH$ on a gold film is 9×10^{-8} [43]. Assuming a constant sticking coefficient, which is valid only at low coverage, estimate the time required to achieve a coverage of 0.01 ML for adsorption from a 5×10^{-3} mol l^{-1} solution. Take the surface density of atoms to be 1×10^{19} m^{-2}.

6.7   Your lab partner has prepared two silicon crystals but has not labelled them. One is hydrogen terminated; the other is terminated with an oxide layer. Propose and explain an experiment you could perform in your kitchen that would distinguish the two.

6.8   Explain the observed trend that C_4 straight-chain amphiphiles generally do not form Langmuir–Blodgett films or SAMs that exhibit a structure that is as well ordered as that of C12 straight-chain amphiphiles.

6.9   Describe what would occur during vertical deposition of a Langmuir–Blodgett film if the barrier of the trough were stationary and a large-surface-area substrate were used.

Further Reading

A. W. Adamson and A. P. Gast, *Physical Chemistry of Surfaces*, 6th edn (John Wiley, New York, 1997).

L. H. Dubois and R. G. Nuzzo "Synthesis, Structure and Properties of Model Organic Surfaces", *Annu. Rev. Phys. Chem.* **43** (1992) 437.

H. Ibach, "The role of surface stress in reconstruction, epitaxial growth and stabilization of mesoscopic structures", *Surf. Sci. Rep.* **29** (1997) 193.

P. Jensen, "Growth of nanostructures by cluster deposition: experiments and simple models", *Rev. Mod. Phys.* **71** (1999) 1695.

D. A. King and D. P. Woodruff (eds), *The Chemical Physics of Solid Surfaces and Heterogeneous Catalysis: Surface Properties of Electronic Materials*, Volume 5 (Elsevier, Amsterdam, 1988).

D. A. King and D. P. Woodruff, (eds), *The Chemical Physics of Solid Surfaces and Heterogeneous Catalysis: Growth and Properties of Ultrathin Epitaxial Layers*, Volume 8 (Elsevier, Amsterdam, 1997).

H. Lüth, *Surfaces and Interfaces of Solid Materials*, 3rd ed (Springer, Berlin, 1995).

S. M. Prokes and K. L. Wang (eds), "Novel Methods of Nanoscale Wire Formation", *MRS Bulletin* **24** (1999).

F. Schreiber, "Structure and growth of self-assembling monolayers", *Prog. Surf. Sci.* **65** (2000) 151.

D. K. Schwartz, "Langmuir–Blodgett film structure", *Surf. Sci. Rep.* **27** (1997) 24.

V. A. Shchukin and D. Bimberg, "Spontaneous ordering of nanostructures on crystal surraces", *Rev. Mod. Phys.* **71** (1999) 1125.

G. A. Somorjai, *Introduction to Surface Chemistry and Catalysis* (John Wiley, New York, 1994).

A. Ulman, *An Introduction to Ultrathin Organic Films from Langmuir–Blodgett to Self-assembly* (Academic Press, Boston, MA, 1991).

A. Ulman, "Formation and structure of self-assembled monolayers", *Chem. Rev.* **96** (1966) 1533.

Z. Zhang and M. G. Lagally, "Atomistic processes in the early stages of thin-film growth", *Science* **276** (1997) 377.

References

[1] A. W. Adamson and A. P. Gast, *Physical Chemistry of Surfaces*, 6th edn (John Wiley, New York, 1997).

[2] J. W. Gibbs, *Collected Works, Volume 1: Thermodynamics* (Longman, London, 1928).

[3] J. Dąbrowski, E. Pehlke and M. Scheffler, *Phys. Rev. B* **49** (1994) 4790.

[4] R. Koch, "Intrinsic stress of epitaxial thin films and surface layer", in *The Chemical Physics of Solid Surfaces and Heterogeneous Catalysis. Growth and Properties of Ultrathin Epitaxial Layers*, Volume 8, eds. D. A. King and D. P. Woodruff (Elsevier, Amsterdam, 1997) p. 448.

[5] G. Wulff, *Z. Kristallogr. Mineral.* **34** (1901) 449.

[6] C. Rottman and M. Wortis, *Phys. Reports* **103** (1984) 503.

[7] M. C. Desjonquères and D. Spanjaard, *Concepts in Surface Physcis*, 2nd edn (Springer, Berlin, 1996).

[8] B. S. Clausen, J. Schiøtz, L. Gråkæk, C. V. Quesen, K. W. Jacobsen, J. K. Nørskov and H. Topsøe, *Topics in Catalysis* **1** (1994) 367.

[9] F. C. Frank and J. H. van der Merwe, *Proc. R. Soc. London, Ser. A* **198** (1949) 205.

[10] I. N. Stranski and L.Kr'stanov, *Sitzungsber. Akad. Wiss. Wien* **146** (1938) 797.

[11] M. Volmer and A. Weber, *Z. Phys. Chem.* **119** (1926) 277.

[12] V. A. Shchukin and D. Bimberg, *Rev. Mod. Phys.* **71** (1999) 1125.

[13] A. P. Alivisatos, *Science* **271** (1996) 933.

[14] A. D. Yoffe, *Adv. Phys.* **42** (1993) 173.

[15] F. M. Ross, R. M. Tromp and M. C. Reuter, *Science* **286** (1999) 1931.

[16] I. Daruka and A.-L. Barabási, *Phys. Rev. Lett.* **79** (1997) 3708.

[17] I. Daruka and A.-L. Barabási, *Phys. Rev. Lett.* **78** (1997) 3027.

[18] D. Jesson, *Phys. Rev. Lett.* **77** (1996) 1330.

[19] W. M. Tong and R. S. Williams, *Annu. Rev. Phys. Chem.* **45** (1994) 401.

[20] G. Rosenfeld, B. Poelsema and G. Comsa, "Epitaxial growth modes far from equilibrium", in *The Chemical Physics of Solid Surfaces and Heterogeneous Catalysis: Growth and Properties of Ultrathin Epitaxial Layers*, Volume 8, eds D. A. King and D. P. Woodruff (Elsevier, Amsterdam, 1997) p. 66.

[21] A. C. Jones, *J. Crystal Growth* **129** (1993) 728.

[22] J. M. Jasinski and S. M. Gates, *Accts. Chem. Res.* **24** (1991) 9.

[23] J. M. Jasinski, B. S. Meyerson and B. A. Scott, *Annu. Rev. Phys. Chem.* **38** (1987) 109.

[24] J. Wintterlin and P. Avouris, *Surf. Sci.* **286** (1993) L529.

[25] J. Wintterlin and P. Avouris, *J. Chem. Phys.* **100** (1994) 687.

[26] H.-G. Rubahn, *Laser Applications in Surface Science and Technology* (John Wiley, Chichester, Sussex 1999).

[27] P. M. Ajayan and T. W. Ebbesen, *Rep. Prog. Phys.* **60** (1997) 1025.

[28] D. S. Bethune, C. H. Klang, M. S. de Vries, G. Gorman, R. Savoy, J. Vazquez and R. Beyers, *Nature* **363** (1993) 605.

[29] S. Iijima and T. Ichihashi, *Nature* **363** (1993) 603.

[30] T. Guo, P. Nikolaev, A. Thess, D. T. Colbert and R. E. Smalley, *Chem. Phys. Lett.* **243** (1995) 49.

[31] S. Fan, M. G. Chapline, N. R. Franklin, T. W. Tombler, A. M. Cassell and H. Dai, *Science* **283** (1999) 512.

[32] J. Kong, H. T. Soh, A. M. Cassell, C. F. Quate and H. Dai, *Nature* **395** (1998) 878.

[33] Y. Q. Zhu, W. B. Hu, W. K. Hsu, M. Terrones, N. Grobert, J. P. Hare, H. W. Kroto, D. R. M. Walton and H. Terrones, *J. Mater. Chem.* **9** (1999) 3173.

[34] K. J. Laidler, *The World of Physical Chemistry* (Oxford University Press, Oxford, 1993).

[35] I. Langmuir, *J. Am. Chem. Soc.* **39** (1917) 1848.

[36] S. Hughes, *Washi: The World of Japanese Paper* (Kodansha International, Tokyo, 1978).

[37] W. C. Bigelow, D. L. Pickett and W. A. Zisman, *J. Colloid Interface Sci.* **1** (1946) 513.

[38] R. G. Nuzzo and D. L. Alara, *J. Am. Chem. Soc.* **105** (1983) 4481.

[39] A. Ulman, *An Introduction to Ultrathin Organic Films from Langmuir–Blodgett to Self-assembly* (Academic Press, Boston, MA, 1991).

[40] A. Ulman, *Chem. Rev.* **96** (1996) 1533.

[41] F. Schreiber, *Prog. Surf. Sci.* **65** (2000) 151.

[42] J. M. Buriak, *Chem. Commun.* (1999) 1051.

[43] L. S. Jung and C. T. Campbell, *Phys. Rev. Lett.* **84** (2000) 5164.

[44] L. H. Dubois and R. G. Nuzzo, *Annu. Rev. Phys. Chem.* **43** (1992) 437.

[45] G. E. Poirier and E. D. Pylant, *Science* **272** (1996) 1145.

[46] M. Schunack, L. Petersen, A. Kühnle, E. Læsgaard, I. Stensgaard, I. Johannsen and F. Besenbacher, *Phys. Rev. Lett.* **86** (2001) 456.

[47] D. L. Patrick, V. J. Cee and T. P. Beebe Jr, *Science* **265** (1994) 231.

[48] K. L. Prime and G. M. Whitesides, *Science* **252** (1991) 1164.

[49] H. Takano, J. R. Kenseth. S.-S. Wong, J. C. O'Brien and M. D. Porter, *Chem. Rev.* **99** (1999) 2845.

[50] K. Kern, H. Niehus, A. Schatz, P. Zeppenfeld, J. George and G. Comsa, *Phys. Rev. Lett.* **6** (1991) 855.

[51] H. Brune, M. Giovannini. K. Bromann and K. Kern, *Nature* **394** (1998) 451.

[52] S. Aggarwal, A. P. Monga, S. R. Perusse, R. Ramesh, V. Ballarotto, E. D. Williams, B. R. Chalamala, Y. Wei and R. H. Reuss, *Science* **287** (2000) 2235.

Appendix A

A.1 Fundamental Constants and Conversion Factors

A.1.1 Fundamental constants

Table A.1 Recommended values of fundamental constants, Source: Peter J. Mohr and Barry N. Taylor, "The fundamental physical constants", *Physics Today* **53**(8) (2000) BG6.

Quantity	Symbol	Value		Units
Boltzmann constant	k_B	1.38066	10^{-23}	$J\,K^{-1}$
Planck constant	h	6.62608	10^{-34}	$J\,s$
	$\hbar = h/2\pi$	1.05457	10^{-34}	$J\,s$
Avogadro constant	N_A	6.02214	10^{23}	mol^{-1}
Bohr radius	a_0	5.29177	10^{-11}	m
Rydberg constant	R_∞	1.09737	10^5	cm^{-1}
Speed of light	c	2.99792458	10^8	$m\,s^{-1}$
Elementary charge	e	1.602176	10^{-19}	C
Gas constant	$R = N_A k_B$	8.31451		$J\,K^{-1}\,mol^{-1}$
Vacuum permittivity	ε_0	8.85419	10^{-12}	$J^{-1}\,C^2\,m^{-1}$
Atomic mass unit	u	1.66054	10^{-27}	kg
Electron mass	m_e	9.10939	10^{-31}	kg
Proton mass	m_p	1.67262	10^{-27}	kg
Neutron mass	m_n	1.67493	10^{-27}	kg

A.1.2 Conversion factors

Table A.2 Energy conversion factor

1 eV	1.602176×10^{-19} J
1 eV/hc	8065.5 cm^{-1}
1 meV/hc	8.0655 cm^{-1}
1 eV/particle	96.485 kJ mol^{-1}
1 kcal mol^{-1}	4.184 kJ mol^{-1}

Table A.3   Thermal energy conversion factors

Thermal Energy	4 K	77 K	298 K
$RT/(\text{kJ mol}^{-1})$	0.03254	0.6264	2.424
k_BT/meV	0.3447	6.635	25.68
$\dfrac{k_BT}{hc}/\text{cm}^{-1}$	2.780	53.52	207.1

Table A.4   Pressure Conversion Factors

1 atm	101325 Pa
1 atm	1013.25 mbar
1 atm	760 torr
1 bar	10^5 Pa
1 torr	1.3332 mbar
1 torr	133.32 Pa

A.1.2.1   The Langmuir: unit of exposure

$$1 \text{ L} = 1 \times 10^{-6} \text{ torr s}$$
$$= 1.3332 \times 10^{-6} \text{ mbar s}$$
$$= 1.3332 \times 10^{-4} \text{ Pa s}$$

A.1.3   Prefixes

Table A.5   Prefixes

a	f	p	n	μ	m	c	d	k	M	G	T
atto	femto	pico	nano	micro	milli	centi	deci	kilo	mega	giga	terra
10^{-18}	10^{-15}	10^{-12}	10^{-9}	10^{-6}	10^{-3}	10^{-2}	10^{-1}	10^{3}	10^{6}	10^{9}	10^{12}

Appendix B

B.1 Abbreviations

1D, 2D, 3D	one-, two-, and three-dimensional
2PPE	two-photon photoemission
8CB	4'-octyl-4-biphenylcarbonitrile
II–VI	compound semiconductors composed of elements from the old groups II and VI, now groups 12 and 16, respectively
III–V	compound semiconductors composed of elements from the old groups III and V, now groups 13 and 15, respectively
AES	Auger electron spectroscopy
AFM	atomic force microscopy [also called scanning force microscopy (SEM)]
AO	atomic orbital
ARUPS	angle-resolved ultraviolet photoemission spectroscopy
ATR	attenuated total reflectance
bcc	body-centred cubic
BET	Brunauer–Emmet–Teller (isotherm)
CA	chemically amplified
CI	configuration interaction
CITS	constant current tunnelling spectroscopy
CTST	conventional transition-state theory
CVD	chemical vapour deposition
DFT	density function theory
DNA	deoxyribonucleic acid
DNQ	diazonaphthoquinone
DRIFTS	diffuse reflectance infrared Fourier transform spectroscopy
ECS	equilibrium crystal shape
EELS	electron energy loss spectroscopy
E–R	Eley–Rideal (reaction mechanism)
ESCA	electron spectroscopy for chemical analysis
fcc	face-centred cubic
FM	Frank–van der Merwe (layer-by-layer growth)
FT	Fourier transform

F–T	Fischer–Tropsch
FWHM	full width at half maximum
GGA	generalized gradient approximation
HA	Hot-atom (reaction mechanism)
HC	hydrocarbon
hcp	hexagonal close packed
HOMO	highest occupied molecular orbital
HOPG	highly oriented pyrolitic graphite
HREELS	high-resolution electron energy loss spectroscopy
HtBDC	hexa-*tert*-butyldecacyclene
IC	integrated circuit
IR	infrared
IRAS	infrared reflection absorption spectroscopy
K-cell	Knudsen cell
k-space	crystal momentum space; $\mathbf{k}$ is the wavevector of an electron or phonon in a lattice
L	Langmuir, a unit of exposure (for definition, see Appendix A)
LB	Langmuir–Blodgett
LED	light-emitting diode
LEED	low-energy electron diffraction
L–H	Langmuir–Hinshelwood (reaction mechanism)
LUMO	lowest unoccupied molecular orbital
MARI	most abundant reaction intermediate
MBE	molecular beam epitaxy
MBRS	molecular beam relaxation spectrometry
MEMS	microelectromechanical systems
MIR	multiple internal reflection, or mid-infrared
ML	monolayer, can be defined either with respect to number of adsorption sites or number of surface atoms
MO	molecular orbital
MPI	multiphoton ionization
MPPE	multiphoton photoemission
MWT	multiwalled nanotube
NO_x or NOX	nitrogen oxides
NSOM	near-field scanning optical microscopy [also called scanning near-field optical microscopy (SNOM)]
PEEM	photoemission electron microscopy
PES	potential energy (hyper)surface
PMMA	poly(methylmethacrylate)
por-Si	porous silicon
PSTM	photon scanning tunnelling microscopy
QMS	quadrupole mass spectrometer
RAIRS	reflection absorption infrared spectroscopy

RDS	rate-determining step
RSH	alkane thiol (or HSRX, where X = tail group)
SAM	self-assembled monolayer
SATP	standard ambient temperature and pressure (298 K and 1 atm)
SEM	scanning electron microscopy
SFM	scanning force microscopy
SK	Stranski–Krastanov (layer-plus-island growth)
SNOM	scanning near-field optical microscopy
SPM	scanning probe microscopy
STM	scanning tunnelling microscopy
STOM	scanning tunnelling optical microscopy
STS	scanning tunnelling spectroscopy
SWT	single-walled nanotube
syn gas	synthetic gas, $H_2 + CO$
TDS	thermal desorption spectrometry
TEM	transmission electron microscopy
TPD	temperature-programmed desorption [also called thermal desorption spectrometry (TDS)]
TPRS	temperature-programmed reaction spectrometry
TS	transition state
TST	transition-state theory
UHV	ultrahigh vacuum, pressure range below 10^{-8} torr (10^{-6} Pa)
UPS	ultraviolet photoelectron spectroscopy
VW	Wolmer–Weber (island growth)
XPS	X-ray photoelectron spectroscopy

Appendix C

C.1 Symbols

a	absorptivity
$\mathbf{a}$	basis vector of a substrate lattice
$\mathbf{a}^*$	reciprocal lattice vector of a substrate
A	Arrhenius pre-exponential factor for a first-order process; absorbance
A_p	peak area
A_s	surface area
$\mathbf{b}$	basis vector of an overlayer
$\mathbf{b}^*$	reciprocal lattice vector of an overlayer
B_v	rotational constant for vibrational level v
c	speed of light; concentration
d	depletion layer width; tip-to-surface separation; island diameter
D	diffusion coefficient
D_0	diffusion pre-exponential factor
$D(E)$	transmission probability in tunnelling
$D(\mathrm{M-A})$	dissociation energy of the M−A bond
E	energy
E_0	In electron energy loss spectroscopy, the energy of the incident electron beam (primary energy)
E_a	generalized activation energy
E_{ads}	adsorption activation energy
E_B	electron binding energy
E_C	energy of the conduction band minimum
E_{des}	desorption activation energy
E_{dif}	diffusion activation energy
E_f	final energy
E_F	Fermi energy
E_g	band gap energy
E_i	initial energy; position of E_F in an intrinsic semiconductor
E_k	kinetic energy
E_n	component of kinetic energy along the surface normal (normal energy)

E_n	energy of an image potential (Rydberg) state of principal quantum number n
E_V	energy of the valence band maximum
E_{vac}	vacuum energy
F_N	force on a cantilever
g	degeneracy
G^s	surface Gibbs energy
$\mathscr{H}$	Hamiltonian operator
H	enthalpy
I	tunnelling current; intensity; moment of inertia
k	rate constant; crystal momentum
k_1	rate constant of forward reaction 1
k_{-1}	rate constant of reverse reaction 1
$\mathbf{k}$	wavevector
K	equilibrium constant
k_N	cantilever force constant
l	path length
L	desorption rate from chamber walls
m	mass
M	molar mass (also called molecular or atomic weight)
$\mathbf{M}$	transformation matrix between $\mathbf{a}$ and $\mathbf{b}$
n	principal quantum number; real part of the index of refraction; number of moles
$\tilde{n}$	complex index of refraction
n_i	intrinsic carrier density
$n(\mathbf{k}, p)$	occupation number of the phonon with wavevector $\mathbf{k}$ in branch p
N_0	number of surface sites or atoms
N_A	density of electrically active acceptor atoms
N_{ads}	number of adsorbates
N_C	effective density of states in the conduction band
N_D	density of electrically active donor atoms
N_{exp}	number of atoms or molecules exposed to (i.e. incident upon) the surface
N_V	effective density of states in the valence band
$N_{v,J}$	number of molecules in vibrational level v and rotational level J
p_b	base (steady-state) pressure
q	partition function
q_{ads}	heat of adsorption for a single molecule or atom
Q_{ads}	canonical partition function for adsorption
r	rate; radial distance; molecular bond length
R	reflectivity of an adsorbate-covered surface
R_0	reflectivity of a clean surface
R_∞	reflectivity of an optically thick sample; Rydberg constant

s	sticking coefficient (also called sticking probability)
$\mathbf{s}$	direction of diffraction electron beam
s_0	initial sticking coefficient
$\mathbf{s}_0$	direction of incident electron beam
s_A	Auger sensitivity factor for element A
S	scattering function; vacuum chamber pumping speed
t	time
T	temperature; transmittance
T_p	temperature at which a temperature-programmed desorption peak maximum occurs
T_s	surface temperature
T_θ	isokinetic temperature
u	displacement
v	velocity
V	voltage; potential; volume
V_B	tunnelling barrier height
V_{fb}	flat band potential
V_{surf}	magnitude of band bending
W	work
x	direction in the plane of the surface
x_A	in quantitative electron spectroscopy, the mole fraction of element A
y	direction in the plane of the surface
Y	Young's modulus
z	direction normal to the surface
Z	atomic number
Z_w	collision rate
α	absorption coefficient; polarizability
δ_{ij}	Kronecker delta function
$\delta\varepsilon_{relax}, \varepsilon_{rel}, \delta\varepsilon_{corr}$	In electron spectroscopy, corrections to the Hartree-Fock energies involving electron relaxation, relativistic effects and electron correlation
Δ	change
$\mathit{\Delta}$	the energy difference between substrate bonds, Φ_{AA}, and the interface bonds, Φ_{AB}, within a heterolayer
ΔE_b	chemical shift
ΔG	Gibbs energy change
ΔH	enthalpy change
$\Delta_{ads}H$	integral heat of adsorption
$\Delta_{ads}H^{diff}$	differential heat of adsorption
$\Delta_{ads}H^{st}$	isoteric heat of adsorption
S	enthalpy change
χ	electron affinity
ε	exposure; lattice mismatch; permitivity

ε_k	orbital energy
$\varepsilon_{\alpha\beta}$	lattice deformation
$\varepsilon(\omega)$	dielectric function
γ	surface tension or energy; angle between basis vectors
η	electrochemical potential
ϕ	work function; azimuthal angle
ϕ_c	contact potential
Φ_{AA}, Φ_{AB}, Φ_{BB}	at the interface between two materials, the energies of A−A, A−B and B−B bonds
Γ	intrinsic (homogeneous) line width
κ	imaginary part of the index of refraction; transmission coefficient
λ	wavelength
λ_m	mean free path in material m
μ	chemical potential; reduced mass; dipole moment
ν	frequency
ν_n	pre-exponential factor for a process of order n
θ	fractional coverage in monolayers
θ_{00}	number of adjacent sites in the presence of lateral interactions
θ_D	Debye temperature
θ_p	coverage at the maximum of a temperature-programmed desorption peak
θ_{req}	functional form of the sticking coefficient dependence on coverage
ϑ	polar angle
$\rho_s(E)$	density of states at energy E
σ	areal density of adsorbates; symmetry number in rotational partition function; differential conductance
σ_0	areal density of surface sites or atoms
σ_*	areal density of empty sites
τ	lifetime
$\tau_{\alpha\beta}(\mathbf{m})$	intrinsic surface stress tensor
ω_0	radial frequency of the fundamental mode of a harmonic oscillator
ω_D	Debye frequency
Ξ	grand canonical partition function
ψ	contact angle; wavefunction
ζ	degree of supersaturation

C.1.1 *Kisliuk model of precursor-mediated adsorption*

f_c	probability of chemisorption from the intrinsic precursor
f_{des}	probability of desorption from the intrinsic precursor
f_{mig}	probability of migration to next site in the intrinsic precursor
f_c'	probability of chemisorption from the extrinsic precursor
f_{des}'	probability of desorption from the extrinsic precursor

| f'_{mig} | probability of migration to the next site in the extrinsic precursor |
| α | trapping coefficient into the precursor |

C.1.2   *Symbols commonly used in subscripts and superscripts*

$\ddagger$	transition state
0 or i	initial
ads	adsorption
des	desorption
dif	diffusion
elec	electronic
f	final
g	gas
in	incident
K	kinetic
max	maximum or saturation value
p	peak
rot	rotational
s	surface
trans	translational
vib	vibrational

C.1.3   *Mathematical conventions*

| $\langle x \rangle$ | mean value of variable x |
| det $\mathbf{M}$ | determinant of matrix $\mathbf{M}$ |

C.1.4   *Chemical notation*

*	surface site
[A]	concentration of compound A
e^-	electron
h^+	hole (the absence of an electron)

C.1.4.1   *Amphiphiles*

| α end | head group (polar end) |
| ω end | tail group (hydrophobic end) |

C.1.4.2   *Vibrational modes*

δ	in-plane bend
γ	out-of-plane bend
ν	stretch
τ	torsion

Appendix D

D.1 Useful Mathematical Expressions

$E = (n + \tfrac{1}{2})\hbar\omega_0$ harmonic oxcillator energy levels

$E_K = \tfrac{1}{2}mv^2$ kinetic energy

$E_J = hcB_v J(J+1)$ rigid rotor energy levels

$f(E) = \left\{ \exp\left[\dfrac{(E - \mu)}{k_B T}\right] + 1 \right\}^{-1}$ Fermi–Dirac distribution

$f(v) = 4\pi \left(\dfrac{M}{2\pi RT}\right)^{3/2} v^2 \exp\left(\dfrac{-Mv^2}{2RT}\right)$ Maxwellian velocity distribution

$k = A \exp\left(\dfrac{E_a}{RT}\right)$ Arrhenius expression

$\langle n(\omega, T)\rangle = \left[\exp\left(\dfrac{\hbar\omega}{k_B T}\right) - 1\right]^{-1}$ Planck distribution

$N_v = N \exp\left(\dfrac{-E_v}{k_B T}\right) = N \exp\left(\dfrac{-hcG_0(v)}{k_B T}\right)$ vibrational Maxwell–Boltzmann distribution

$N_{vJ} = N_v \dfrac{hcB_v}{k_B T}(2J + 1)\exp\left(\dfrac{-E_J}{k_B T}\right)$ rotational Maxwell–Boltzmann distribution

$q = \displaystyle\sum_{i=1}^{\infty} g_i \exp\left(\dfrac{-E_i}{k_B T}\right)$ general partition function

$q_{trans} = \displaystyle\prod_i \dfrac{(2\pi m k_B T)^{1/2}}{h}$ translational partition function in i dimensions

$q_{rot} = \dfrac{8\pi^2 I k_B T}{\sigma h^2}$ rotational partition function, linear molecule

$q_{rot} = \dfrac{8\pi^2 (8\pi^3 I_A I_B I_C)^{1/2}(k_B T)^{3/2}}{\sigma h^3}$ rotational partition function, nonlinear molecule

$q_{vib} = \displaystyle\prod_i \left[1 - \exp\left(\dfrac{hv_i}{k_B T}\right)\right]^{-1}$ vibrational partition function

$$\langle x^2 \rangle^{1/2} = (4Dt)^{1/2} = \left[\frac{4D_0}{A} \exp\left(\frac{E_{\text{ads}} - E_{\text{dif}}}{RT}\right) \right]^{1/2}$$

mean square displacement on a uniform two-dimensional potential energy surface

$$Z_{\text{w}} = \frac{N_a p}{(2\pi MRT)^{1/2}}$$

$M = $ molar mass

$$Z_{\text{w}} = \frac{p}{(2\pi m k_{\text{B}} T)^{1/2}}$$

$m = $ particle mass

$$Z_{\text{w}} = 3.51 \times 10^{22} \text{ cm}^{-2}\text{s}^{-1} \frac{p}{(MT)^{1/2}}$$

Z_{w} in cm^{-2} s^{-1}, M in g mol^{-1}, p in torr, T in K

$$Z_{\text{w}} = 2.63 \times 10^{24} \text{ m}^{-2}\text{s}^{-1} \frac{p}{(MT)^{1/2}}$$

Z_{w} in m^{-2} s^{-1}, M in g mol^{-1}, p in Pa

$$\Delta G = \Delta H - T\Delta S$$

definition of change in Gibbs energy

$$\varepsilon = Z_{\text{w}} t$$

definition of exposure

$$\lambda = \frac{h}{mv} = \frac{h}{p}$$

de Broglie relationship

$$\mu = \frac{m_1 m_2}{m_1 + m_2}$$

reduced mass

$$\theta = \frac{p}{p + p_0(T)}$$

thermodynamic expression of the Langmuir isotherm for molecular adsorption

$$p_0(T) = \left(\frac{2\pi k_{\text{B}} T}{h^2}\right)^{3/2} k_{\text{B}} T \exp\left(\frac{-q_{\text{ads}}}{k_{\text{B}} T}\right)$$

$$\theta = \frac{pK}{1 + pK}$$

kinetic expression of the Langmuir isotherm for molecular adsorption

$$K = \frac{k_{\text{ads}}}{k_{\text{des}}}$$

$$\sigma = s\varepsilon$$

coverage resulting from exposure, ε

Index